INFORMATION THEORY AND CODING

BASICS AND PRACTICES

S Veluswamy

ME (IISc.), FIETE, MSEE,

Additional Director (Retd.)

DRDO, India

Former Professor & Head

Department of ECE, MVJ College of Engineering

Bangalore, India

Publishing Globally

NEW AGE INTERNATIONAL (P) LIMITED, PUBLISHERS

New Delhi • Bangalore • Chennai • Cochin • Guwahati

Hyderabad • Kolkata • Lucknow • Mumbai

Visit us at **www.newagepublishers.com**

Published by New Age International (P) Ltd., Publishers
First Edition: 2014

BRANCHES

- **Bangalore** 37/10, 8th Cross (Near Hanuman Temple), Azad Nagar, Chamarajpet Bangalore-560 018, Tel.: (080) 26756823, Telefax: 26756820 **E-mail: bangalore@newagepublishers.com**
- **Chennai** 26, Damodaran Street, T. Nagar, Chennai-600 017, Tel.: (044) 24353401 Telefax: 24351463, **E-mail: chennai@newagepublishers.com**
- **Cochin** CC-39/1016, Carrier Station Road, Ernakulam South, Cochin-682 016 Tel.: (0484) 2377004, Telefax: 4051303 **E-mail: cochin@newagepublishers.com**
- **Guwahati** Hemsen Complex, Mohd. Shah Road, Paltan Bazar, Near Starline Hotel Guwahati-781 008, Tel.: (0361) 2513881, Telefax: 2543669 **E-mail: guwahati@newagepublishers.com**
- **Hyderabad** 105, 1st Floor, Madhiray Kaveri Tower, 3-2-19, Azam Jahi Road Nimboliadda, Hyderabad-500 027, Tel.: (040) 24652456 Telefax: 24652457, **E-mail: hyderabad@newagepublishers.com**
- **Kolkata** RDB Chambers (Formerly Lotus Cinema) 106A, 1st Floor S.N. Banerjee Road, Kolkata-700 014, Tel.: (033) 22273773 Telefax: 22275247, **E-mail: kolkata@newagepublishers.com**
- **Lucknow** 16-A, Jopling Road, Lucknow-226 001, Tel.: (0522) 2209578, 4045297 Telefax: 2204098, **E-mail: lucknow@newagepublishers.com**
- **Mumbai** 142C, Victor House, Ground Floor, N.M. Joshi Marg, Lower Parel Mumbai-400 013, Tel.: (022) 24927869, Telefax: 24915415 **E-mail: mumbai@newagepublishers.com**
- **New Delhi** 22, Golden House, Daryaganj, New Delhi-110 002 Tel.: (011) 23262368, 23262370, Telefax: 43551305 **E-mail: sales@newagepublishers.com**

ISBN: 978-81-224-3672-3

₹ 199.00

C-13-11-7215

Printed in India at Ajit Printing Press, Delhi.
Typeset at In House, Delhi.

PUBLISHING GLOBALLY
NEW AGE INTERNATIONAL (P) LIMITED, PUBLISHERS
7/30 A, Daryaganj, New Delhi - 110002
Visit us at **www.newagepublishers.com**

INFORMATION THEORY AND CODING

BASICS AND PRACTICES

Dedicated to

Claude Elwood Shannon
(1916–2001)
"Father of Information Theory"

&

Richard Wesley Hamming
(1915–1998)
"Pioneer in Error Control Coding"

Preface

Presently we are living in the so called information age which brought a fast evolution of technology in daily life, as well as of educational life style; the Information Age has allowed rapid global communications and networking to shape modern society and made the world into a global village. The Information Age, also commonly known as the Computer Age or Digital Age, is an idea that the current age will be characterized by the ability of individuals to transfer information freely, and to have instant access to information that would have been difficult or impossible to find previously. The idea is linked to the concept of a digital age or digital revolution, and carries the consequences of a shift from traditional industry that the industrial revolution brought through industrialization, to an economy based on the manipulation of information, i.e., an information society. The Information Age formed by capitalizing on the computer microminiaturization advances, with a transition spanning from the advent of the personal computer in the late 1970s to the Internet's reaching a critical mass in the early 1990s, and the adoption of such technology by the public in the two decades after 1990.

The major seed for this growth was the introduction of the concepts namely, Information theory and Coding theory. However it was considered impractical to have a cost effective communication system using these concepts, because of lack of technology in those days. As the technology (both hardware and software) improved in the information age these subjects enjoy a growth in interest. Many books are available on this subject, but with emphasis primarily on the mathematics and research aspects. Any system designer wants the concepts to be put in practice. Any subject understood physically, supported by mathematical background and practical aspects will be lost longer in the minds of the student. This book is written with this in mind. The gaps between theory and practice are brought in many cases. The complexity of system in implementing a concept is also brought out. What value of a parameter in an algorithm is viable in practice is also explained in detail for many aspects. The subject of information theory and coding requires good mathematical background, which created hesitation to students and practicing engineers (both hardware and software) who are interested in these techniques to realize a practical system. One of the purposes in this book is to present the complex, mathematical oriented subject in a way that it can be understood and practiced only with a minimum mathematical background. The approach is to explain the subject material with minimum mathematical rigor, and with an emphasis both on theory and practice. All the algebra required for understanding the information theory and coding techniques is presented in the book.

The close relation between the subject and implementation techniques is stressed wherever required. The design of elementary circuits to implement an encoder – decoder pair and modification of these circuits to reduce the number of elements are also broughtout. Even though the elementary circuits like shift registers can be implemented in software, the modification ideas will simplify the software programme. Codes are not of much interest if no decoding algorithm or decoder is possible or available. This book concentrates on both basics and practices.

This book began with the materials taught to students and practiced in the industry, by the author, for many years and is written with better understanding of the needs of the students and the practising engineers in the subject. This book meets the requirements of senior undergraduate students of electronics and communication engineering, telecommunication engineering branch of Indian universities and practising engineers in the industry. This book also will meet the interests of graduate students of physics and mathematics. The book contains eight chapters. The concepts are examplified with worked out examples to make them very clear. The origin of concepts are mentioned appropriately. A number of exercise problems are given for practise. In order to motivate the students for further studies important references are also included. A brief of the chapters are as follows:

Chapter 1 starts with an introduction to the subject of Information Theory and Coding. Then it talks about the measure of information and their units. Average information content of symbols in long independent and dependent sequences (*Entropy*); computations of entropy and information rate are explained in detail. Different kind of information sources like, zero memory, Markov sources, extended sources and the information conveyed by these sources are explained elaborately.

Chapter 2 deals with Source Coding Techniques. Principles of source coding, properties of different source codes; code construction procedures, code efficiency and redundancy are explained in detail. Shannon's Noiseless coding theorem; Shannon's encoding algorithm, Shannon – Fano encoding algorithm for binary as well as ternary sources are explained elaborately. Huffman's minimum redundancy coding techniques for binary, ternary and quarternay alphabets, their properties and decoding procedures are explained vividly. Concept of compression ratio is also explained.

Chapter 3 takes up the study on the different kinds of channels through which the information is to be transmitted. Discrete Communication channels, their representation using Transition Probability diagram, Channel Matrix and Joint Probability Matrix are explained in depth. Concepts of Entropy Function, Mutual Information, Channel Capacity, Channel Efficiency and Redundancy are explained in a step by step manner with examples. Different practically encountered channels like Binary Symmetric Channel (BSC), Binary Erasure Channel (BEC), deterministic channel, Noiseless channel, Cascaded channel and their capacity are explained in detail. The concept of Continuous channels, their entropy and how to maximize their entropy are dealt in detail. Channel capacity (Shannon-Hartley) theorem, the capacity of Band limited channels with Additive White Gaussian Noise (AWGN), channel capacity trade off with different parameters like noise and bandwidth; Shannon's limit of channel capacity are explined in detail with graphs showing the band width limited and power limited regimes.

Chapter 4 starts with an introduction to Error Control Coding (Channel Coding) techniques like Forward Error Correction (FEC) and Automatic Repeat request (ARQ) schemes, their merits and demerits; the types of channels, types of noise and types of errors encountered in a practical communication system. Next a simplified model of a coded communication system is considered and the two important types of FEC schemes namely the Linear block codes and Convolutional codes are introduced. The concepts of Code mapping principle, code distance and Hamming Distance are explained. Next an elaborate study on the Linear block codes is given in this chapter. Principles of Generator matrix, systematic format of codewords, Encoding and Decoding procedures and circuitries; error detection and error correction capabilities, probability of erroneous decoding, simultaneous error detection and error correction capabilities of the code are explained in detail with different options and many examples. The principle of standard array, table look-up decoding procedure, Perfect Code, Hamming bound and related concepts are brought out with examples.

Chapter 5 deals with another kind of linear error control code namely Binary Cyclic code. The definition of cyclic codes, their algebraic structure and properties which simplify encoding and error detection are explained. The concept of generator polynomial, systematic format of cyclic codes and the steps involved in obtaining systematic cyclic codes are brought out. The use of generator matrix for generation of cyclic codes and Parity check matrix in the decoding of cyclic codes are explained clearly with worked out examples. The Linear switching circuits consisting of shift registers and logic circuits, for polynomial multiplication, division and simultaneous multiplication and division and their modifications are well explained with examples. This will help the designers to implement the encoder- decoder both in hardware and software efficiently. Error detection and correction procedures like standard array decoding are illustrated with examples. The special decoding technique using Megitt decoder is explained. The random and burst error detection and correction capabilities of the cyclic codes are brought out clearly. The properties of the generator polynomial are summarized at the end.

Chapter 6 concentrates on the study about the important class of error control codes namely the Convolutional codes. The chapter starts with definition of convolutional codes and encoding procedure using time domain approach. Different parameters of the encoder like memory order, constraint length, fractional rate loss etc., are explained clearly with examples. The generator matrix method of generating convolution codes is also brought out. Next transform domain approach of encoding convolutional codes using transfer function matrix are explained clearly. Systematic convolutional codes and their advantages; catastrophic codes and their deleterious effects, measures to avoid catastrophic codes are also explained. The concepts of state diagram and code tree diagram using the structural properties of convolutional codes are introduced. The procedures for drawing these diagrams and using them for encoding, decoding and using the modified state diagram for minimum distance computation of the code are brought out with clear examples. The different distance properties, like minimum free distance, Column Distance Function (CDF) are explained. Maximum Likelihood Decoding of Convolutional codes using Viterbi Algorithm, decoding of Convolutional Codes using Minimum Distance criteria of paths in Trellis Diagram and Sequential decoding procedure using code tree diagram are dealt in depth with clear examples. The implementation issues of Viterbi algorithm is discussed elaborately and the limitations are brought out. The dependence of both the Viterbi algorithm and sequenctial decoding on the memory order of the encoder are well explained. Applications of Convolutional codes in the real world situation are summarized at the end of the chapter.

Chapter 7 deals with the coding procedures which use advanced techniques and modification of the error control codes discussed in chapters 4 through 6. First a discussion on the well-known multiple error correcting perfect code, namely the Golay code is carried out. Its capabilities and decoding procedures like error trapping and table look-up methods are explained. Next the Bose – Chaudhuri – Hocquenghem (BCH) codes which are large class of powerful multiple error correcting Cyclic Codes that occupies an important place in the theory and practice of channel coding is considered. The definition of BCH codes and the encoding procedure for a BCH non-binary code is explained clearly. Different decoding algorithms for BCH codes, suggested by different researchers and their complexity are mentioned. Next an important sub-class of non-binary, q-array BCH codes, called the Reed–Solomon (RS) code, also known as maximum distance separable codes, which are known for their powerful burst error correcting capability, is taken up. The encoding of RS codes, starting from generation of Galios field, is explained in length with an example. The steps involved in RS code decoding, including the implementation of the Massey-Berlekamp algorithm for error location polynomial computation and the error correction procedure are explained in depth with the help of programme steps and flow chart. Next shortened cyclic codes, burst error correcting codes, burst and random error correcting codes are explained in detail. Interleaved codes, Concatenated codes, product codes, Binary repetition codes are the other topics covered in detail. Different modifications, like expanding, puncturing, shortening, expurgating etc., to remodel a known code to fit a specific application are explained with examples. The modified procedure of Massey and Belekamp algorithm for error and erasure correction is explained in depth with the help of flow chart. Finally the chapter ends with a discussion on "Block codes versus

Convolutional codes", wherein we bring out the situations in which block codes can be used, where convolutional codes can be used and where a mix of both can be used with the benefit of doing so. A comparison study on coded and uncoded systems is done with examples. The applications of the error control coding are also tabulated. In this chapter practical aspects are covered more. Some of the materials may be beyond the syllabus coverage in undergraduate level. However it gives a completion of the discussion on error control codes, with the present day industry situation, and for the students to quench their limitless thirst of knowledge in this field.

Appendix A gives those basic concepts of modern algebra that is required for understanding the theory of information and coding, to take up design and implementation work in this area.

The author dedicates this book for the frontiers of communication like Claude E Shannon, RW Hamming, Golay, Bose – Chaudhuri – Hocquenghem, Reed-Solomon etc., for their pioneering work and opportunity for people to take up study and research in this field.

The author acknowledges all those who motivated him, directly or indirectly, to write this book. The author would like to extend his special thanks to his parents, wife Rani and children Deepa, SVS Kumar and Anupallavi for their continued support and constant encouragement for completing this task. Thanks also goes to publisher for bringing out this book.

S. Veluswamy

Contents

CHAPTER

1

INFORMATION THEORY

1.0 Introduction

Information theory can be thought of as a branch of applied mathematics and electrical communication engineering involving the quantification of information. Information theory is not to be confused with Information Science. Information theory was developed by Claude E. Shannon to find fundamental limits on signal processing operations such as compressing the information (data) and on reliably storing and communicating the information (data). Since its origin it has developed to find applications in many other areas, including statistical inference, natural language processing, cryptography, neurobiological networks, the evolution and function of molecular codes, model selection in ecology, thermal physics, quantum computing, piracy detection and other forms of data analysis.

The meaning of word "***information***" in information theory is "***message***" or "***intelligence***". An important measure of information is known as ***entropy***, which is usually expressed by the average number of bits needed to store or communicate one symbol in a message. Entropy quantifies the uncertainty involved in predicting the value of a random variable. For example, specifying the outcome of a fair coin flip (two equally likely outcomes) provides less information (lower entropy) than specifying the outcome from drawing a card from a deck of 52 cards (52 equally likely outcomes).

Applications of fundamental topics of information theory include lossless data compression (*e.g.*, ZIP files), lossy data compression (*e.g.*, MP3s and JPGs), and channel coding (*e.g.*, for Digital Subscriber Line (DSL)). The field of information theory involves mathematics, statistics, computer science, physics, neurobiology, and electrical communication engineering. Its impact has been crucial to the success of many important missions to deep space, the invention of the compact disc, the feasibility of mobile phones, the development of the Internet, the study of linguistics and of human perception, the understanding of black holes (a region of space-time from which nothing, not even light, can escape), and numerous other fields. Important sub-fields of information theory are source coding, channel coding, algorithmic complexity theory, algorithmic information theory, information-theoretic security, and measures of information.

The main concepts of information theory can be understood by considering the most widespread means of human communication: language. Two important aspects of a language are as follows: First, the most common words *(e.g.*, "a", "the") should be shorter than less common words (*e.g*., "benefit", "generation", "mediocre"), so that sentences will not be too long. Such a trade-off in word length is analogous to *data compression* and is the essential aspect of ***source coding***. Second, if part of a sentence is unheard or misheard due to noise (may be from a passing car) the listener should still be able to glean the meaning of the underlying message. Such robustness is as essential for an electronic communication system as it is for a language; properly building such robustness into communications is done by ***channel coding***. *Source coding* and *channel coding* are the fundamental concerns of information theory.

Information theory is generally considered to have been founded in 1948 by Claude E Shannon in his seminal work, "A Mathematical Theory of Communication". The central paradigm of classical information theory is the engineering problem of the transmission of information over a noisy channel. The most fundamental results of this theory are Shannon's source coding theorem, which establishes that, on average, the number of *bits* needed to represent the result of an uncertain event is given by its entropy; and Shannon's noisy-channel coding theorem, which states that *reliable* communication is possible over *noisy* channels provided that the rate of communication is below a certain threshold, called the channel capacity. The channel capacity can be approached in practice by using appropriate encoding and decoding systems.

Information theory is a broad and deep mathematical theory, with equally broad and deep applications, amongst which is the vital field of coding theory. Coding theory is concerned with finding explicit methods, called *codes*, of increasing the efficiency and reducing the net error rate of data communication over a noisy channel to near the limit that Shannon proved is the maximum possible for that channel. These codes can be roughly subdivided into data compression (source coding) and error-correction (channel coding) techniques. In the latter case, it took many years to find the methods Shannon's work proved were possible. A third class of information theory codes are cryptographic algorithms (both codes and ciphers). Concepts, methods and results from coding theory and information theory are widely used in cryptography and cryptanalysis. Information theory is also used in information retrieval, intelligence gathering, gambling, statistics, and even in musical composition.

In Shannon's revolutionary and ground breaking paper, the work for which had been substantially completed at Bell Labs by the end of 1944, Shannon for the first time introduced the qualitative and quantitative model of communication as a statistical process underlying information theory, opening with the assertion that *"The fundamental problem of communication is that of reproducing at one point, either exactly or approximately, a message selected at another point."* With it came the ideas of

- the information entropy and redundancy of a source, and its relevance through the source coding theorem;
- the mutual information, and the channel capacity of a noisy channel, including the promise of perfect loss-free communication given by the noisy-channel coding theorem;
- the practical result of the Shannon-Hartley law for the channel capacity of a Gaussian channel; as well as
- the bit—a new way of seeing the most fundamental unit of information.

1.1 Generic Information System

In order to understand the concepts of information theory let us consider a general information system as shown in Fig. 1.1 below:

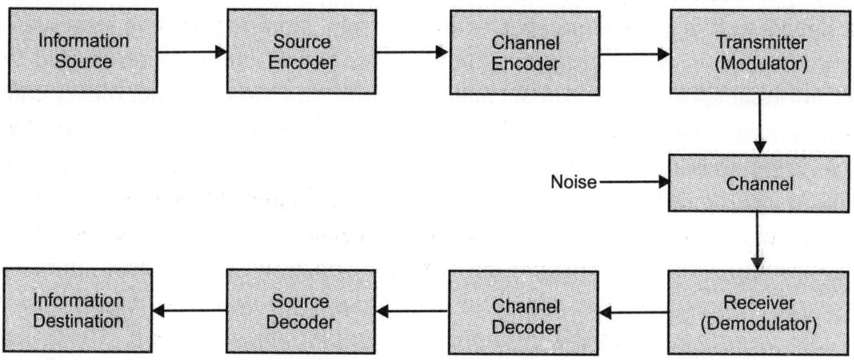

Fig. 1.1. Generic Information System

The information source is the origin of the message. The message can be an electrical signal such as a voltage, current or power or an audio message or video message such as television/picture/music message or a digital/discrete message from a computer or any digital source. In general an information source can be classified as an analog information source or a discrete information source. Analog information sources, like a microphone actuated by speech or a video camera produce one or more continuous amplitude electrical signals with respect to time. The discrete information source such as key board of a digital handheld terminal or a computer produces a sequence of discrete symbols or letters. The analog information can be converted into discrete information by analog to digital conversion methods.

The important parameters of a discrete information source are *source alphabet, symbol rate, probability of occurrence of the source alphabet* and the *inter-dependence of symbols* in a sequence of a message.

The examples of source alphabet are the alpha numeric characters (26 English alphabets, numerals and special characters like comma, full stop etc.) produced by a key board of a digital system. The symbol rate is the rate at which the source alphabet is produced by the information source. That is, if the device produces the alphabet at a speed of 100 characters/sec, then the symbol rate is 100 symbols/sec.

When we are using a language, for example, English, some alphabet (like 'E') will appear more frequently than some other alphabet (like 'Z') and some alphabet (like 'U') will always appear immediately after some alphabet (like 'Q'). These structural properties of the alphabet (symbols) in a sequence can be characterized by the probability of occurrence of the individual symbols and also by the conditional probabilities of occurrence of symbols (*i.e*, the inter-dependence of symbols).

The source encoder transforms the source output into a sequence of binary digits (0's and 1's) called the information sequence. The source encoder is ideally designed so that (1) the number of bits per unit time required to represent the source output is minimized, and (2) the source output can be reconstructed from the information sequence without any ambiguity. At source encoder binary coding is preferred because of its high transmission efficiency over the channel. Other coding methods are the ternary and quaternary coding. The simplest way of coding is to assign a fixed length binary code word to each message symbol in the input sequence. But this procedure is not efficient in general and is suitable and efficient only for the sources which produce symbols with equal probabilities. In most of the practical cases the symbols occur with unequal probabilities. In such cases the source encoder assigns variable length code words to these symbols. The different methods of source coding and important parameters of a source encoder namely the code word length, average data rate, coding efficiency and redundancy of coding are discussed in Chapter 2.

The channel coding provides the required robustness to the information while being transmitted over the channel in the presence of noise which produces error. That is the channel coding should provide the communication system the capability of detecting and correcting the channel induced error and deliver error free information at the receiver. The channel encoder (which performs the channel coding) transforms the information sequence coming from the source encoder output into a discrete encoded sequence. In most cases the encoded sequence is also a binary sequence, although in some applications non-binary coding have been used. There are two types of channel coding techniques in common use, namely the block codes and convolutional codes.

For the purpose of encoding the information sequence into encoded code sequence the channel encoder of the block code, divides the information sequence into message blocks of fixed length (say 'k' bits). Then each message block is independently transformed into an 'n' bit word called the code word or code vector. This is done by adding (n-k) redundant bits normally called the parity bits, to the k-bit message, according to certain rules of channel coding, based on the 'k' bit message of the present time unit. Therefore, corresponding to 2^k different possible messages there are 2^k possible code words at the channel encoder output. This set of 2^k code words of length 'n' is called an ***(n, k) block code***. The ratio $R = (k/n)$ is called the ***code rate*** and can be interpreted as the number of information bits entering into the encoder per transmitted symbol. The error detecting and correcting capability of the code used depends on the amount of redundancy (parity bits) used.

The convolutional code Encoder also accepts a k-bit message block and produces an encoded sequence (code word) of n-symbol block. However, each encoded block depends not only on the corresponding k-bit message block at the same time unit, but also on 'm' previous message blocks. Hence, the encoder has a memory of order 'm'. The set of code words produced by a k-input, n-output encoder of memory order 'm' is called a **(n, k, m) *Convolutional Code***. The ratio $R = k/n$ is called the code rate. Here for a fixed value of R, reliable communication can be achieved by proper use of the memory. How to use the memory to achieve a required level of reliable communication is the major problem in designing the encoder. The design and implementation of channel encoder and decoder pairs to combat the noisy environment in which code words must be transmitted is one of the major topics of this book. The detailed study on the subject of channel coding is dealt in Chapters 4 through 7 of this book.

Discrete symbols are not suitable for transmission over physical channel. So, the ***'Modulator'*** unit of the transmitter transforms each of the channel encoder output symbols into a waveform of appropriate period which is suitable for transmission over the physical channel. This waveform enters the channel and is corrupted by noise.

The communication channel provides the (electrical) connectivity between the information source and the destination. The channel can be a pair of wires or a telephone cable or free space over which the information signal is transmitted or radiated. Due to the limited availability of the channel bandwidth the information signal gets distorted (both in amplitude and in phase) as it travels through the channel. In addition, the signal power also decreases due to attenuation of the channel. Further the signal is corrupted by the unwanted and unpredictable electrical disturbances called the ***noise***. When binary information is transmitted the effect of the noise is to convert the binary '0's into '1's and binary '1's into '0's. The signals are said to be corrupted by noise and errors are induced in the received signal. A detailed study on the communication channels is provided in

Chapter 3. The main aim of a communication system is to combat the effects of the noise and provide a reliable error free communication as much as possible.

The ***'demodulator'*** processes each received waveform during the waveform period and produces an output that may be discrete. The sequence of demodulator outputs corresponding to the channel encoder output is called the received sequence.

The ***'channel decoder'*** transforms the received sequence into a binary sequence called the estimated sequence. The channel decoding is based on the channel encoding algorithm and the noise characteristics of the channel so as to minimize the probability of decoding error. The estimated sequence will be a replica of the source encoder output, even though the noise might have caused some decoding error.

The ***'source decoder'*** transforms the estimated sequence into an estimate of the source output and delivers the estimate to the destination. When the source is continuous this involves digital to analog conversion. In a perfectly designed system, the estimate will be a faithful reproduction of the source output except when the channel is very noisy.

In the forthcoming sections of this chapter, we will see some of the important aspects of Shannon's work on the information theory.

1.2 Measure of Information

The measure of information from an event depends on our expectation from that event. That is the more unexpected an event, the more information is obtained when that event occurs. For example, in a classroom, if a regular late coming student come to class in time, in a particular day, it creates a surprise and lot more information can be gained.

The **Information Content** or the **Amount of Information or Self-Information** from an event or message can be expressed or defined as

$$\mathbf{I(x) = \log (1/(p(x)) = -\log p(x)} \tag{1.1}$$

where $p(x)$ = probability of occurrence of the event 'x'.

We see that **I(x)** is a dimensionless number. By convention, in general the "**unit**" assigned is ***"bit"***; the unit bit is used when the base of the logarithm is 2. For example, when $p = 1/4$, $\mathbf{I(x) = \log_2(4) = 2}$ **bits**. The use of the base 2 is convenient when base band modulation techniques like binary PCM is used.

However, the unit of **I(x)** depends on the base for the expression log [ ], in equation (1.1). Accordingly when natural logarithmic base is used the unit is "***nat***" and when the base is 10 the unit is "***Hartley*** or ***Decit***". So we can express **I(x)** as

$$\mathbf{I(x) = \{-\log_2 p(x)\}\ bit,} \tag{1.2}$$

$$\mathbf{= \{-\log_e p(x)\} = \{-\ln p(x)\}\ nat,}$$

$$\mathbf{= \{-\log_{10} p(x)\}\ Hartley\ or\ Decit}$$

In general $\mathbf{I(x) = -\log_D p(x)}$ **D- ary unit.**

The information measure or amount of information or self-information can be expressed by any monotonically decreasing function of the probability 'p'. However, the function '$-\log(p)$' is more convenient for the following reasons/ properties:

(*a*) The information content of a message increases as $p(x)$ decreases and *vice versa,*

(*b*) **I(x)** is a continuous function of $p(x)$,

(*c*) The total information content of two or more independent messages or events is the sum of individual information.

The property (*c*) can be illustrated with the following example: Suppose two independent events 'x' and 'y' have probability of occurrence as $p(x)$ and $p(y)$ then the joint probability of their occurrence $= p = p(x).\ p(y)$ and the information conveyed by their occurrence is

$$\mathbf{I(x, y) = -\log(p) = -\log[p(x\ and\ y)] = -\log[p(x).\ p(y)]}$$

$$\mathbf{= [-\log p(x)] + [-\log p(y)]}$$

$$\mathbf{= I(x) + I(y)}$$

In general $I = -\Sigma \log[p(i)]$, where 'i' the different events. For example, consider the two news messages **Indian cricket team lifts the world cup** and **the Indian Hockey team wins over their rival.** Both are independent news and the total amount of information obtained is equal to the sum of information obtained from both the messages.

In order to understand the concept of amount of information consider the amount of information conveyed by a binary source emitting binary digits with equal probability, that is $p(0) = p(1) = 1/2$, then the amount of information conveyed by every binary digit emitted by the source is $\mathbf{I = \log_2(2) = 1\ bit.}$

Suppose binary '0' occurs with a probability $1/4$ and binary '1' occurs with probability of $3/4$, then binary '0' conveys an amount of information $= \log_2(4) = 2$ bits and binary '1' conveys an amount of information $= \log_2(4/3) = 0.42$ bits.

Suppose the source emits M equally likely and independent messages and $\mathbf{M = 2^N}$, where **N** is an integer, then the amount of information in each message is $\mathbf{I = \log_2 M = \log_2 (2^N) = N\ bits.}$

Consider the definite event "Sun rises in the East everyday", for which $p(x) = 1$, then the amount of information in 'x' is **I(x) = $\log_2(1)$ = 0 bits**. When $p(x)$ decreases from 1 to 0, **I(x)** increases monotonically from 0 to infinity. Thus, the more unexpected an event, the more information is obtained when that event occurs.

Thus, from the above discussions the following can be inferred:

- The amount of information is inversely proportional to the probability of occurrence of the event.
- The lowest possible information is "zero" which occurs when a "certain" event, with a probability of occurrence $p = 1$, takes place.
- Maximum information can be obtained from an event having least probability of occurrence.
- When independent events occur then the total information from all the events is equal to the sum of individual information.

1.2.1. Relation between Different Units of Information

From the definition of information

$$I = \{\log_2(1/p)\} \text{ bit}$$

$$= \{\log_e(1/p)\} \text{nat}$$

$$= \{\log_{10}(1/p)\} \text{ Hartley}$$

We have 1 Hartley = $\{I/[\log_{10}(1/p)]\} = \{\log_e(1/p)\}$ nats/$[\log_{10}(1/p)]$

$= [-\log_e p/-\log_{10} p]$ nats

$= [\log_p(10)/\log_p(e)]$ nats [that is $\log_a b = 1/(\log_b a)$]

Therefore **1 Hartley** = $\log_e(10)$ nats = **2.303 nats**

Similarly **1 Hartley** = $\log_2(10)$ bits = $[1/\log_{10}2]$ bits = **3.32 bits**

and **1 nat** = $\log_2(e)$ bits = $[1/\log_e(2)$ bits$]$ = **1.443 bits.**

Note: The following base conversions can be used for calculation purposes:

$\log_{10}2 = 0.30103$, $\log_2 10 = 3.3219$

$\log_{10}e = 0.43429$, $\log_e 10 = 2.30259$

$\log_e 2 = 0.69315$, $\log_2 e = 1.44270$

$\log_a X = \log_b X \,.\, \log_a(b)$;

for example $\log_2 X = (\log_{10} X).(\log_2 10)$

$= 3.322(\log_{10} X).$

or $\log_a X = (\log_b X)/\log_b(a)$;

for example $\log_2 X = (\log_{10} X)/(\log_{10} 2)$

$= 3.322(\log_{10} X)$.

As an example let us consider a source S, producing messages $M = \{m_1, m_2, m_3\}$ with $p = \{1/2, 1/4, 1/4\}$, then the message content of each message is

$$I(m_1) = \log(1/p_1) = \log(2) = 1 \text{ bit}$$
$$= (1/3.32) \text{ Hartley}$$
$$= (1/1.443) \text{ nats}$$
$$I(m_2) = \log(1/p_2) = \log(4) = 2 \text{ bits}$$
$$= (2/3.32) \text{ Hartley}$$
$$= (2/1.443) \text{ nats}$$
$$I(m_3) = \log(1/p_3) = \log(4) = 2 \text{ bits}$$
$$= (2/3.32) \text{ Hartley}$$
$$= (2/1.443) \text{ nats.}$$

1.3 Average Information Content of Symbols in Long Independent Sequences: (Entropy)

Now let us assume that the information or message is obtained from an information source. Let us consider a discrete information source producing a sequence of symbols or messages from a fixed set of alphabet, $M = \{m_1, m_2, m_3, \ldots, m_i, \ldots, m_n\}$. Let these symbols be statistically independent of each other and their probability of occurrence do not dependent on each other. So we say that such a source does not have any memory. Such a source is called a **"Memoryless" or "Zero memory" Source** or more precisely a **Discrete Memoryless Source**.

Let us consider a discrete zero memory source. **The average amount of information conveyed by the source, per source symbol or the rate of actual information is defined as the '*Entropy*' of the source**. The entropy of a source is denoted as **H(M)**. Entropy is also known as **measure of unpredictability**.

1.3.1. Computation of Entropy

Let us consider a zero memory discrete information source 'S' producing messages from its alphabet $M = \{m_1, m_2, m_3, \ldots, m_i, \ldots, m_n\}$. Let the probabilities of emitting these alphabet be $p\{m\} = \{p_1, p_2, p_3, \ldots, p_i, \ldots, p_n\}$, and $\sum p_i = 1$. Let us assume that N messages are produced, ($N \to$ infinity).

Then the symbol or message 'm_i' is produced Np_i times and each occurrence produces an information $\{-\log p_i\}$. Hence the total information from 'Np_i' occurrences of the messages $m_i = \{-Np_i \cdot \log p_i\}$. Therefore the total information conveyed by all the N messages is given by

$$I_{total} = -N \sum \{p_i \log [p_i]\} \text{ bits , for } i = 1 \text{ to } n \text{ and } \sum p_i = 1$$

Hence, the Entropy of the source, that is the average information per source symbol, is given by

$$H(M) = I_{total}/N = 1/N\,(-N \sum \{p_i \log [p_i]\}) \text{ bits/message , for } \sum p_i = 1.$$

$$H(M) = -\sum \{p_i \log [p_i]\} \text{ bits/message for } \sum p_i = 1.$$

$$\mathbf{H(M) = \sum \{p_i \log [1/p_i]\} \text{ bits/message}} \text{ for } \sum p_i = 1 \quad (1.3)$$

$$\mathbf{H(M) = \sum \{p_i I\,(m_i)\} \text{ bits/message}} \text{ for } \sum p_i = 1.$$

Thus, the Entropy is associated with the source and the information is associated with the message. H(M) is many times referred as **Marginal Entropy** and $p(m_i)$ as **Marginal probability**. The unit of Entropy will be **bits/message** or **nats/message** or **Hartleys/message** depending on the base of the logarithm used. Also the definition of entropy given in equation (1.3) is based on time-averaging and hence holds good for ensemble averages provided the source is ergodic (for which the ensemble and time averages are identical) . The Entropy is also referred as **"average uncertainty"** or **"average amount of surprise"**.

To understand the concept of Entropy let us consider the following examples:

***Example* 1.1:** Consider a binary source with source alphabet M = $\{m_1, m_2\}$ with probabilities $\{m\}$ = (1/256, 255/256},

Then the entropy $H(M) = \sum \{p_i \log [1/p_i]\}$ for $i = 1$ and 2.

$$= (1/256) \log_2 (256) + (255/256) \log_2 (255/256)$$

$$= \mathbf{0.037 \text{ bits/message.}}$$

Therefore, the entropy is very small and is relatively very easy to guess whether m_1 or m_2 will occur.

***Example* 1.2:** Let M′ = $\{m_3, m_4\}$ with $p\ \{m\} = \{7/16, 9/16\}$,

Then the entropy $H(M) = \sum \{p_i \log [1/p_i]\}$ for $i = 1$ and 2.

$$= (7/16) \log_2 (16/7) + (9/16) \log_2 (16/9)$$

$$= \mathbf{0.989 \text{ bits/message.}}$$

In this case, it is not easy to decide which symbol was produced.

***Example* 1.3:** Let M″ = $\{m_5, m_6\}$ with $p\ \{m\} = \{1/2, 1/2\}$,

Then the entropy $H(M) = \sum \{p_i \log [1/p_i]\}$ for $i = 1$ and 2.

$$= (1/2) \log_2 (2) + (1/2) \log_2 (2)$$

$$= \mathbf{1 \text{ bit/message.}}$$

In this case, the uncertainty is maximum for the binary source; it is not possible to decide which symbol was produced.

From the above examples it is clear about the dependence of Entropy of a source on the probabilities of occurrence of the message symbols.

1.3.2. Properties of Entropy

The entropy, given by expression in equation (1.3) has the following significant properties:

(*a*) ***Lower Bound*:** The Entropy of a source has a lowest value H(M) = 0, when the probability of occurrence of all its symbol except one, are zero.

(*b*) ***Upper Bound*:** The Maximum value of the entropy of a source is obtained when all the symbols produced by the source are equi-probable or equally-likely.

The entropy of a source producing 'n' symbols as given by equation (1.3) is

$$H(M) = -\Sigma \{p_i \log [p_i]\} \text{ for } i = 1 \text{ to } n, \text{ and } \Sigma p_i = 1.$$

$$= -\Sigma \{(1/n) \log [1/n]\} = -n. \{(1/n) \log [1/n]\} \text{ bits/message}$$

$$\mathbf{H(M)_{max} = \log(n) \text{ bits/message symbol}} \quad (1.4)$$

The concept can be made clear by considering the entropies in 'nats'.

$$\log_e(n) - H(M) = \Sigma p_i.\log_e[n] + \Sigma p_i. \log_e [p_i] \text{ for } p_i \neq p_j \text{ and } \Sigma p_i = 1.$$

$$= \Sigma p_i. \log_e [np_i]$$

Considering the inequality,

$$\log_e [1/n] \geq (1-n),$$

We have $\log_e(n) - H(M) \geq \Sigma p_i . [1 - (1/np_i)]$ for $i = 1$ to n;

$$\geq \{\Sigma p_i - (1/n) \Sigma(p_i/p_i)\} \geq (1-1)$$

$$\geq 0$$

and hence $\mathbf{\log_e(n) = H(M)}$ iff $p_i = p_j$

For a binary source which produces (0, 1) with probabilities (p, $1 - p$),

$$H(M) = p \log (1/p) + (1 - p) \log [1/(1 - p)]$$

$$= H(p).$$

H(p) is plotted in Fig. (1.2) below. It is seen from figure that H(p) is a continuous

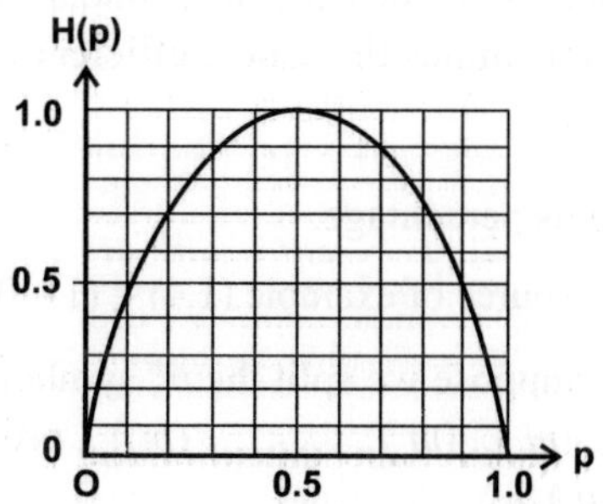

Fig. 1.2. Entropy of Binary Source

function of 'p' and is maximum with H(p) = 1 bit , only when $p = (1 - p) = 0.5$. Also H(p) = 0 for $p = 0$ and $p = 1$. That is ***the entropy of a source is zero both for a least likely symbol as well as for a most certain symbol.***

For a 'n' symbol source, the maximum Entropy $H(M)_{max}$ increases only as log(n) , and for any other $p_i \neq p_j$, the entropy of the source is less. This is illustrated in the following example:

***Example* 1.4:** A source produces four symbols m_1, m_2, m_3, m_4 with p {M} = {1/2, 1/4, 1/8, 1/8} then the entropy of the source is given by

$$H(M) = -\sum p_i \cdot \log_e [p_i] = -[1/2 \log (1/2) + 1/4 \log (1/4) + 1/8 \log (1/8) + 1/8 \log (1/8)]$$

$$= 1.75 \text{ bits/message}$$

If however all symbols were equally likely, *i.e.*, $p(i) = 1/4$, then maximum $H(M) = \log_2(4) = 2$ bits/message.

(*c*) ***Source Efficiency*** **(η_s):** The source efficiency of a source is defined as the ratio of its entropy **H(M)** to the maximum source efficiency $\mathbf{H(M)_{max}}$, that is the ratio of actual rate of information to the maximum possible rate of information and is given by the expression

$$\eta_s = \{H(M) / H(M)_{max}\} \quad (1.5)$$

and can be expressed as percentage.

The efficiency of the source of example (1.4) = (1.75/2) = 0.875 or 87.5 %.

(*d*) ***Source Redundancy*** **($R_{\eta s}$)** : Many times message sources produce symbols that are not absolutely required to convey information. For example, in English language it is not necessary to write the letter 'U' following the letter 'Q' (Quizzes, Queen). The redundancy in English text is estimated to be 50%. That is in the long run half the symbols are not necessary. However, redundancy is required in many applications. For example, in digital communication it is required to include redundancy (parity bits) to achieve error detection and error correction.

When redundancy increases, the actual entropy of the source is reduced. This reduction in entropy is used as the measure of redundancy of a source. The source redundancy is defined as one minus the source efficiency and is expressed as

$$R_{\eta s} = (1 - \eta_s) \quad (1.6)$$

and also can be expressed as percentage.

The redundancy of the source of example (1.4) = (1 – 0.875) = 0.125 or 12.5 %.

(*e*) ***Additive Property*** : Suppose we split the nth symbol m_n into 's' sub-symbols such that $m_n = (m_n)_1, (m_n)_2, (m_n)_3, \ldots\ldots (m_n)_s$, produced with probabilities $(p_n)_1, (p_n)_2, \ldots\ldots, (p_n)_s$,

Such that

$$p_n = (p_n)_1 + (p_n)_2 + \ldots\ldots\ldots + (p_n)_s = \Sigma\,(p_n)_j \text{ for } j = 1 \text{ to } s \tag{1.7}$$

Then the entropy due to the split symbol is $H'(M) = H(p_1, p_2, p_3, \ldots, p_{n-1}, \ldots, (p_n)_1, (p_n)_2, \ldots (p_n)_s)$

$$= \Sigma\,\{p_i \log[1/p_i]\} + \Sigma\,\{(p_n)_q \log[1/(p_n)_q]\}$$

for $i = 1$ to $(n-1)$ and

$q = 1$ to s

$$H'(M) = \Sigma\,\{p_i \log[1/p_i]\} - p_n \log[1/p_n] + \Sigma\,\{(p_n)_q \log[1/(p_n)_q]\}$$

for $i = 1$ to n and $q = 1$ to s

$$= \Sigma\,\{p_i.\log[1/p_i]\} - \Sigma(p_n)_q \log[1/p_n] + \Sigma\,\{(p_n)_q \log[1/(p_n)_q]\}$$

for $i = 1$ to n and $q = 1$ to s using equation (1.7)

$$= \Sigma\,\{p_i.\log[1/p_i]\} + \Sigma(p_n)_q \log[1/(p_n)_q - \log[1/p_n]]$$

$$= \Sigma\,\{p_i.\log[1/p_i]\} + (p_n)\,\Sigma\,(p_n)_q/(p_n)\,\{\log[(p_n)/(p_n)_q]\}$$

for $i = 1$ to n and $q = 1$ to s

$\therefore$ $H'(M) = H(M)$ + a positive quantity since $(p_n)_q \leq p_n$ for all values of 'q'.

$\therefore$ $H'(M) \geq H(M)$

That is splitting of symbols into sub-symbols cannot decrease the entropy. In other words, the entropy is unaltered by the arrangement of events as long as the probabilities of individual events do not change.

1.4 Information Rate

In any communication system, it is desirable to transfer as high a message at as high a speed possible. There may be sources with equal entropies but some may produce more messages per unit time than the other. Therefore, it is not sufficient to describe the entropy of a source alone and it is required to specify the ***"Entropy Rate"*** or ***"Information Rate"***. Considering a source producing message at a rate of 'r' messages per second, the entropy rate or the information rate, R, of the source can be defined as

$$\mathbf{R = r.\,H(M) \text{ bits per second (or 'bps')}} \tag{1.8}$$

That is we are taking here the time element into consideration for measuring the information. That is the information rate is the amount of information produced by a source per unit of time.

***Example* 1.5:** Let us compute the information rate of a source producing a synchronization pattern consisting of 16 symbols, in blocks of four, at a rate of 1000 blocks per second. The first symbol in each block is always the same; the remaining three are filled by any of the 16 symbols with equal probability.

Since, all the symbols are of equal probability, the Entropy of source will be maximum and is given by

$$H = H_{max} = \log_2(16) = 4 \text{ bits/symbol.}$$

Since, the symbols are produced in blocks of four, the total entropy is given by

$$H_T = H_1 + H_2 + H_3 + H_4$$

where $H_1 = 0$, as the first symbol is the same always (that is a definite event) and $H_2 = H_3 = H_4 = 4$ bits/symbol as the second, third and fourth symbols are filled by any of the 16 symbols with equal probability.

$$\therefore \qquad H_T = 0 + 4 + 4 + 4 = 12 \text{ bits/block}$$

Hence, the ***Information rate*** **(R) = r.H$_T$** = (1000 block/sec). (12 bits/block) = 12000 bits/sec.

***Example* 1.6:** Let us assume that a continuously varying analog signal is band limited to 1000 Hz and is sampled at Nyquist Rate. Let the samples are quantized into 4 levels. The quantization levels (or symbols) can be independent and occur with probabilities $p_1 = p_4 = 3/8$ and $p_2 = p_3 = 1/8$. Then the information rate of source can be computed as follows:

$$H(M) = \Sigma p_i . \log [1/p_i] \text{ for } i = 1 \text{ to } 4$$

$$= 2.[(1/8) \log (8)] + 2.[(3/8) \log (8/3)] = 1.8113 \text{ bits/symbol}$$

As the signal is sampled at Nyquist rate, the symbol rate, $r = 2B = 2 \times 1000$ = 2000 symbols/sec.

Therefore, the Information Rate = R = r.H(M)

$$= 2000 \times 1.8113 = 3622.6 \text{ bits/sec.}$$

1.5 Extension of Zero Memory Source

In many communication systems, it is necessary to transform one form of signal into another form for efficient transmission within the available resources (like bandwidth). This is called encoding of signal from one form into another form. For example, in a QPSK modulation scheme for bandwidth efficiency, the information from a binary source S = {0, 1} is grouped as {00, 01, 10, 11} = {Q_1, Q_2, Q_3, Q_4}. Now each grouped symbol, Q_i, consisting of two binary digits, represents one of the 4 different phases of a carrier wave. That is two symbols (bits) of the source S represents one of the phase (new symbol) of the 4 different unique phases.

In information theory, the source producing the symbols {00, 01, 10, 11} is called the second extension, denoted by S^2, of the basic binary source S = {0, 1}. The number of symbols produced by the 2nd extension of a binary source will be = (number of basic source symbols)^2 = 2^2 = 4. In general the number of symbols in the nth extension, S^n, of a binary source S will be equal to (number of basic source symbols)$^n = 2^n$. If S is a zero memory source then its n^{th}extension, S^n, will also be a zero memory source.

In order to have more understanding on the concept of the source extension, let us consider a binary source S, producing symbols m_1 and m_2 with probabilities p_1 and p_2 respectively such that $p_1 + p_2 = 1$.

Then the second extension S^2 of this binary source will have $2^2 = 4$ symbols given by

$m_1 m_1$ with probabilities $p_1\, p_1 = (p_1)^2$

$m_1 m_2$ with probabilities $p_1\, p_2 = (p_1 p_2)$

$m_2 m_1$ with probabilities $p_2\, p_1 = (p_1 p_2)$

and $m_2 m_2$ with probabilities $p_2\, p_2 = (p_2)^2$

The sum of the probabilities of occurrences of the symbols of the source S^2 is

$$= (p_1)^2 + 2\,(p_1 p_2) + (p_2)^2 = (p_1 + p_2)^2 = 1.$$

The entropy of binary source S is given by

$$H(S) = \Sigma p_i \cdot \log[1/p_i] = p_1 \log(1/p_1) + p_2 \log(1/p_2) \text{ for } i = 1 \text{ to } 2 \qquad (1.9)$$

The entropy of the 2nd extended source S^2 is given by

$$\begin{aligned}
H(S^2) &= \Sigma\, p_j \cdot \log[1/p_j] \text{ for } j = 1 \text{ to } 4 \\
&= (p_1)^2 \log[1/(p_1)^2] + (p_1 p_2) \log(1/p_1 p_2) + (p_1 p_2) \log(1/p_1 p_2) + (p_2)^2 \\
&\quad \log[1/(p_2)^2] \\
&= 2\,(p_1)^2 \log[1/p_1] + 2\,(p_1 p_2) \log(1/p_1 p_2) + 2\,(p_2)^2 \log[1/p_2] \\
&= 2\,(p_1)^2 \log[1/p_1] + 2\,(p_1 p_2) \log(1/p_1) + 2\,(p_1 p_2) \log(1/p_2) + \\
&\quad 2\,(p_2)^2 \log[1/p_2] \\
&= 2\,(p_1)\,(p_1 + p_2) \log[1/p_1] + 2\,(p_2)(p_1 + p_2) \log[1/p_2] \\
&= 2\,\{p_1 \log[1/p_1] + p_2 \log[1/p_2]\} \text{ for } p_1 + p_2 = 1. \\
&= 2\,H(S) \text{ using equation (1.9)}
\end{aligned}$$

Therefore $H(S^2) = 2\,H(S)$,

That is the entropy of the 2nd extended source is twice that of the basic source.

In the same way, the 3rd extension of a basic binary source will have $2^3 = 8$ symbols given by

$m_1\, m_1\, m_1$ with probabilities $p_1\, p_1\, p_1 = (p_1)^3$

$m_1\, m_1\, m_2$ with probabilities $= (p_1)^2 p_2$

$m_1 m_2 m_1$ with probabilities $= (p_1)^2 p_2$

$m_1 m_2 m_2$ with probabilities $= (p_1)(p_2)^2$

$m_2 m_1 m_1$ with probabilities $= (p_1)^2 p_2$

$m_2 m_1 m_2$ with probabilities $= (p_1)(p_2)^2$

$m_2 m_2 m_1$ with probabilities $= (p_1)(p_2)^2$

$m_2 m_2 m_2$ with probabilities $= (p_2)^3$

The sum of the probabilities of occurrences of the symbols of the 3rd extended binary source S^3 is

$$= (p_1)^3 + 3(p_1)^2 p_2 + 3p_1(p_2)^2 + (p_2)^3 = (p_1 + p_2)^3 = 1.$$

And the entropy of this 3rd extended binary source can be similarly derived to be

$$\mathbf{H(S^3) = 3\ H(S)} \tag{1.10}$$

So in general, **the nth extended binary source S^n will have 2^n symbols and entropy**

$$\mathbf{H(S^n) = n\ H(S)} \tag{1.11}$$

Proof: Let us consider the source $S = [s_1, s_2, s_3, \ldots, s_q]$ with probability $P = \{p_1, p_2, p_3, \ldots, p_q\}$. The nth extension of this source, S^n, will have q^n symbols, namely $\{\sigma_1, \sigma_2, \sigma_3, \ldots, \sigma_q{}^n\}$. The probabilities of each of these σ_i [for i to q^n] is given by

$$\mathbf{P(\sigma_i) = P(S_{i_1})\, P(S_{i_2})\, P(S_{i_3}) \ldots P(S_{i_n}),}$$

$$\therefore \sum_{s^n} \mathbf{P(\sigma_i) = \Sigma P(S_{i_1})\, P(S_{i_2})\, P(S_{i_3}) \ldots P((S_{i_n}), \text{ for } i = 1 \text{ to } q^n}$$

$$\mathbf{= \Sigma P(S_{i_1}).\ \Sigma P(S_{i_2}).\ \Sigma P(S_{i_3}) \ldots \Sigma P(S_{i_n})}$$

$$\mathbf{\text{for } i_1, i_2, i_3, \ldots i_n = 1 \text{ to } q^n}$$

$$\mathbf{= 1.\ 1.\ 1 \ldots 1}$$

$$\therefore \sum_{s^n} \mathbf{P(\sigma_i) = 1} \tag{1.12}$$

That is the sum of all probabilities of the nth extended source is equal to 1.

The entropy of the nth extended source is given by

$$\mathbf{H(S^n)} = \sum_{s^n} \mathbf{P(\sigma_i) \log[1/P(\sigma_i)]}$$

$$= \sum_{s^n} \mathbf{P(\sigma_i) \log[1/\ P(S_{i_1})\, P(S_{i_2})\, P(S_{i_3}) \ldots P(S_{i_n})]}$$

$$= \sum_{s^n} \mathbf{P(\sigma_i) \log[1/\ P(S_{i_1})] +} \sum_{s^n} \mathbf{P(\sigma_i) \log[1/\ P(S_{i_2})]}$$

$$+ \ldots + \sum_{s^n} \mathbf{P(\sigma_i) \log[1/\ P(S_{i_n})]} \tag{1.13}$$

Let us consider the term

$$\sum_{s^n} P(\sigma_i)\log[1/P(S_{i_1})] = \sum_{s^n} P(S_{i_1})\,P(S_{i_2})\,P(S_{i_3})\ldots P(S_{i_n})\log[1/P(S_{i_1})]$$

$$= \sum_{s^n} P(S_{i_1})\log[1/P(S_{i_1})]\,\{\Sigma P(S_{i_2}).\Sigma P(S_{i_3})\ldots \Sigma P(S_{i_n})\}$$

for $i_1, i_2, i_3, \ldots\ldots\ldots, i_n = 1$ to q^n

$$= H(S).\ 1.\ 1.\ 1\ldots\ldots 1$$

$$\therefore \sum_{s^n} P(\sigma_i)\log[1/P(S_{i_1})] = H(S)$$

Similarly

$$\sum_{s^n} P(\sigma_i)\log[1/P(S_{i_1})] = \sum_{s^n} P(\sigma_i)\log[1/P(S_{i_2})]$$

$$= \{\sum_{s^n} P(\sigma_i)\log[1/P(S_{i_n})]\} = H(S) \qquad (1.14)$$

Using equation (1.14) in equation (1.13), we get

$$H(S^n) = H(S) + H(S) + \ldots\ldots\ldots\ldots\ldots\ldots\ldots + H(S)\ \text{'}n\text{' terms}$$

$$\therefore\ H(S^n) = n\,H(S)$$

1.6 Average Information Content of Symbols in Long Dependent Sequences

We started our discussions on the average information of symbols in long independent sequences, under section (1.3), and considered symbols which are statistically independent of each other and their probability of occurrence do not dependent on each other. We defined sources which produce such sequences as memoryless or zero memory sources. We have also derived expressions for the average information (entropy) and information rate for such sources. However, in reality most of the sources are not memoryless. They produce symbols or sequences that are statistically dependent. For example, in communication systems, the parity symbols added to the information are dependent on the information symbols. Another, well known example is the English text which is fairly predictable. Even if we don't know exactly what is going to come next, we can be fairly certain that, for example, there will be many more *e*'s than *z*'s, or that the combination '*qu*' will be much more common than any other combination with a '*q*' in it and the combination 'th' will be more common than any of them or the occurrence of a consonant implies that the following letter will most probably a vowel and so on. In other words, the entropy of such sources will be fairly low when compared to the zero memory sources. There are statistical models available for such sources with finite memory which produce sequences with interdependent symbols.

1.7 'Mark-off' Statistical Model for Information Source (Sources with Finite Memory)

The real life sources with memory produce symbols which are interdependent. That is the emission of a symbol depends on certain number of previously produced symbols. Specifically the emission of a source symbol s_1 depends on a finite number **'r'** of the previously produced symbols. A source which produces symbols in this manner is known as an **"r^{th} order Mark-off or Markov Source", S_r.** The text in any language is an excellent example of such sources. Since the occurrence of a symbol from this source depends on the occurrences of previous symbols, the probability of a symbol depends on the probability of occurrences of those previous symbols also. Hence, Markov sources are generally specified by a set of conditional probabilities.

Markov sources are commonly used in **communication theory**, as a model of a transmitter. Markov sources also occur in **natural language processing** where they are used to represent hidden meaning in a text. Markov source is also a good example of a redundant source. In a Markov source the occurrence of a symbol x_i in the zero-th position s_0 of a message depends on the previous r symbols $\{s_1, s_2, s_3, \ldots, s_r\}$. The probability of occurrence of x_i is given by the following conditional probabilities:

$$\mathbf{P}\,(\mathbf{x_i} / \mathbf{s_1}, \mathbf{s_2}, \mathbf{s_3}, \ldots, \mathbf{s_r}) \tag{1.15}$$

where x_i is the symbol in the position $\mathbf{s_0}$ and each **s** has a m-symbol alphabet $\{\mathbf{x_1}, \mathbf{x_2}, \mathbf{x_3}, \ldots, \mathbf{x_m}\}$. As the probability of occurrence of $\mathbf{x_i}$ that is $\mathbf{P(x_i)}$ depends on the probability of occurrences of the previous **r** symbols, the transitional probabilities specified in equation (1.15) may be represented by a state diagram. The state diagram will have $\mathbf{m^r}$ possible states previous to $\mathbf{s_0}$ and the behavior of the Markov source can be predicted from this state diagram.

The sequence $(\mathbf{s_1}, \mathbf{s_2}, \mathbf{s_3}, \ldots, \mathbf{s_r})$ represents the state at the beginning of the $(r + 1)^{th}$ symbol interval. The word "state" is used here to know the previous history or residual influence as in modern control theory or sequential switching theory. State diagram is a graphical representation of the Markov source in which the states are denoted by nodes and the transition from state to state is represented by directed line segments. The transitional probabilities and the symbols produced are marked along the line segments of the diagram. The concepts are illustrated in the following example (1.7):

***Example* 1.7:** Let us consider a 2nd order Markov source having four states. The four states may represent the binary digits A = 00, B = 01, C = 10 and D = 11. Let the transitional probabilities of this source $\mathbf{p}\,(\mathbf{x_i}/\mathbf{s_1}, \mathbf{s_2})$, are

$p\,(0/00) = p\,(1/11) = 0.6$

$p\,(1/00) = p\,(0/11) = 0.4$

$p\,(0/01) = p\,(0/10) = p\,(1/01) = p\,(1/10) = 0.5.$

The state diagram of this source is given in Fig. 1.3.

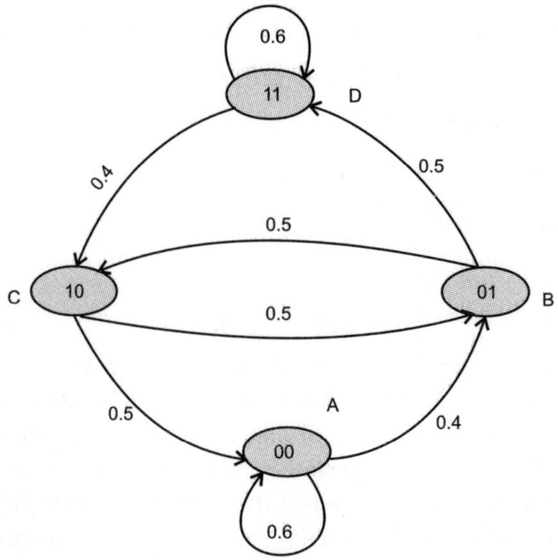

Fig. 1.3. State Diagram of a Second Order Markov Source

From the state diagram, the state equations can be written and the probabilities of the four different states, known as ***the stationary distribution of the states***, can be computed as follows:

The state equations are

$$p(\mathrm{A}) = 0.6\,p(\mathrm{A}) + 0.5\,p(\mathrm{C}) \tag{1.16}$$

$$p(\mathrm{B}) = 0.4\,p(\mathrm{A}) + 0.5\,p(\mathrm{C}) \tag{1.17}$$

$$p(\mathrm{C}) = 0.5\,p(\mathrm{B}) + 0.4\,p(\mathrm{D}) \tag{1.18}$$

$$p(\mathrm{D}) = 0.6\,p(\mathrm{D}) + 0.5\,p(\mathrm{B}) \tag{1.19}$$

From equation, (1.16) we can write that

$$\boldsymbol{p(\mathrm{A}) = 1.25\,p(\mathrm{C})} \tag{1.20}$$

From equations (1.17) and (1.20)

$$\boldsymbol{p(\mathrm{B}) = (0.4)(1.25)\,p(\mathrm{C}) + 0.5\,p(\mathrm{C})}$$

$$\boldsymbol{= p(\mathrm{C})} \tag{1.21}$$

From equations (1.19) and (1.21)

$$\boldsymbol{p(\mathrm{D}) = (0.6)\,p(\mathrm{D}) + 0.5\,p(\mathrm{B})}$$

$$\boldsymbol{= 1.25\,p(\mathrm{C})} \tag{1.22}$$

We know that

$$\boldsymbol{p(\mathrm{A}) + p(\mathrm{B}) + p(\mathrm{C}) + p(\mathrm{D}) = 1} \tag{1.23}$$

Therefore using (1.20), (1.21), (1.22) and (1.23), we have

$$1.25\, p(\mathrm{C}) + p(\mathrm{C}) + p(\mathrm{C}) + 1.25\, p(\mathrm{C}) = 1$$

and $p(\mathrm{C}) = 1/4.5 = 2/9$

$$p(\mathrm{B}) = p(\mathrm{C}) = 2/9$$

$$p(\mathrm{A}) = 1.25\, p(\mathrm{C}) = (5/18)$$

$$p(\mathrm{D}) = 1.25\, p(\mathrm{C}) = (5/18)$$

That is $p(\mathrm{A}) = p(\mathrm{D}) = 5/18$ and $p(\mathrm{C}) = p(\mathrm{B}) = 2/9$.

So for the given transition probabilities at a given time the above source/system can be at the state A or D with a probability of 5/18 and at state B or C with a probability of 2/9. If the source is in state A, then it can produce the symbol A with probability 0.6 and returns to state A or it may produce the symbol B with probability 0.4 and remains at B. When the source is at state B it may produce symbol C with probability 0.5 and return back to B with same probability 0.5 or it may produce symbol D with probability 0.5 and remain at state D. Other state transitions take place similarly as seen in the state diagram. Based on the above probabilities a typical message from the source can be as given below which satisfies the given conditions.

….. BDCB | DDDDDCBCAAAAABCBCBDDDDDCB | DDDD….

The state transitions and symbol generation can be represented using a **"Tree Diagram"**. A tree diagram is a graph in which the nodes represent the states of the source and branches represent the transitions. The tree diagrams help in the easy computation of the probabilities of producing various symbols of the message by a given source.

***Example* 1.8:** To understand the concept of using a tree diagram let us consider the first order Markov source producing three equally likely symbols A, B and C with probabilities of occurrence $p(\mathrm{A}) = p(\mathrm{B}) = p(\mathrm{C}) = 1/3$. Being the first order source, the probability of occurrence of a symbol depends on that particular symbol and the immediate previous symbol. That is the inter symbol dependence exists only for one symbol duration only. Since, the previously produced symbols can be either A, B or C the status of the source can be represented by three states one for each symbol and each state can be denoted by A, B or C.

Let the transition probabilities of source be

$$p(\mathrm{A}\rightarrow\mathrm{A}) = p(\mathrm{B}\rightarrow\mathrm{B}) = p(\mathrm{C}\rightarrow\mathrm{C}) = 0.5$$

$$p(\mathrm{A}\rightarrow\mathrm{B}) = p(\mathrm{B}\rightarrow\mathrm{A}) = p(\mathrm{A}\rightarrow\mathrm{C}) = p(\mathrm{C}\rightarrow\mathrm{A}) = p(\mathrm{C}\rightarrow\mathrm{B}) = p(\mathrm{B}\rightarrow\mathrm{C}) = 0.25.$$

The Tree diagram for this source, representing the initial state (first state at the end of first symbol duration and the second state at the end of second symbol duration) is shown in Figure (1.4). From the diagram we can see that the sequence AB can be produced by either one of the transitions $A \rightarrow A \rightarrow B$ or $B \rightarrow A \rightarrow B$ or $C \rightarrow A \rightarrow B$. Therefore, the probability of source producing the two symbol sequence AB is given by

$$p(AB) = p(A \rightarrow A \rightarrow B) \text{ or } p(B \rightarrow A \rightarrow B) \text{ or } p(C \rightarrow A \rightarrow B)$$

As the three transition paths are mutually exclusive, that is disjoint; we can write the probability of producing the sequence AB as

$$\begin{aligned} p(AB) &= p(A \rightarrow A \rightarrow B) + p(B \rightarrow A \rightarrow B) + p(C \rightarrow A \rightarrow B) \\ &= (1/3)(1/2)(1/4) + (1/3)(1/4)(1/4) + (1/3)(1/4)(1/4) = (1/12) \end{aligned}$$

The same can be obtained using the tree diagram shown in Figure (1.4). The dashed path(s) in the Tree diagram gives the possible ways of producing the sequence 'AB'. So, in general, the probability of source producing a particular sequence is equal to the sum of the products of the probabilities on the Tree diagram along all the paths that gives the particular interested sequence.

Similarly, the probability of the other 'two symbol' sequences can be computed as below:

$$\begin{aligned}
p(AC) &= (1/3)(1/2)(1/4) + (1/3)(1/4)(1/4) + (1/3)(1/4)(1/4) = (1/12) \\
p(AA) &= (1/3)(1/2)(1/2) + (1/3)(1/4)(1/2) + (1/3)(1/4)(1/2) = (1/6) \\
p(BA) &= (1/3)(1/2)(1/4) + (1/3)(1/4)(1/4) + (1/3)(1/4)(1/4) = (1/12) \\
p(BB) &= (1/3)(1/2)(1/2) + (1/3)(1/4)(1/2) + (1/3)(1/4)(1/2) = (1/6) \\
p(BC) &= (1/3)(1/2)(1/4) + (1/3)(1/4)(1/4) + (1/3)(1/4)(1/4) = (1/12) \\
p(CA) &= (1/3)(1/2)(1/4) + (1/3)(1/4)(1/4) + (1/3)(1/4)(1/4) = (1/12) \\
p(CB) &= (1/3)(1/2)(1/4) + (1/3)(1/4)(1/4) + (1/3)(1/4)(1/4) = (1/12) \\
p(CC) &= (1/3)(1/2)(1/2) + (1/3)(1/4)(1/2) + (1/3)(1/4)(1/2) = (1/6)
\end{aligned}$$

The above can be easily traced through the Tree diagram also.

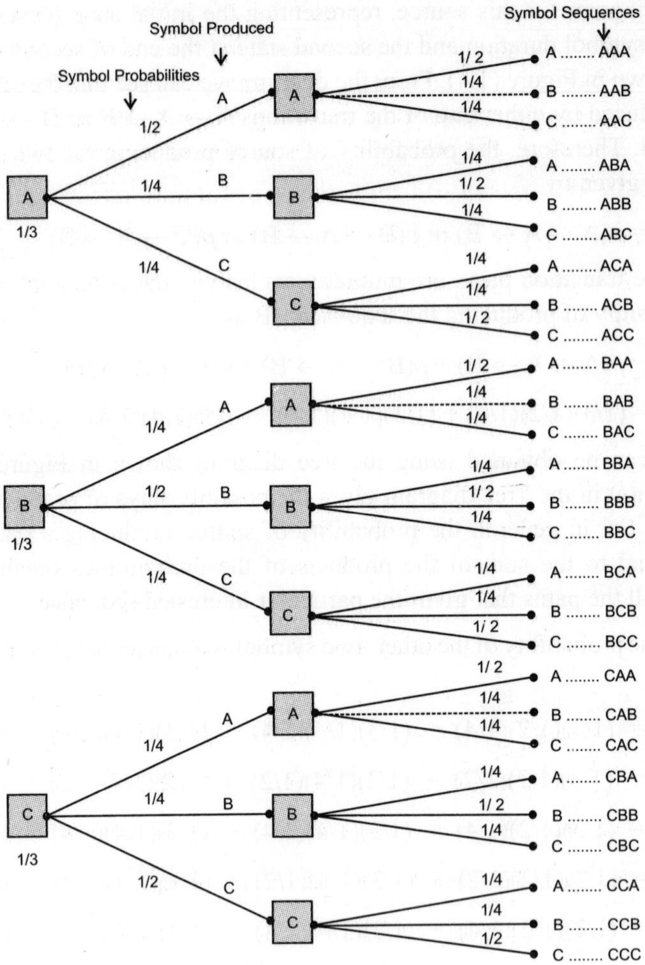

Fig. 1.4. Tree Diagram of the First Order Markov Source

1.8 Entropy of a Mark-off (Markov) Source

The entropy of a Markov source is defined as the weighted average of the symbols produced from each state. The entropy of state 'i' is defined as the average information content of the symbols produced from the ith state and is given by

$$H_i = \sum_{j=1}^{n} p_{ij} \log(1/p_{ij}) \text{ bits/symbol} \tag{1.24}$$

Then the Entropy of the source is the average of the entropy of each state and is given by

$$H = \sum_{i=1}^{n} p_i H_i$$

$$= \sum_{i=1}^{n} p_i \{\sum_{j=1}^{n} p_{ij} \log(1/p_{ij})\} \text{ bits/symbol} \quad (1.25)$$

where p_i is the probability that the source is in the state 'i'.

1.9 Information Rate of a Mark-off (Markov) Source

The average information rate R_s for the source is defined as

$$R_s = r_s H \text{ bits/sec} \quad (1.26)$$

where r_s is the number of state transitions per second, that is the symbol rate of the source.

***Example* 1.9:** Let us find out the entropy of the Markov source of example (1.8)

The probability of each source is $p(A) = p(B) = p(C) = 1/3$ and the transition probabilities are

$$p(A\to A) = p(B\to B) = p(C\to C) = 0.5 = p$$

$$p(A\to B) = p(B\to A) = p(A\to C) = p(C\to A) = p(C\to B) = p(B\to C)$$
$$= 0.25 = p/2$$

The entropy of each state of the source is

$$H(A) = H(B) = H(C) = p \log (1/p) + [(1-p)/2] \log [1/(p/2)] + \{[(1-p)/2] \log[1/(p/2)]\}$$

$$= p \log (1/p) + (1-p) \log(2 / p)$$

Then the Entropy of the source $H(S) = p(A)H(A) + p(B)H(B) + p(C)H(C)$

$$= (1/3) [3 \{p \log (1/p) + [(1-p)] \log [2/p]\}]$$

That is $H(S) = p.\log(1/p) + (1-p). \log [2/(1-p)]$

$$= -p.\log(p) - (1-p) \{\log[(1-p)/2]\}$$

$$= -p.\log(p) - (1-p) \{\log(1-p) - \log(2)\}$$

$$= -p.\log(p) - (1-p) \{\log(1-p) - 1\}$$

$$= -p.\log(p) - (1-p) \log(1-p) + (1-p) \textbf{ bits/symbol}$$

For $p = 0$, $H(S) = 1$ bit/symbol

$p = 1$, $H(S) = 0$ bit/symbol

For the given example $p = 1/2$, hence $H(A) = H(B) = H(C) = p \log (1/p) + (1-p) \log [2/p]$

$$= (1/2) \log (2) + (1/2) \log (4) = 1/2 + 1$$

$$= 1.5 \text{ bits / symbol.}$$

and $H(S) = -p.\log(p) - (1-p)\log(1-p) + (1-p)$

$= (1/2)\log(2) + (1/2)\log(2) + (1/2)$

$= 1.5$ bits / symbol.

1.10 Extension of a Markov Source and Adjoint Source

Like a zero memory source the Markov sources can also be extended. An rth order Markov source can be extended to a Source S^m, with q^m symbols. However, it is seen that the symbols of S^m have reduced inter symbol influence. Thus the mth extension of source S is a μth order Markov source, where $\mu = r/m$ and an rth order Markov source can be converted into the first order Markov source by making $m = r$. The extended source can be obtained completely by specifying appropriate conditional probabilities. Similar to the entropy of the extended zero memory sources, the entropy of the extended first order Markov source S^m is given by

$$H(S^m) = m.H(S) \tag{1.27}$$

Consider a rth order Markov source with source alphabet $S = \{s_1, s_2, s_3, \ldots, s_q\}$. For the first order Markov source, the set of states is the same as the set of source symbols and the probabilities of states namely, $\{p_1, p_2, p_3, \ldots, p_q\}$, gives the unconditional probabilities of the source symbols directly.

An Adjoint source, Š, can be defined as a zero memory source that has the same source alphabet, S, and the first order probabilities $\{p_1, p_2, p_3, \ldots, p_q\}$. Although S and Š have identical probabilities it is seen that

$$H(S) \leq H(\check{S}) \tag{1.28}$$

or $H(\check{S}) - H(S) \geq 0$, independent of '*m*'. Using equations (1.27) and (1.28), it can be shown that

$$H(\check{S}^m) = m\,H(S) + [H(\check{S}) - H(S)] \tag{1.29}$$

Using equations (1.28) and (1.29), we can write $H(\check{S}^m) = m\,H(S) + \varepsilon$, where $\varepsilon \geq 0$.

or $[H(\check{S}^m)/m] = H(S) + (\varepsilon / m)$

and when $m \to \infty$ $[H(\check{S}m)/m] = H(S)$ (1.30)

That is for large values of '*m*', in a Markov source the inter symbol dependence becomes less significant. Hence, considering the sequence σ_k with probabilities $p(\sigma_k)$ and letting

$$G_m = (1/m)\{\Sigma\, p(\sigma_k)\log[1/p(\sigma_k)]\} \tag{1.31}$$

Then $\mathbf{G_m}$ **is a monotonically decreasing function of 'm'** and

When $m \to \infty$ $G_m = H(S)$ bits/sec (1.32)

***Example* 1.10 :** Let us compute the values of G_1, G_2, G_3 for the source of Example (1.8).

From equation (1.31), for m = 1, that is for the first order symbols, the source is a zero memory source and Š produces symbols with equal probability.

Therefore $G_1 = H(\check{S}) = (1/3)\log(3) + (1/3)\log(3) + (1/3)\log(3)$

$$\mathbf{G_1 = H(\check{S}) = 1.585 \text{ bits/symbol.}}$$

We can see from the value of H(S) computed in Example (1.9), $\mathbf{G_1 > H(S)}$.

Using the values of second order symbol probabilities computed in example (1.8) we can calculate the entropy of second order symbols and the value of G_2, as follows:

$$p(AA) = p(BB) = p(CC) = (1/6)$$

$$p(AB) = p(AC) = p(BA) = p(BC) = p(CA) = p(CB) = (1/12)$$

Therefore, the entropy of second order symbols, that is the entropy of states at the end of second symbol interval is given by

$$H(\check{S}^2) = 3\,[(1/6)\log(6)] + 6\,[(1/12)\log(12)]$$

$$H(\check{S}_2) = 3.085 \text{ bits/symbol.}$$

From equation (1.31) for $m = 2$,

$$G_2 = (1/2)\,\{\Sigma\, p(2_k)\log[1/p(2_k)]\} = (1/2)\,H(\check{S}^2) = (1/2)\,3.085 \text{ bits/symbol}$$

$\therefore$ $\mathbf{G_2 = 1.5425}$ **bits/symbol**, which is also **greater than H(S).**

The third order symbol probabilities of source can be traced from the tree diagram shown in Fig. (1.4), and the entropy of the states at the end of the third symbol interval can be computed as follows:

$$H(\check{S}^3) = 3\,[(1/12)\log(12)] + 12\,[(1/24)\log(24)] + 12\,[(1/48)\log(48)]$$

$$= 4.585 \text{ bits/symbol}$$

$$\therefore\ G_3 = \{H(\check{S}^3)/3\} = (4.585/3)$$

i.e, $\mathbf{G_3 = (1.5285)}$ **bits/symbol.**

Thus, we have seen that $G_1 > G_2 > G_3 > H(S)$.

***Example* 1.11:**

(*a*) A Television picture is composed of approximately 3,00,000 basic picture elements. Each of these elements can assume ten distinguishable brightness levels with equal probability. Find the information content of the Television picture frame.

(*b*) An announcer describes the above picture frame orally in 1000 words out of his vocabulary of 10,000 words, each having equal probability of their occurrence in description.

(*i*) Determine the amount of information broadcast by the announcer in describing the picture.

(*ii*) Can we say that the announcer is doing justice to the picture in 1000 words, i.e is he exaggerating or underrating the reality?

(*c*) If 60 frames are generated per second, what is the rate of information generated by this source (in bits/sec).

Solution:

(*a*) Number of picture elements = 3,00,000

Number of distinct brightness levels (with equal probability) = 10.

Information in each picture $I = -\log(p_i) = -\log(1/10)$

$= \log_{10}(10)$ Hartley = **1 Hartley.**

Therefore, information content in a Television picture frame = Information in total number of elements in each frame = 1 × 3,00,000 = 3,00,000 Hartleys.

(*b*) (*i*) Number of words used to describe the picture = 1000.

Number of words in the vocabulary = 10,000.

Therefore, probability of each word = (1/10,000)

Therefore information content in the 1000 words used by the announcer $= \log(1/p) \times 1000 = \log_{10}(10^4) \times 1000 =$ **4000 Hartleys.**

(*ii*) The actual information content of the picture frame is 300,000 Hartleys [from (a)], whereas the information conveyed by the announcer is only 4000 Hartleys.

Therefore, the announcer is not doing the justification to the picture. i.e he is underrating the reality.

(*c*) The rate at which the frames are generated = r = 60 frames/second.

Therefore, information rate = r. H

= Number of frames/second × Number of bits/frame.

$= \{60 \times \sum p_i \log[1/p_i]\}$ Hartleys/sec, for $i = 1$ to 300,000.

$= 60\,[(1/10)\log_{10}(10) \times 300{,}000]$ Hartleys/sec.

= 60 × 30,000 Hartleys/sec

= 60 × 30,000 × 3.32 = 5976000 bits/sec

≈ **6 Mbps.**

PROBLEMS

1.1. The Morse code is composed of 'dots' and 'dashes'. Assume that a dash is 3 times as long as a dot and has one-third of the probability of occurrence.

(*a*) What is the amount of information contained in a dot and in a dash?

(*b*) Compute the entropy of the dot-dash code.

(*c*) Find the 'average rate of information' if the period of the dot is 10 m-seconds; and the time between each symbol is also 10 m-seconds.

1.2. Consider a deck of cards consisting of 52 cards.

(*a*) If the card drawn from the deck is a diamond, how much information does this event convey?

(*b*) How much information it contains if the card drawn is a King?

(*c*) How much information is conveyed if the card drawn is a diamond King?

(*d*) Is the information obtained in (c) is the sum of information conveyed in (a) and (b).

1.3. Let X represents the outcome of a single roll of a fair dice. What is the entropy of X?

1.4. A fair coin is tossed repeatedly.

Let X = {event of getting 3 heads out of 5 trials} and

Y = {event of getting 5 heads out of 8 trials}

(*a*) Find the self information of the two events.

(*b*) Show which event conveys less information.

1.5. An information source produces 128 symbols, 16 of which occur with a probability of 1/32 and the remaining occur with a probability of 1/224. The source produces symbols at a rate of 1000 symbols per second. Assuming that the symbols are produced independently, find the average information rate of this source.

1.6. Let S_1 and S_2 are two zero memory sources. S_1 produces '*n*' symbols with probabilities p_i, for $i = 1, 2, \ldots, n$ and S_2 produces '*m*' symbols with probabilities q_j, for $j = 1, 2, \ldots\ldots, m$. Let $S(\alpha)$ be a new source such that $S(\alpha) = S_1 + S_2$, called a 'mixture' and has $(n + m)$ symbols. Also the first '*n*' symbols of $S(\alpha)$ has probabilities αp_i for $i = 1, 2, \ldots\ldots, n$ and the last '*m*' symbols has probabilities $\bar{\alpha} q_j$, for $j = 1, 2, \ldots\ldots, m$, where $\bar{\alpha} = (1-\alpha)$.

(*a*) Show that the entropy of the new source is $H[S(\alpha)] = \alpha H[S_1] + \bar{\alpha} H[S_2] + H(\alpha)$.

(*b*) Express α_0 in terms of $H[S_1]$ and $H[S_2]$.

(*c*) Show that $H[S(\alpha)]_{max} = H[S(\alpha_0)] = H[S_2] + \log(1 + 2^{H[S_1] - H[S_2]})$.

(*d*) Suppose $m = n$, show that $H[S_1] \leq \sum_{k=1}^{n} p_k \log(1/q_k)$.

1.7. The state diagram of a Mark-off source is shown in figure below. For $q_1 = q_2 = 2/3$

(*a*) Find the stationary (state probability) distribution.

(*b*) Find the entropy of the states.

(*c*) Find the entropy of the source.

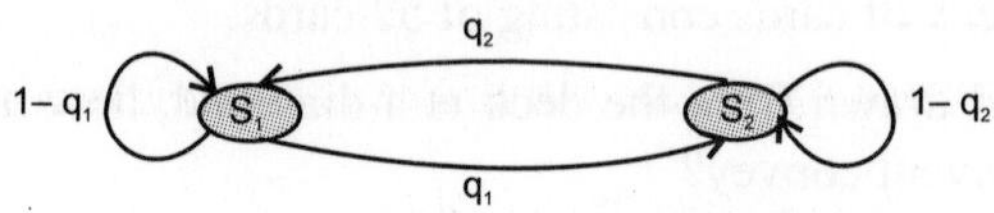

Mark-off Source of Problem 1.7

1.8. For the state diagram of the Mark-off source of problem 1.7, let $q_1 = 3/5$ and $q_2 = 1/6$.

(*a*) Find the entropy H of the source

(*b*) Find G_1, G_2 and G_3. Also verify that $G_1 > G_2 > G_3 > H$.

1.9. Consider the first order Mark-off model shown in figure below.

(*a*) Find the stationary (state probability) distribution.

(*b*) Find the entropy of the states.

(*c*) Find the entropy of the source.

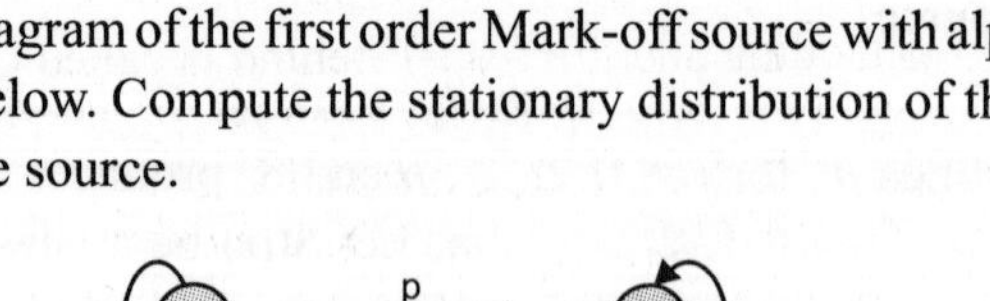

Mark-off Source of Problem 1.9

1.10. The state diagram of the first order Mark-off source with alphabet $S = \{0, 1, 2\}$ is shown below. Compute the stationary distribution of the states, H(S) and $H(S^2)$ of the source.

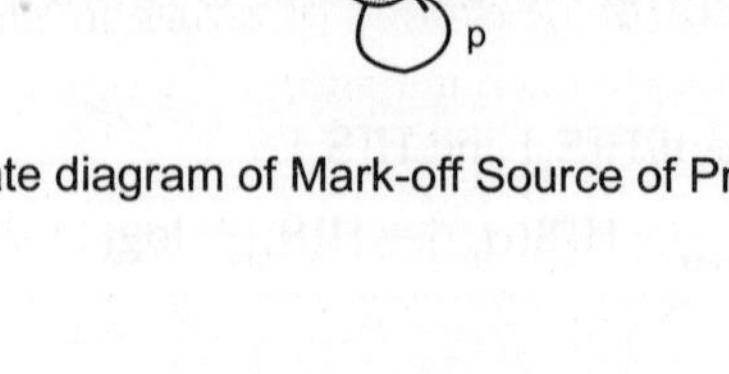

State diagram of Mark-off Source of Problem 1.10

CHAPTER 2

SOURCE CODING

2.0 Introduction to Source Encoding

The purpose of any system, particularly a communication system, will be to transfer information as efficiently as possible. That is the rate of information transfer should be as high as possible. In digital communication systems in which the information is transmitted as sequences of bits, the number of bits required to transmit a source symbol should be as minimum as possible. That is it is required to compress the data from a source for efficient transmission. This practice is found everyday on the Internet where the common Zip data compression is used to reduce the network load and make files smaller. Also removal of all data that superfluous the need of transmitter is required which decreases the bandwidth required for the transmission process. For example, in present day cell phone SMS (Short Message Service) sending we use all short ways of message sending which increases the efficiency. Consider the following SMS message,

"Gm u shd b abl 2 re.d tis ev.ntho. sevrl l.t.rs r mis.ng "

which completely conveys the meaning, "Good morning, you should be able to read this even though several letters are missing". So there are several methods available for efficient transfer of information. This method of transforming one form of information to another form, for the purpose of efficient information transfer is referred to as Coding and the theory behind is called Coding theory.

Coding theory is the study of the properties of codes and their fitness for a specific application. Codes are used for data compression, cryptography, error-correction and recently for network coding. Codes are studied by various scientific disciplines such as information theory, electrical engineering, communication, mathematics, and computer science for the purpose of designing efficient and reliable data transmission methods. This typically involves the removal of redundancy and the correction (or detection) of errors in the transmitted data. There are essentially two aspects to Coding theory:

1. Data compression (or, *source coding*).
2. Error correction (or, *channel coding*).

These two aspects may be studied in combination. The first aspect, source encoding compresses the data for efficient transmission. The second, channel encoding, adds extra data bits to make the transmission of data more robust to disturbances present on the transmission channel. The ordinary user may not be aware of many applications using channel coding. A typical music CD uses the Reed-Solomon code to correct for scratches and dust. In this application the transmission channel is the CD itself. Cell phones also use coding techniques to correct for the fading and noise of high frequency radio transmission. Data modems, telephone transmissions, satellite communication etc., all employ advanced modern channel coding techniques like Turbo codes and LDPC (Low Density Parity Check) codes to get the bits through. In this chapter, we will take up the study on source coding. We will study the basic principle of source coding, their properties and different source coding methods.

2.1 Principles of Source Coding

The aim of source coding is to take the source data and make it smaller/shorter. Basically source codes try to reduce the redundancy present in the source, and represent the source with fewer bits that carry more information. Since, the amount of information from a source is represented by the entropy of a source, the minimum number of bits per second required to represent a source should be equal to the entropy of the source. *In one of the three theorems introduced by Claude E. Shannon for digital communications in 1948, Shannon relates entropy to the minimum number of bits per second required to represent a source without much loss (or distortion).* That is any source coding scheme can, only, achieve a maximum rate that is equivalent to the entropy of the source. Therefore, in principle a source coding or data compression scheme can be achieved by assigning short descriptions to the most frequent outcomes of the data source and long description to the less frequent outcomes. That is a source symbol which is having large probability of occurrence will be represented by less number of bits and the symbols that have low probability of occurrence will be represented by large number of bits. In other words, the number of bits of a code which represents a source symbol is inversely proportional to the probability of occurrence of the symbol. *Hence, the data compression scheme or the source coding technique which explicitly tries to minimize the average length of messages according to a particular assumed probability model is known as* **"Entropy Coding"**.

A formal definition for source coding can be stated as, *Source Coding is the process of transforming the m-ary source alphabets* $S = \{s_1, s_2, s_3, \ldots, s_m\}$ *to a convenient form, say, binary or D-ary digit or code alphabets* $C = \{c_1, c_2, c_3, \ldots, c_d\}$ *to match the channel for efficient transmission.* The purpose of coding is to transform or map all the possible source alphabet of S into sequences of code alphabets in C. Any sequence of alphabets in C is called a "**code word**" and the number of alphabets in a code word is called "**word length**".

***Example* 2.1:** c_1, $c_1\, c_3\, c_4$, $c_3\, c_5\, c_7\, c_8$, $c_1\, c_2\, c_3\, c_3\, c_4$ are code words having word lengths 1, 3, 4 and 5 respectively.

The unit or block in the communication system (as shown in Fig. 2.1) that performs this transformation is called a "**Source Encoder**". In binary coding, the code alphabet will contain only two symbols, namely, C = {0, 1}. Hence, a Binary Source encoder will have only two states and the process of transmission becomes much easier.

Fig. 2.1. Source Encoding

***Note*:** For cases where the source is analog the Source Encoder can be an Analog to Digital Converter and there are powerful compression algorithms that specify all the steps from sampling, quantization, and binary representation. These are referred to as waveform coders. JPEG, MPEG, vocoders are a few examples for image, video, and voice coders, respectively. However, in this chapter, we will take up the study on the discrete information sources.

2.2 Properties of Codes used for Source Coding

In order to avoid ambiguity in decoding the source encoder outputs, to have a one-to-one correspondence between the source alphabet and the code alphabet, as well as, to minimize the time for transmission of information, the codes should have the following properties:

2.2.1. Block Code

The first property a code should have is that, a code should transform each of the source alphabets in S into a "finite sequence" or "code word" using the symbols in the code alphabet C. Such a code is called a Block Code. *That is a Block Code map each of the source symbols in to a finite sequence called the code word. A code can be a "fixed length code" or a "variable length code".*

***Example* 2.2:** Let the source alphabets be S = $\{s_1, s_2, s_3, s_4\}$ and code alphabet C = {0, 1}. Then the corresponding block code can be C_1 ={0, 11, 10, 11}. That is the sequences 0, 11, 10, 11 are the code words for source symbols s_1, s_2, s_3, s_4 respectively.

2.2.2. Non-Singular Code

In example 2.2, we can see that the source alphabets s_2 and s_4 are represented by the same code word, namely "11". That is from the code words we cannot distinguish the corresponding source symbols. In this case, the decoding will be ambiguous. When the codes are non-distinguishable on a simple inspection the code set is called "Singular Code". For non-ambiguous decoding the code set should be distinct and easily distinguishable. That is each source symbol should be represented by a unique code word. Accordingly it is defined that a block code is said to be **"non-singular"** iff all the code words are **"distinct"** and easily **"distinguishable"** from one another.

***Example* 2.3:** The example 2.2 can be modified with the code as $C_1 = \{0, 00, 10, 11\}$. Here each source symbol is uniquely represented by distinct code alphabets and can be decoded without any ambiguity.

An in depth study reveals that even though the codes of example (2.3) appears to be non-singular, while transmitting continuous symbols there may be problems in decoding. For example, if the transmitted sequence is 0011, it may be decoded as $s_1\ s_1\ s_4$ or $s_2\ s_4$. Therefore, there is an ambiguity about the code. Even though the code is non singular when the single symbols are transformed and decoded the code becomes singular for continuous transmission of symbols. That is non-singularity property can guarantee an unambiguous description of a single source symbol only. When we want to transmit a sequence of symbols, one way to ensure unambiguous decoding is by adding a special symbol ("comma") between any two code words. But this is an inefficient use of the special symbol and there are better ways to solve this problem.

2.2.3. Uniquely Decodable Codes

The problem described in the previous section can be solved by using uniquely decodable codes. A non-singular code is uniquely decodable if every code word in a sequence of code words can be uniquely identified or distinguished.

Let us consider the source alphabet $S = \{s_1, s_2, s_3, s_4\}$ and encoded using binary code alphabet $C = \{0, 1\}$. A possible block code can be as given in table below:

Table 2.1: Code Table

Source Symbols	Code C_1
s_1	00
s_2	01
s_3	10
s_4	11

For making the Uniquely Decodable codes concept clear, let us consider the second extension of the Code C_1 of the above table. The extended code is given in table below:

Table 2.2 : Second Extension Code Table for Code C_1

Source Symbols	Code C_1'	Source Symbols	Code C_1'
$s_1 s_1$	0000	$s_3 s_1$	1000
$s_1 s_2$	0001	$s_3 s_2$	1001
$s_1 s_3$	0010	$s_3 s_3$	1010
$s_1 s_4$	0011	$s_3 s_4$	1011
$s_2 s_1$	0100	$s_4 s_1$	1100
$s_2 s_2$	0101	$s_4 s_2$	1101
$s_2 s_3$	0110	$s_4 s_3$	1110
$s_2 s_4$	0111	$s_4 s_4$	1111

From the Table 2.2, it is clear that code C_1' is non-singular and uniquely decodable. Similarly 3rd, 4th, 5th, ... , n^{th} extensions of code C_1' will also be non-singular and uniquely decodable.

Therefore, it can be defined that a *block code is said to be **"uniquely decodable"** iff the nth extension code-words is non-singular for every finite value of 'n'.*

In other words, any encoded string in a uniquely decodable code has only one possible source string producing it. Accordingly the code of Table (2.1) is uniquely decodable, whereas the code given in example (2.3) is not uniquely decodable.

2.2.4. Instantaneous Codes

An ***instantaneous code*** is a uniquely decodable code in which the end of a code word can be identified independently and immediately without the help of the succeeding code word or symbol. Using this code, a symbol can be decoded as soon as we come to the end of the corresponding code word. As a result the decoding is instantaneous and there is no time delay in decoding the received symbol.

The instantaneous code is also known as a ***Prefix code*** (prefix-free code). A code is called a prefix code if no code word is a prefix of any other code word. This property of the code is known prefix property. An instantaneous code is also referred as an ***Irreducible code***. That is a code word cannot be further reduced to get another code word.

Prefix of a code: Consider the code word C = $\{c_1, c_2, c_3, \ldots\ldots, c_D\}$. Then the sequence $c_1, c_2, c_3, \ldots\ldots, c_j$ for all $j \leq$ D are called the "prefixes" of the code 'C'.

The instantaneous and prefix property of a source code can be illustrated using the following example.

Example **2.4:** Consider the following three uniquely decodable codes A, B, C shown in Table (2.3) below:

Table 2.3: Three Uniquely Decodable Codes.

Source Symbols	Code A	Code B	Code C
s_1	00	0	0
s_2	01	10	01
s_3	10	110	011
s_4	11	1110	0111

Code A is the simplest possible uniquely decodable fixed length non-singular code. When code A is used, the decoding can be done immediately after the two code symbols are received, without waiting for any further symbols; that is without referring to the succeeding symbols. Therefore, there is no time delay in decoding. Hence, Code A is an instantaneous code.

Code B is also an instantaneous, uniquely decodable code like Code A. However, it has a special feature that in any of the code word the symbol '0' indicates the end of the code word. Therefore, using this property an imaginary comma can be included at the receiver whenever the symbol '0' is received and the decoding of the code word can be completed. For this reason code B is also called as a **"Comma Code" or "Self-Punctuating Code"**.

When we analyze the structure of Code C, it is clear that even though it is non-singular and uniquely decodable, it is not instantaneous. That is decoding cannot be done word by word independently. We see that every code word starts with the symbol '0' and the code lengths are different. Therefore, the decoding of a received code word can start only after receiving the starting '0' symbol of the next succeeding code word. For example, if we receive the code word '01', we cannot decode it as 's_2' until we receive the next code symbol. If the next received code symbol is '0' then the previous '01'can be decoded as 's_2'. Instead if the next received code symbol is '1' then we are still unable to decode it as 's_3', because the first three symbols of 's_4' is also '011'. Therefore, we have to wait further till we receive the symbol '0' which indicates the start of a code word. That is to decode a received code word we have to wait for the arrival of the succeeding code word. So, there is a definite time delay before a received code word can be decoded. Hence, Code C is not an instantaneous code.

Further we can see in Code C, that the code word for source symbol 's_1', that is '0', is a prefix of code word for 's_2'; and the code word for source symbol 's_2', that is '01', is a prefix of code word for 's_3'; and the code word for source symbol 's_3', that is '011', is a prefix of code word for 's_4'. Because of this we are not in a position to decode the code words as soon as they are received; we have to wait till the start symbol of the next code word is received.

Hence, a necessary and sufficient condition for a source code to be 'instantaneous' is that ***"no complete code word be a prefix of some other code word"***.

This definition of an instantaneous code word enables us the immediate decoding of a received code word. Once a received sequence of symbols matches with a code word the decoding can be done without waiting for further symbols. Since, by definition it is not a prefix of any other code word, the decoding is unique and there will not be any waste of time in decoding.

2.2.5. Optimal Codes

The next property that a source code should have is the optimality of the code length. That is the number of bits used in a code word to represent a source output should be optimum.

An instantaneous code is defined as an **"optimal code"**, if it has "minimum expected or average length", for a specified set of probability of occurrence of the source symbols.

In such codes, as we have mentioned earlier, source symbols with large probability of occurrence will be represented by shorter code words and the symbols that have low probability of occurrence will be represented by longer code words.

***Example* 2.5:** Suppose that a source symbol s_i with probability p_i is transformed into a code word of length l_i and a source symbol s_j with probability p_j is transformed into a code word of length l_j. If $p_i > p_j$ then let $l_i < l_j$. Then the average length of two code words is given by

$$L_1 = p_i l_i + p_j l_j \tag{2.1}$$

In case we interchange the code words, that is transform the symbol 's_i' into the code word of length l_j and symbol s_j into code word of length l_i, then the average length of the code words becomes

$$L_2 = p_i l_j + p_j l_i \tag{2.2}$$

Then $L_2 - L_1 = p_i(l_j - l_i) + p_j(l_i - l_j)$

$$= (p_i - p_j)\,(l_j - l_i) \tag{2.3}$$

Since $(p_i > p_j)$ and $(l_i < l_j)$ it is clear that $(L_2 - L_1)$ is positive and L_1 is shorter than L_2 which is the requirement for optimal codes.

The properties of Source codes are summarized in the form of tree diagram and circle diagram of Figs. 2.2. and 2.3. respectively as given below:

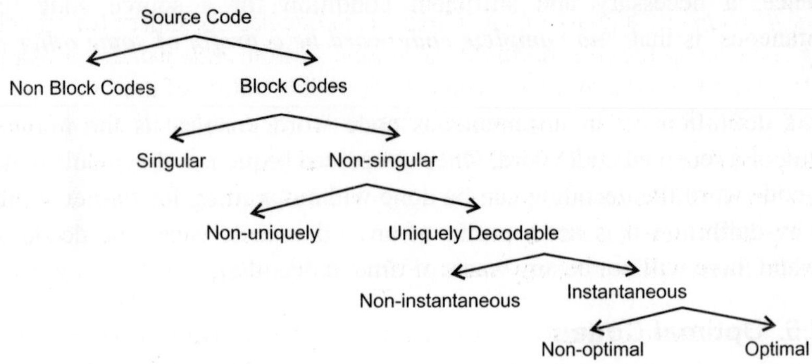

Fig. 2.2. Properties of Source Codes—Tree Codes

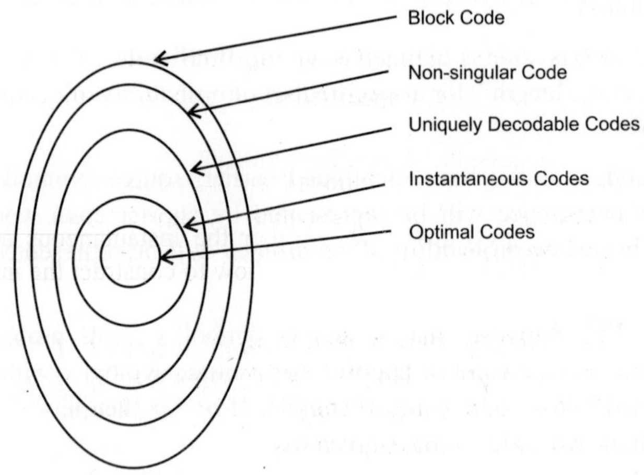

Fig. 2.3. Properties of Source Codes—Circle Diagram

2.3 Kraft Inequality

From the above discussions, it is clear that the aim of any source coding is to construct instantaneous codes of minimum average length to describe a given source. The necessary and sufficient condition for the construction of a code is specified by the **"Kraft Inequality"**.

Theorem: For any instantaneous code over an alphabet of size D, the code lengths $l_1, l_2, \ldots\ldots, l_m$ must satisfy the inequality

$$\sum_{i=1}^{m} D^{(-l_i)} \leq 1 \tag{2.4}$$

where '*m*' is the size of the source alphabet. Equation (2.4) is called the **"Kraft Inequality"** (stated by Kraft in 1949).

Conversely, given a set of code word lengths that satisfy this inequality, there exists an instantaneous code with these word lengths.

Proof*:** Consider a D-ary tree in which each node has D children (or branches). Let the branches of the tree represent the symbols of the ***code word. For example, the D branches from the root represent the D possible values of the first symbol of the code word. Then each code word is represented by a leaf on the ***tree***. The path from the root traces out the symbols of the code word.

The prefix condition on the code words implies that no code word is an ancestor (prefix) of any other code word on the tree. Hence, each code word eliminates its descendants as possible code words. Let l_{max} be the length of the longest code word of the set of code words. Consider all nodes of the tree at level l_{max}. Some of them are code words, some are descendants of codewords, and some are neither. A code word at level l_i has $D^{l_{max}-l_i}$ descendants at level l_{max}. Each of these descendant set must be disjoint. Also, the total number of nodes in these sets must be less than or equal to $D^{l_{max}}$. Hence, summing over all code words, we

have $\sum D^{l_{\max}-l_i} \le D^{l_{\max}}$

or $\sum D^{(-l_i)} \le 1$, which is the **"Kraft Inequality"**.

However this inequality just tells whether the instantaneous code we are seeking will exist or not; and it does not show how to construct the instantaneous code.

Conversely, given any set of code words with lengths l_1, l_2,......, l_m which satisfy the Kraft Inequality, we can always construct a D-ary tree as shown in Fig. 2.4.

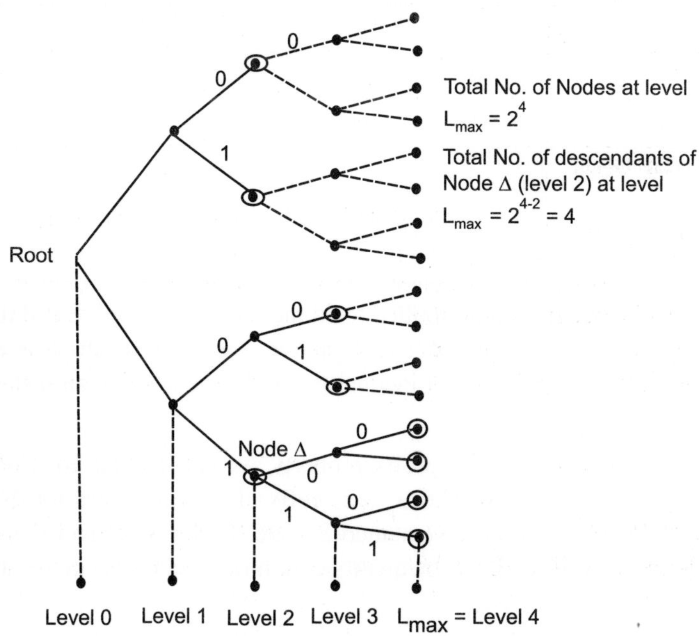

Fig. 2.4. D-ary Tree Satisfying Kraft Inequality

Using this tree a prefix/instantaneous code can be constructed. The procedure is as follows: Label the first node (lexicographically) of length l_1, as code word 1, and move its descendants from the tree. Then label the first remaining node of length l_2 as code word 2, etc. Processing this way, we construct a prefix code with the specified lengths $l_1, l_2, \ldots\ldots, l_m$.

For Binary codes we have D = 2 so that the Kraft inequality of equation (2.4) becomes

$$\sum_{i=1}^{m} 2^{(-l_i)} \leq 1 \tag{2.5}$$

The different aspects of Kraft Inequality are illustrated by the following example (2.6):

***Example* 2.6:** Let us consider a six symbol source encoded into binary codes as shown in the Table (2.4) below:

Table 2.4: Codes to Describe Kraft Inequality Property

Source Symbols	Code A	Code B	Code C	Code D	Code E
s_1	00	0	0	0	0
s_2	01	1000	10	1000	10
s_3	10	1100	110	1110	110
s_4	110	1110	1110	111	1110
s_5	1110	1101	11110	1011	11110
s_6	1111	1111	11111	1100	1111
Average Code length $= \sum_{k=1}^{6} 2^{(-l_k)} \rightarrow$	1	(13/16) < 1	1	(7/8) < 1	(33/32) > 1

The Kraft Inequality test is applied to all the codes A through E and the results are given in the row 8 of the Table (2.4) above. From the results it is clear that Code E does not satisfy the Kraft inequality and hence it is not an instantaneous code.

When we apply the test for prefix property, we can see that for Code D, the complete code word of the symbol s_4 is a prefix of the code word for the symbol s_3. Hence, Code D is not an instantaneous code. Codes A, B and C satisfy the prefix property, as well as, Kraft Inequality conditions and hence are instantaneous codes.

Individual code tree or decision tree can be drawn for each of the above codes. The decision tree for the code C is shown in Fig 2.5. The tree has an initial state and six terminal states corresponding to the six source symbols s_1 to s_6. A node in the tree represents a decision point.

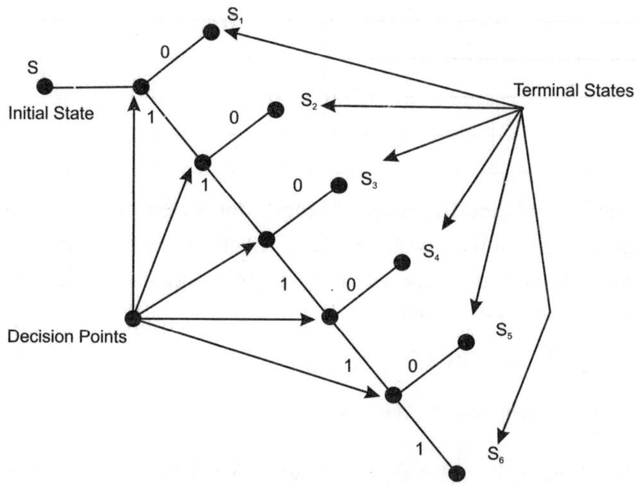

Fig. 2.5. Code (Decision) Tree for Code C of Table (2.4)

The decoding at the receiver end starts always from the initial state in the code tree. If the first received bit is a '0' the decoder moves to the terminal state s_1, instead if it is a binary '1' the decoder moves to the second decision point. Next if the second received bit is a binary '1' the decoder moves to the third decision point; instead if the second received bit is a binary '0' then the decoder moves to the terminal point s_2. This process continues. Once each terminal state is reached and the symbol is identified, the decoder is reset to the initial state before the bit corresponding to the next symbol starts arriving. It is to be noted here that each bit of the received sequence is examined only once.

***Example* 2.7:** In order to understand the working of the decoder let us assume that the received sequence (for Code C in Table 2.4) is "0 110 10 11110 111110.....". Using the decision tree of Fig. 2.5, the first bit is decoded as s_1, the next three bits as s_3, the next two bits as s_2, next five bits as s_5, next five bits as s_6 and the last bit as s_1. After the decoding of each symbol, the decoder is reset to the initial state.

***Example* 2.8:** Now let us find out whether an instantaneous code can exist for a given set of code word lengths with an example. Let the code alphabets be C = {0,1,2}. The number of code words with different word length for three different codes is given in Table (2.5).

Table 2.5: Code Table for Example (2.8)

Number of Code Words with length l_i			Code Length l_i
Code 1	Code 2	Code 3	
2	2	1	1
1	2	4	2
2	2	6	3
4	3	0	4
1	1	0	5

Code 1 consists of 2 code words of length one (ternary) digit each, one code word of length 2 digit each, 2 code words of length 3 digits each, 4 code words of length 4 digits each and one code word of length 5 digits each. Let us see whether this code satisfies the Kraft inequality.

For C = {0, 1, 2}, we have D = 3.

The Kraft inequality for 'Code 1' is

$$\sum_{i=1}^{m} D^{(-l_i)} = 2 \times 3^{-1} + 1 \times 3^{-2} + 2 \times 3^{-3} + 4 \times 3^{-4} + 1 \times 3^{-5} = 0.90535 < 1$$

That is 'Code 1' satisfies the Kraft inequality (as per equation 2.4); and hence can be used to construct an instantaneous code.

The Kraft inequality for 'Code 2' is

$$\sum_{i=1}^{m} D^{(-l_i)} = 2 \times 3^{-1} + 2\times3^{-2} + 2\times3^{-3} + 3 \times 3^{-4} + 1 \times 3^{-5} = 1.0041 > 1$$

That is 'Code 2'does not satisfy the Kraft inequality (as per equation 2.4); and hence it cannot be used to construct an instantaneous code.

The Kraft inequality for 'Code 3' is $\sum_{i=1}^{m} D^{(-l_i)} = 1 \times 3^{-1} + 4 \times 3^{-2} + 6 \times 3^{-3} = 1$

That is 'Code 3' satisfies the Kraft inequality (as per equation 2.4); with equality sign and hence it can be used to construct an instantaneous code with the code word lengths specified.

2.4 Instantaneous Code Construction Procedure

By now we are in a position to construct an instantaneous code for a given set of code word lengths, by applying Kraft Inequality condition and prefix property.

***Example* 2.9:** Let us construct an instantaneous code for a source with alphabets S ={s_1, s_2, s_3, s_4, s_5}. We can assign '0' to the symbol s_1. Therefore 's_1'→ 0.

Hence, all other symbols can be represented by a '1' in the first position of the code words. As '1'at the first position of the code word represents all other symbols we can assign s_2 →10. Similarly we can assign s_3 → 110. This is because we cannot assign '11' to s_3 since there is no other combination of symbols left for s_4 and s_5, which will satisfy the prefix property. Similarly we can assign s_4 → 1110 and s_5 → 1111.

All the codes assigned are arranged in the Table (2.6). Another way of assignment is shown in Table (2.7)

Table 2.6: Code Table for Example 2.9.

Source symbols	Code assigned
s_1	0
s_2	10
s_3	110
s_4	1110
s_5	1111

Table 2.7: Code Table for Example 2.9

Source symbols	Code assigned
s_1	00
s_2	01
s_3	10
s_4	110
s_5	111

2.5 Code Efficiency and Redundancy

In section 2.1, while discussing about the principles of source coding we have mentioned that Claude E. Shannon related entropy of a source to the minimum number of bits per second required to represent a source and that source coding scheme can only achieve a maximum rate that is equivalent to the entropy of the source. However it is found, all the practical source coding schemes produces code words with average length L, which is ≥ the entropy of the source H(M).

The "average length" of the code word produced by a source encoder is given by

$$L = \sum_{i=1}^{m} p_i l_i \text{ binits/message symbol} \tag{2.6}$$

where p_i is the probabilities of occurrences of the source symbols s_i, l_i is the length of the code words representing the source symbols s_i and binits is the short form of Binary Digits representing the word length.

The Entropy of the source as per equation (1.3) is

$$H(M) = \sum_{i=1}^{m} p_i \log(1/p_i) \text{ bits/message symbols} \tag{2.7}$$

For practical systems it is always seen that

$$\mathbf{L \geq H(M)} \text{ for binary code} \quad (2.8)$$

$$\text{and } \mathbf{L \geq H_D(M)} \text{ for D-ary codes} \quad (2.9)$$

where $H_D(M)$ is the entropy in D-ary units/message symbols given by,

$$\mathbf{H_D(M) = \{H(M)/log_2 D\}} \quad (2.10)$$

where H(M) is measured in bits/message symbol and 'D' is the number of different symbols used in code alphabet.

The **Coding Efficiency** is defined as the ratio of the Entropy of the Source to the average length of the code words used to represent the source symbols by the source encoder and is given by

$$\eta_c = \{H(M)/L\} \text{ for binary codes} \quad (2.11)$$

$$\text{and } \eta_c = \{H_D(M)/L\} \text{ for D-ary codes} \quad (2.12)$$

Since L, the average code length of the source encoder, is always greater than or equal to H(M), the entropy of the source, the coding efficiency is always less than or equal to unity. That is $\eta_c \leq 100$ % and the **Coding Redundancy** denoted by $\mathbf{R}_{\eta_c}$ is defined as

$$\mathbf{R}_{\eta_c} = 1 - \eta_c \quad (2.13)$$

***Example* 2.10:** Let a source produce message symbols $M = \{s_1, s_2, s_3, s_4, s_5\}$ with probabilities P = {1/2, 1/4, 1/8, 1/16, 1/16}

$$H(M) = (1/2) \log(2) + (1/4) \log(4) + (1/8)\log(8) + 2 \times (1/16) \log (16)$$
$$= 1.875 \text{ bits/symbol.}$$

For the code of Table (2.6) the average length L using equation (2.6) is

$$L = \sum_{i=1}^{5} p_i l_i = p_1 l_1 + p_2 l_2 + p_3 l_3 + p_4 l_4 + p_5 l_5$$
$$= (1/2) \times 1 + (1/4) \times 2 + (1/8) \times 3 + 2 \times (1/16) \times 4$$
$$= 1.875 \text{ binits/message symbol.}$$

Using equation (2.11) the coding efficiency = η_c = H(M)/L = 1.875/1.875 =1

$$= 100 \text{ \%.}$$

And coding redundancy $R_{\eta_c} = 1 - \eta_c = 0$

$$= 0 \text{ \%.}$$

Similarly the average length for the code of Table (2.7) is

$$L = \sum_{i=1}^{5} p_i l_i = p_1 l_1 + p_2 l_2 + p_3 l_3 + p_4 l_4 + p_5 l_5$$
$$= (1/2) \times 2 + (1/4) \times 2 + (1/8) \times 2 + 2 \times (1/16) \times 3$$
$$= 2.125 \text{ binits/message symbol}$$

The efficiency of this code = η_c = H(M)/L = 1.875/2.125 = 0.8824

That is η_c = 88.24 %.

And coding redundancy $R_{\eta_c} = 1 - \eta_c = 1 - 0.8824 = 0.1176$

R_{η_c} = 11.76 %.

From this example, it is clear that the code of Table (2.7) is less efficient and poor way of encoding the source symbols when compared to the code of Table (2.6). Therefore, by proper encoding the coding efficiency can be increased and the redundancy can be reduced.

2.6 Shannon's First Theorem – Noiseless Coding Theorem

In information theory, **Shannon's source coding theorem** (or **noiseless coding theorem**) establishes the limits to possible data compression, and the operational meaning of the Shannon entropy. The **source coding theorem** shows that it is not possible to compress the data such that the code rate (average number of bits per symbol) is less than the Shannon entropy of the source, otherwise information will be lost. However, it is possible to get the code rate arbitrarily close to the Shannon entropy, with negligible probability of loss. The **source coding theorem for symbol codes** places an upper and a lower bound on the minimal possible expected (average) length of code words as a function of the entropy of the information source and the size of the code word length. **The first part of theorem says that no matter how we compress, we will need to send at least 'nH(M)' bits for a message of 'n' symbols.**

2.6.1. Shannon's Statement

For an information source of symbol alphabets M which are encoded into a D-ary code, the entropy of the source H(M) and the average code length L are related as

$$[H(M)/\log_2(D)] \leq L \leq [H(M)/\log_2(D)] + 1 \qquad (2.14)$$

***Proof*:** As per Shannon the length l_i of a code word is given by

$$l_i = \log_D(1/p_i) \qquad (2.15)$$

This is as per the concept that a source symbol with larger probability will be represented by a shorter code word and a source symbol with smaller probability will be represented by a longer code word. l_i calculated as per equation (2.15) may not be an integer always. In order to have l_i an integer we can have

$$\log_D(1/p_i) \leq l_i \leq \log_D(1/p_i) + 1 \qquad (2.16)$$

Equation (2.16) also satisfies Kraft inequality, since the left inequality gives $(1/p_i) \leq (D)^{l_i}$, or $(p_i) \geq (D)^{-l_i}$ from which we can write

$$\sum_{i=1}^{m} p_i = 1 \geq \sum_{i=1}^{m} (D)^{-l_i}$$

Also as we have $\log_D(1/p_i) = [\log(1/p_i)/\log(D)]$, equation (2.16) can be written as

$$[\log(1/p_i)/\log(D)] \leq l_i \leq [\log(1/p_i)/\log(D)] + 1 \qquad (2.17)$$

Multiplying equation (2.16) by p_i and summing for all values of 'i' we have

$$\sum_{i=1}^{m} [p_i \log_D(1/p_i)/\log(D)] \leq \sum_{i=1}^{m} p_i l_i \leq \sum_{i=1}^{m} [p_i \log_D(1/p_i)/\log(D)] + \sum_{i=1}^{m} p_i$$

Or $$[H(M)/\log_2(D)] \leq L \leq [H(M)/\log_2(D)] + 1 \qquad (2.17a)$$

From equation (2.10) we have $H_D(M) = \{H(M)/\log_2 D\}$; substituting this in the above equation

$$H_D(M) \leq L \leq H_D(M) + 1 \qquad (2.18)$$

Therefore, $L \geq H_D(M)$ (2.19)

Equation (2.19) is the same as equation (2.9) and hence equation (2.9) is proved.

In order to have better efficiency normally we use the nth extension of the source alphabets and the new average code word length for this extension can be denoted by L_n. Since equation (2.14) or (2.17a) holds good for a zero memory source, it is valid for the extended source also. Hence we have

$$[H(M^n)/\log(D)] \leq L_n \leq [H(M^n)/\log(D)] + 1 \qquad (2.20)$$

Since, $H(M^n) = n\,H(M)$ equation (2.20) becomes

$$[H(M)/\log(D)] \leq (L_n/n) \leq [H(M)/\log(D)] + (1/n) \qquad (2.20a)$$

or $$[H_D(M)] \leq (L_n/n) \leq [H_D(M)] + (1/n) \qquad (2.21)$$

Therefore, when 'n' tends to infinity, we have

$$\lim_{n\to\infty}\left(\frac{L_n}{n}\right) = [H(M)/\log(D)] = [H_D(M)] \qquad (2.22)$$

When L is the average length of the code word for the basic source, for the extended source the ratio $(L_n / n) \neq L$ and in general it is found that

$$(L_n/n) \leq L \qquad (2.23)$$

Equation (2.21) is called ***Shannon's first theorem*** or ***the "Noiseless Coding theorem"***. This is called so because we have not considered the effects of channel noise. This is because the importance here is to study about the most efficient source encoding techniques for maximum coding efficiency. Even though the theorem is proved for zero memory sources, it is equally valid for finite memory that is Markov sources also.

To summarize the **Shannon's first theorem** or **Shannon's fundamental theorem** or the "**Noiseless Coding theorem**" can be stated as "*for a code alphabet with 'D' symbols and source alphabet of 'm' symbols, the average length of code words can be as close as the entropy of the source, $H_D(M)$, by increasing the extension*".

In other words, "by increasing the value of '*n*' in equation (2.22) the value (L_n/n) can be reduced to the maximum extend as we require, but cannot be below the value of $H_D(M)$". That is by increasing '*n*', the coding efficiency ($\eta_c = \{H_D(M)/L\}$, and for n^{th} extension $\eta_c = \{H_D(M)/(L_n/n)\}$) can be increased to the maximum extend (however to a value $\leq 100\%$). That is when '*n*' is increased to infinity the code rate reaches the maximum value equal to the entropy of the source.

In order to understand the concept let us consider some more examples. Shannon's lossless source coding theorem in fact is based on the concept of block coding. Let us consider a special information source in which the alphabet consists of only two letters or symbols *i.e* S = {*a*, *b*}. An nth order block code is just a mapping which assigns to each block of '*n*' consecutive symbols a sequence of bits of varying length. The following illustrate this concept.

***Example* 2.11:**

1. **First-Order Block Code:** Each symbol is mapped to a single bit.

B_1	$p(B_1)$	Code word
A	0.5	0
B	0.5	1

An example:

$$\left\{\begin{array}{l}\text{Original Data}: \underline{a}\ \underline{a}\ \underline{a}\ \underline{a}\ \underline{a}\ \underline{a}\ \underline{a}\ \underline{b}\ \underline{b}\ \underline{b}\ \underline{b}\ \underline{b}\ \underline{b}\ \underline{b}\ \underline{b}\ \underline{b}\ \underline{b}\ \underline{b}\ \underline{b}\ \underline{a}\ \underline{a}\ \underline{a}\ \underline{a} \\ \text{Encoded data}: 0\,0\,0\,0\,0\,0\,0\,1\ 1\ 1\ 1\ 1\ 1\ 1\ 1\ 1\ 1\ 1\ 1\ 0\ 0\ 0\ 0\end{array}\right\}$$

Here 24 bits are used to represent 24 symbols --- an average of **1 bit/symbol**.

2. **Second-Order Block Code**: Pairs of symbols are mapped to one, two, or three bits.

B_2	$p(B_2)$	Code word
aa	0.45	0
ab	0.45	10
ab	0.05	110
ba	0.05	111

***An example*:**

Original Data: aa aa aa ab bb bb bb bb bb bb aa aa

Encoded data: 0 0 0 110 10 10 10 10 10 10 0 0

Here 20 bits are used to represent 24 symbols --- an average of **0.83 bits/ symbol.**

3. **Third-Order Block Code**: Triplets of symbols are mapped to bit sequence of lengths one through six.

B_3	$p(B_3)$	Code word
aaa	0.405	0
bbb	0.405	10
aab	0.045	1100
abb	0.045	1101
bba	0.045	1110
baa	0.045	11110
aba	0.005	111110
bab	0.005	111111

***An example*:**

- **Original Data**: aaa aaa abb bbb bbb bbb bba aaa

  Encoded data: 0 0 1101 10 10 10 1110 0

Here 17 bits are used to represent 24 symbols --- an average of **0.68 bits/ symbol**.

Note :

- The rates (R) or the average length of the code in bits per symbol shown in the above tables are calculated from

  R = (1/*n*) Σp(B_n) *l*B(n) bits/symbol

  Where *l*B(n) is the length of the code word for block B_n
- The higher the order, the lower the rate (i.e., better compression).

2.6.2. The Gap between Theory and Practice

Most of the times we know that it is not possible to achieve practically, whatever we derive based on theory. Increasing the value of '*n*' for the purpose of increasing the coding efficiency results in increased coding complexity of the encoder, as large number of source symbols are to be handled and in the increased time required for encoding and transmitting the symbols. As per as the noiseless coding is concerned, we will see in this section, what are the gaps between theory and practice.

- The theory holds when the block length n approaches infinity. In real-time source encoding or compression, the compression algorithm must wait for n consecutive source samples before it can begin the compression. When n is large, this wait (or delay) may be too long. For example, in real-time speech compression, the speech signal is sampled at 8000 samples/second. If n is, say, 4000, then the compression delay is half a second. In a two-way conversation, this long delay may not be desirable.
- The theory does not take into consideration the complexities associated with the compression and decompression operations. Typically, as the block length n increases, the complexities also increase. Often, the rate of increase is exponential in n.
- The theory assumes that the statistical property of the source is known. In practice, this information may not be available.
- The theory assumes that there is no error in the compressed bit stream. In practice, due to noise in the communication channel or imperfection in the storage medium, there are errors in the compressed bit stream.

However, all of these problems have been successfully solved by researchers in the field over a period of time.

2.7 Source Coding Techniques

In the preceding sections, we have taken up the study on the properties of the codes, bounds on the code lengths and the Shannon's first fundamental theorem. In this section, we will take up the study on certain important source coding techniques.

2.7.1. Shannon's Encoding Algorithm

The Shannon's encoding algorithm is a basic and important source coding algorithm and is easy to use and generates fairly efficient codes. Even though it is an easy algorithm it is tedious when done manually. There are five encoding steps. These steps are illustrated below:

1. The given source symbols are arranged in the order of decreasing probability of occurrences.

   $S = \{s_1, s_2, s_3, \ldots\ldots, s_m\}$ with probabilities

   $P = \{p_1, p_2, p_3, \ldots\ldots, p_m\}$; and $p_1 \geq p_2 \geq p_3 \ldots\ldots \geq p_m$

2. The following sequences are computed

   $a_1 = 0$

   $a_2 = p_1 + 0 = p_1 + a_1,$

   $a_3 = p_2 + p_1 = p_2 + a_2,$

   $a_4 = p_3 + p_2 + p_1 = p_3 + a_3$

....

....

$$a_{m+1} = p_m + p_{m-1} + \dots\dots\dots\dots + p_2 + p_1 = p_m + a_m$$

3. A set of integers, l_i, which are the smallest integer solution of the inequalities are determined.

   $2^{l_i} p_i \geq 1$ for $i = 1, 2, 3, \dots\dots\dots, m$

4. The decimal numbers 'a_i' is expanded in binary form to l_i places.
5. The decimal point is removed to get the required code.

***Example* 2.12:** Encode the following source symbols using Shannon's (binary) encoding algorithm. **S = {s_1, s_2, s_3, s_4}** with probabilities **P = {(2/5), (3/10), (1/5), (1/10)}**; also compute the coding efficiency and redundancy.

Solution:

***Step* 1:** $p_1 \geq p_2 \geq p_3 \dots\dots \geq p_m \rightarrow (2/5) > (3/10) > (1/5) > (1/10) \rightarrow (4/10) > (3/10) > (2/10) > (1/10)$

***Step* 2:** Sequences of a's are computed.

$a_1 = 0$
$a_2 = (4/10)$
$a_3 = (4/10) + (3/10) = (7/10)$
$a_4 = (7/10) + (2/10) = (9/10)$
$a_5 = (9/10) + (1/10) = 1.0$

***Step* 3:** The smallest integer value of l_i are determined

(*i*) $2^{l_i} p_i \geq 1$; $2^{l_1} \geq (1/p_1)$; $2^{l_1} \geq (10/4) \geq 2.5$; i.e $l_1 = 2$

(*ii*) $2^{l_2} \geq (1/p_2)$; $2^{l_2} \geq (10/3) \geq 3.3$; i.e $l_2 = 2$

(*iii*) $2^{l_3} \geq (1/p_3)$; $2^{l_3} \geq (10/2) \geq 5.0$; i.e $l_3 = 3$

(*iv*) $2^{l_4} \geq (1/p_4)$; $2^{l_4} \geq (10/1) \geq 10.0$; i.e $l_4 = 4$

***Step* 4:** The decimal numbers 'a_i' are expanded in binary form to 'l_i' places.

$\mathbf{a_1 = 0}$

$\mathbf{a_2 = (0.4)_{10}}$; $0.4 \times 2 = 0.8$ with carry '0'

$0.8 \times 2 = 0.6$ with carry '1'

$0.6 \times 2 = 0.2$ with carry '1'. Therefore $\mathbf{a_2 = (0.4)_{10} = (0.011)_2}$

Similarly

$a_3 = (0.7)_{10} = (0.10110)_2$
$a_4 = (0.9)_{10} = (0.111001)_2$

***Step* 5:** The codes are

$a_1 = (0)_2$ and $l_1 = 2$; $s_1 \rightarrow 00$
$a_2 = (0.011)_2$ and $l_2 = 2$; $s_2 \rightarrow 01$

$a_3 = (0.1011)_2$ and $l_3 = 3$; $s_3 \rightarrow 101$

$a_4 = (0.111001)_2$ and $l_4 = 4$; $s_4 \rightarrow 1110$

The average length of code is computed using equation (2.6)

$$L = \sum_{i=1}^{m} p_i l_i \text{ binits/message symbol}$$

$$= 2 \times (2/5) + 2 \times (3/10) + 3 \times (1/5) + 4 \times (1/10)$$

$$= 2.4 \text{ binits/message symbol}$$

The entropy of source as per equation (2.7) is

$$H(M) = \sum_{i=1}^{5} p_i \log(1/p_i) \text{ bits/message symbols}$$

$$= (2/5)\log(5/2) + (3/10)\log(10/3) + (1/5)\log(5) + (1/10)\log(10)$$

$$= 1.84644 \text{ bits/symbol.}$$

Therefore, coding efficiency using equation (2.11) is

$\eta_c = \{H(M)/L\} = (1.84644/2.4) = 0.76935$; i.e $\eta_c =$ **76.935 %**

Redundancy $R_{\eta_c} = 1 - \eta_c = 1 - 0.76935 = 0.23065$; i.e $R_{\eta_c} =$ **23.065 %**

The results can be put in the form of a table

Table 2.8: Code Table for Example (2.12)

Source Symbol	p_i	Binary Code	Code length 'l_i' in binits
s_1	2/5	00	2
s_2	3/10	01	2
s_3	1/5	101	3
s_4	1/10	1110	4

From the table, we can see that the code is meeting the prefix property, i.e no code word is a prefix of any other code word and hence the code is an instantaneous code.

***Example* 2.13:** Consider a source with source alphabets **S = {s$_1$, s$_2$, s$_3$}** with probabilities **P = {0.5, 0.3, 0.2}**. Let us apply Shannon's binary encoding algorithm and compute the coding efficiency and redundancy. Also let us apply the same technique for the second order extension of this source and see the improvement in the code efficiency.

Solution: (i)

***Step* 1:** **S = {s$_1$, s$_2$, s$_3$}**with probabilities **P = {p$_1$, p$_2$, p$_3$} = {0.5, 0.3, 0.2}**;

***Step* 2:** The sequences of "*a*'*s*" are

$a_1 = 0$
$a_2 = 0.5$
$a_3 = 0.5 + 0.3 = 0.8$
$a_4 = 0.8 + 0.2 = 1.0$

***Step* 3:** The smallest integer values of code lengths 'l_i'are

(*i*) $2^{l_1}p_i \geq 1$; $2^{l_1} \geq (1/p_1)$; ; $2^{l_1} \geq (1/0.5) \geq 2$; i.e $l_1 = 1$ binit

(*ii*) $2^{l_2} \geq (1/p_2)$; $2^{l_2} \geq (1/0.3) \geq 3.33$; i.e $l_2 = 2$ binits

(*iii*) $2^{l_3} \geq (1/p_3)$; $2^{l_3} \geq (1/0.2) \geq 5.0$; i.e $l_3 = 3$ binits

***Step* 4:** The decimal numbers 'a_i' are expanded in binary form to l_i places.

$a_1 = 0$
$a_2 = (0.5)_{10}$; $0.5 \times 2 = 0.0$ with carry '1'

Therefore $a_2 = (0.5)_{10} = (0.1)_2$

Similarly

$a_3 = (0.8)_{10} = (0.110....)_2$

***Step* 5:** The codes are

$a_1 = (0)_2$ and $l_1 = 1$; $s_1 \rightarrow 0$; $a_2 = (0.1)_2$ and $l_2 = 2$; $s_2 \rightarrow 10$
$a_3 = (0.110...)_2$ and $l_3 = 3$; $s_3 \rightarrow 110$

Average length $\quad L = \sum_{i=1}^{m} p_i l_i$ binits/message symbol

$= 1\times (0.5) + 2 \times (0.3) + 3 \times (0.2)$

$=$ **1.7 binits/message symbol**

The entropy of source is

$$H(M) = \sum_{i=1}^{3} p_i \log (1/p_i) \text{ bits/message symbols}$$

$= (0.5)\log(2) + (0.3)\log(10/3) + (0.2)\log(5) =$ **1.4855 bits/symbol.**

Therefore,

Coding efficiency is $\eta_c = \{H(M)/L\} = (1.4855/1.7) = 0.8738$; i.e $\eta_c = 87.38\%$

Redundancy is $R_{\eta_c} = 1 - \eta_c = 1 - 0.8738 = 0.1262$; i.e $R_{\eta_c} =$ **12.62%**

(ii) Second Extension:

The 2nd extension of basic source will have $3^2 = 9$ symbols. The symbols and their probabilities are given by

$$\left.\begin{matrix} S_1S_1 & S_1S_2 & S_1S_3 & S_2S_1 & S_2S_2 & S_2S_3 & S_3S_1 & S_3S_2 & S_3S_3 \\ 0.25 & 0.15 & 0.10 & 0.15 & 0.09 & 0.06 & 0.10 & 0.06 & 0.04 \end{matrix}\right\}$$

***Step* 1:** For 2nd extension source, the symbols in the decreasing order of probability

S_1S_1	S_1S_2	S_2S_1	S_1S_3	S_3S_1	S_2S_2	S_2S_3	S_3S_2	S_3S_3
0.25	0.15	0.15	0.10	0.10	0.09	0.06	0.06	0.04
p_1	p_2	p_3	p_4	p_5	p_6	p_7	p_8	p_9

***Step* 2:** The sequences of "*a's*" are

$a_1 = 0;$
$a_2 = 0.25;$
$a_3 = 0.4;$
$a_4 = 0.55;$
$a_5 = 0.65$;
$a_6 = 0.75;$
$a_7 = 0.84;$
$a_8 = 0.90;$
$a_9 = 0.96;$
$a_{10} = 1.$

***Step* 3:** The smallest integer values of code word lengths, 'l_i' are

(*i*) $2^{l_1} \geq (1/p_1)$; $2^{l_1} \geq (1/0.25) \geq 4$; i.e $l_1 = 2$ binits
(*ii*) $2^{l_2} \geq (1/p_2)$; $2^{l_2} \geq (1/0.15) \geq 6.67$; i.e $l_2 = 3$ binits
(*iii*) $2^{l_3} \geq (1/p_3)$; $2^{l_3} \geq (1/0.15) \geq 6.67$; i.e $l_3 = 3$ binits
(*iv*) $2^{l_4} \geq (1/p_4)$; $2^{l_4} \geq (1/0.10) \geq 10.0$; i.e $l_4 = 4$ binits
(*v*) $2^{l_5} \geq (1/p_5)$; $2^{l_5} \geq (1/0.10) \geq 10.0$; i.e $l_5 = 4$ binits
(*vi*) $2^{l_6} \geq (1/p_6)$; $2^{l_6} \geq (1/0.09) \geq 11.11$; i.e $l_6 = 4$ binits
(*vii*) $2^{l_7} \geq (1/p_7)$; $2^{l_7} \geq (1/0.06) \geq 16.67$; i.e $l_7 = 5$ binits
(*viii*) $2^{l_8} \geq (1/p_8)$; $2^{l_8} \geq (1/0.06) \geq 16.67$; i.e $l_8 = 5$ binits
(*ix*) $2^{l_9} \geq (1/p_9)$; $2^{l_9} \geq (1/0.04) \geq 25.00$; i.e $l_9 = 5$ binits

***Step* 4:** The decimal numbers 'a_i' are expanded in binary form to l_i places.

$a_1 = (0)_{10}$.

Therefore, $a_1 = (0)_2$

$a_2 = (0.25)_{10}$; $0.25 \times 2 = 0.5$ with carry '0'
$0.50 \times 2 = 0.0$ with carry '1'

Therefore, $a_2 = (0.25)_{10} = (0.01)_2$

Similarly

$a_3 = (0.40)_{10} = (0.011....)_2$
$a_4 = (0.55)_{10} = (0.1000...)_2$

$a_5 = (0.65)_{10} = (0.1010...)_2$
$a_6 = (0.75)_{10} = (0.11)_2$
$a_7 = (0.84)_{10} = (0.11010..)_2$
$a_8 = (0.90)_{10} = (0.11100..)_2$
$a_9 = (0.96)_{10} = (0.11110..)_2$

***Step* 5:** The codes are

$a_1 = (0)_2$ and $l_1 = 2; s_1s_1 \rightarrow 00$
$a_2 = (0.01)_2$ and $l_2 = 3; s_1s_2 \rightarrow 010$
$a_3 = (0.011..)_2$ and $l_3 = 3; s_2s_1 \rightarrow 011$
$a_4 = (0.1000..)_2$ and $l_4 = 4; s_1s_3 \rightarrow 1000$
$a_5 = (0.1010..)_2$ and $l_5 = 4; s_3s_1 \rightarrow 1010$
$a_6 = (0.11)_2$ and $l_6 = 4; s_2s_2 \rightarrow 1100$
$a_7 = (0.11010..)_2$ and $l_7 = 5; s_2s_3 \rightarrow 11010$
$a_8 = (0.11100..)_2$ and $l_8 = 5; s_3s_2 \rightarrow 11100$
$a_9 = (0.11110..)_2$ and $l_9 = 5; s_3s_3 \rightarrow 11110$

The average length L_2 is computed as

$$L_2 = \sum_{i=1}^{9} p_i l_i$$

$$= 2 \times (0.25) + 3 \times (0.15) + 3 \times (0.15) + 4 \times (0.1) + 4 \times (0.1) + 4 \times (0.09) + 5 \times (0.06) + 5 \times (0.06) + 5 \times (0.04)$$

= **3.36 binits/symbol**

The entropy of second extended source is $\mathbf{H(M^2) = 2\ H(M)}$

$= \mathbf{2 \times 1.4855 = 2.971}$ **bits/symbol.**

% Code Efficiency = $\eta_c^{(2)} = \{H(M^2)/L_2\} \times 100 = (2.971/3.36) \times 100 = 88.42$ **%**

% Code redundancy = $1 - \eta_c^{(2)} = 11.58$ **%**

Code efficiency improves by (88.42 – 87.38) = 1.04%. Also computing (L_n/n) for $n = 2$ we have ($L_2/2$) = (3.36/2) = **1.68 binits/symbol.**

We have **L = 1.7 binits/symbol** for basic source. Therefore, $(L_2/2) < L$ which satisfies equation (2.23). We can also compute L_3 and show that $(L_3/3) < (L_2/2) < L$ and also show that the code efficiency goes on improving and when 'n' tends to infinity the code efficiency reaches 100%.

2.8 Shannon-Fano Encoding Algorithm

One of the well-known and basic techniques for producing more efficient and compact code with less redundancy is the Shannon-Fano algorithm. The different steps involved are as follows:

1. The message symbols are arranged in the decreasing order of probability.
2. Symbols are partitioned into two groups such that the sum of the probabilities of the symbols in each group is the same. i.e $\sum p(m_i)$ in each group = 1/2 .

3. Symbols in group one are designated by binary '1' and in the group two are designed by binary '0'. [It can be done other way also *i.e.* group one by '0' and group two by '1']. This digit forms the starting code symbols of the code.
4. Group one is divided into sub-groups such that the sum of the probabilities of the symbols in each sub-group is approximately the same.
5. All the symbols of the first sub-group are designed by '1'and second sub-group by a '0'.
6. Group two is now sub-divided into two more sub-groups and step 5 is repeated.
7. This procedure is repeated till further sub-division is not possible.

While making the partitions, in step two, symbols with higher probability should be made to correspond to a code with shorter word length. For convenience, a code tree can be constructed and codes can be read directly.

***Example* 2.14:** This example illustrates the steps of the Shannon-Fano algorithm clearly. Let there be 8 possible source symbols. The symbols, their probabilities, different steps of code allocation, the assigned code and the complement of the code are given in the Table (2.9) below.

Table 2.9: Code Table Using Shannon-Fano Algorithm

Symbols	P_i	No. of Selections					Code	Complement of the Code
		1	2	3	4	5		
s_1	1/2	1					1	0
s_2	1/4	0	1				01	10
s_3	1/16	0	0	1	1		0011	1100
s_4	1/16	0	0	1	0		0010	1101
s_5	1/32	0	0	0	1	1	00011	11100
s_6	1/32	0	0	0	1	0	00010	11101
s_7	1/32	0	0	0	0	1	00001	11110
s_8	1/32	0	0	0	0	0	00000	11111

In this example five partitioning/sub-grouping are required. Symbol s_1 is represented by a single bit '1', s_2 by two bit code '10' etc., up to s_8 whose code is a 5 bit word '11111'.The code tree diagram for this example is given in Fig. 2.6.

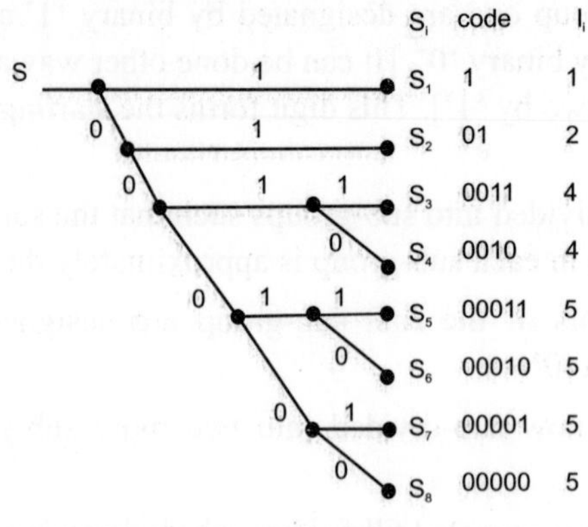

Fig. 2.6. Code Tree for Shannon-Fano Algorithm of Example (2.14).

The following properties of the code are clear from the table.

(*a*) The average length of this code is given by

$$L = \sum_{i=1}^{8} p_i l_i = (1/2) \times 1 + (1/4) \times 2 + (1/16) \times 4 + (1/16) \times 4$$

$$+ 4 \times (1/32) \times 5$$

$$= \mathbf{2.125\ bits/symbol}$$

(*b*) The average information per symbol, i.e. the Entropy of the source is given by

$$H(M) = \sum_{i=1}^{8} p_i \log(1/p_i) \text{ bits/message symbols}$$

$$= (1/2)\log(2) + (1/4)\log(4) + 2 \times (1/16)\log(16) + 4 \times (1/32)\log(32)$$

$$= 1/2 + 1/2 + 1/2 + 5/8$$

$$= \mathbf{2.125\ bits/symbol}$$

$$= \mathbf{L}$$

(*c*) **Coding efficiency** is $\eta_c = \{H(M)/L\} = (2.125/2.125) = 1$; i.e $\eta_c =$ **100 %**

Redundancy is $R_{\eta_c} = 1 - \eta_c = 1 - 1 = 0$; i.e., $R_{\eta_c} =$ **0 %**

(*d*) In this example the coding efficiency is 100%. The reason for this can be obtained by analyzing the expression for coding efficiency, that is Coding efficiency is $\eta_c = \{H(M)/L\} = \{\sum_{i=1}^{m} p_i \log(1/p_i) / \sum_{i=1}^{m} p_i l_i\}$

In this expression if 'l_i = **log(1/p$_i$)**' then the numerator becomes the denominator and hence the coding efficiency is 100%. In this example we can see that for every symbol this condition of 'l_i = **log(1/p$_i$)**' is satisfied and hence the coding efficiency is 100%. In fact **Black** has suggested the expression for the length of the code word as $l_i \approx \log(1/p_i)$.

(*e*) In a long message sequence derived out of binary codes,

the number of occurrences of '1' is $\sum_{i=1}^{m}$ [*Number* of '1's in the code for 's_i'] p_i,

the number of occurrences of '0' is $\sum_{i=1}^{m}$[*Number* of '0's in the code for 's_i'] p_i,

The probability of '1'in a code word = $p(1) = (1/L)\sum_{i=1}^{m}$[*Number* of '1's in the code for 's_i'] p_i,

The probability of '0' in a code word = $p(0) = (1/L)\sum_{i=1}^{m}$ [*Number* of '0's in the code for 's_i'] p_i

From the code Table (2.9) of the example (2.13) we have

$p(1) = (1/2.125)\{1\times(1/2) + 1\times(1/4) + 2\times(1/16) + 1\times(1/16) + 2\times(1/32) + 1\times(1/32) + 1\times(1/32)\} = 0.5$

$p(0) = (1/2.125)\{1\times(1/4) + 2\times(1/16) + 3\times(1/16) + 3\times(1/32) + 4\times(1/32) + 4\times(1/32) + 5\times(1/32)\} = 0.5$

(*f*) The coder output has maximum entropy in terms of binary digits.

(*g*) If the 8 symbols were directly coded into binary digits, then the codes would be {000, 001, 010, 011,....., 110, 111}, that is each symbol represented by 3 bits, that is $\Sigma p_i l_i = 3$ bits/symbol which is greater than the entropy H(M) of the source.

(*h*) The symbol with highest probability is to be represented by a shortest code word.

A required coding efficiency can be obtained by considering different extended versions of a basic source code. Now let us consider the following example to construct a code with a required coding efficiency.

***Example* 2.15:** Consider a Source emitting two symbols 's_1' and 's_2' with probabilities 0.05 and 0.95 respectively. Let the coding efficiency required for source coder is at least 65%.

Table 2.10: Code Table for the Source of Example (2.15)

Symbol	p_i	Code	l_i
s_1	0.05	0	1
s_2	0.95	1	1

The average length of this code is $L = \sum_{i=1}^{2} \text{“}p_i\text{”}l_i = (0.05)\times1+(0.95)\times1$

$= \mathbf{1\ binit/symbol}$

The Entropy of the source is $H(M) = \sum_{i=1}^{2} \text{“}p_i\text{”}\log(1/p_i)$bits/message symbol

$= (0.05)\log(1/0.05) + (0.95)\log(1/0.95) = \mathbf{0.2864\ bits/symbol}$

Coding efficiency $\eta_c = \{H(M)/L\} = (0.2864/1) = 0.2864$ or 28.64 % which is less than the required coding efficiency 65%.

Now let us consider the second extension of the given source.

Table 2.11: Code Table for the 2nd Extension of the Source of Example (2.15)

Symbols	P_i	No. of Selections			Code	Code Length 'l_i'
		1	2	3		
$s_2\,s_2$	0.9025	1			1	1
$s_2\,s_1$	0.0475	0	1		01	2
$s_1\,s_2$	0.0475	0	0	1	001	3
$s_1\,s_1$	0.0025	0	0	0	000	3

The average length of the code for the 2nd extension of the source is

$$L_2 = \sum_{i=1}^{4} \text{“}p_i\text{”}l_i = (0.9025)\times1+(0.0475)\times2 + (0.0475)\times3 + (0.0025)\times3$$

$= \mathbf{1.1475\ binit/symbol}$

The Entropy of the 2nd extension of source is

$H(M^2) = 2\,H(M) = (2)(0.2864) = \mathbf{0.5728}$ bits/message symbol

Coding efficiency of the 2nd extended source $= \eta_c^{(2)} = \{2H(M)/L_2\}$

$= (0.5728/1.1475)$

$= 0.4992$ or 49.92%;

which is less than the required coding efficiency 65%. Therefore let us consider the 3rd extension of the basic source.

The average length of the code for the 3rd extension of the source is

$$L_3 = \sum_{i=1}^{8} \text{“}p_i\text{”}l_i = (0.857375)\times1+(0.045125)\times3\times3 +$$

$$(0.002375)\times3\times5+(0.000125)\times5$$

$= \mathbf{1.29975\ binit/symbol}$

Table 2.12: Code Table for the 3rd Extension of the Source of Example (2.15)

Symbol	P_i	No. of Selections					Code	Code Length 'l_i'
		1	2	3	4	5		
$s_2s_2s_2$	0.857375	1					1	1
$s_2s_2s_1$	0.045125	0	1	1			011	3
$s_2s_1s_2$	0.045125	0	1	0			010	3
$s_1s_2s_2$	0.045125	0	0	1			001	3
$s_2s_1s_1$	0.002375	0	0	0	1	1	00011	5
$s_1s_2s_1$	0.002375	0	0	0	1	0	00010	5
$s_1s_1s_2$	0.002375	0	0	0	0	1	00001	5
$s_1s_1s_1$	0.000125	0	0	0	0	0	00000	5

The Entropy of the 3rd extension of source is

$H(M^3) = 3\ H(M) = (3)(0.2864) =$ **0.8592 bits/message symbol**

Coding efficiency of the 3rd extended source $= \eta_c^{(3)} = \{3H(M)/L_3\}$

$= (0.8592/1.29975)$

$=$ **0.6611 or 66.11%;**

That is the coding efficiency of the 3rd extension of the basic source of example (2.14) is more than 65% and hence is the required code.

2.9 Shannon-Fano Ternary Code

The different steps involved in generating the ternary code are obtained by modifying the steps for the Shannon-Fano encoding algorithm for the binary code given in section 2.8. They are listed below:

1. The message symbols are arranged in the decreasing order of probability.
2. Symbols are partitioned into the groups such that the sum of the probabilities of the symbols in each group is approximately the same. i.e $\Sigma p(m_i)$ in each group $\approx 1/3$.
3. Symbols in group one are designated by a '2', in the second group two are designed by a '1'and the third group by a '0'.
4. Group one is divided into three more sub-groups such that the sum of the probabilities of the symbols in each sub-group is approximately the same.
5. All the symbols of the first sub-group is designated by a '2', second sub-group by a '1' and third sub-group by a '0'.

6. Groups 2 and 3 are sub-divided into three more sub-groups each and step 5 is repeated.
7. This procedure is repeated till further sub-division is not possible.

The following example illustrates the steps of the Shannon-Fano algorithm for ternary code:

***Example* 2.16:** Let the source symbols be **S = {s_1, s_2, s_3, s_4, s_5, s_6, s_7}**with probabilities **P = {0.3, 0.3, 0.12, 0.12, 0.06, 0.06, 0.04}** and the code symbols **D = {0, 1, 2}**. The code table is given below

Table 2.13: Code Table for the Source of Example (2.16)

Symbol	P_i	No. of Selections			Code	Code Length 'l_i'
		1	2	3		
s_1	0.3	2			2	1
s_2	0.3	1			1	1
s_3	0.12	0	2		02	2
s_4	0.12	0	1		01	2
s_5	0.06	0	0	2	002	3
s_6	0.06	0	0	1	001	3
s_7	0.04	0	0	0	000	3

The average length of this code is given by

$$L = \sum_{i=1}^{7} \text{“}p_i\text{”} l_i$$

$$= (0.3)\times1\times2+(0.12)\times2\times2+(0.06)\times3\times2+(0.04)\times3$$

$$= \mathbf{1.56\ trinits/symbol}$$

The Entropy of the source is given by

$$H(M) = \sum_{i=1}^{7} p_i \log(1/p_i) \text{ bits/message symbol}$$

$$= [(0.3)\log(1/0.3)]\times2 + [(0.12)\log(1/0.12)]\times2+[(0.06)\log(1/0.06)]\times2 +(0.04)\log(1/0.04)$$

$$= \mathbf{2.4491\ bits/message\ symbol.}$$

The entropy in D-ary units as per equation (2.10) is

$$H_D(M) = \{H(M)/\log_2 D\}$$

For D = 3 we have $H_3(M) = \{H(M)/\log_2 3\} = (2.4491/\log_2 3)$

$$= 1.5452 \text{ ternary units/message symbol}$$

Therefore ternary **Coding efficiency** is given by

$$\eta_c = \{H_3(M)/L\} = (1.5452/1.56) = 0.9905;$$

i.e., $\eta_c = \mathbf{99.05\ \%}$

Redundancy is $R_{\eta_c} = 1 - \eta_c = 1 - 0.9905 = 0.0095;$

i.e., $R_{\eta_c} = 0.95\ \%$

2.10 Compact Code: Huffman's Minimum Redundancy Code

When each source symbol or message is separately encoded in to a code word of unique combination of '1's and '0's with minimum average code length, the resulting code for the messages is called a compact code. On receiving any code word, the receiver can immediately decode or identify the source symbol/message. In the area of data transmission, it is obvious that, the average number of bits per character (source symbol) transmitted should be minimum for an efficient system so that the redundancy associated will also be minimum. Variable length codes are one of the choices. The primary difficulty associated with variable-length code words is that the rate at which bits are presented to the transmission channel will fluctuate, depending on the relative frequencies of the source symbols. This requires buffering between the source and the channel. However advances in technology have both overcome this difficulty and contributed to the appeal of variable-length codes. Current data networks allocate communication resources to sources on the basis of need and provide buffering as part of the system. These systems require significant amounts of protocol, and fixed-length codes are quite inefficient for applications such as packet headers. In addition, communication costs are beginning to dominate storage and processing costs, so that variable-length coding schemes which reduce communication costs is attractive even if they are complex. For these reasons, one could expect to see even greater use of variable-length coding in the future.

David A. Huffman has suggested a simple coding scheme for the construction of compact optimal codes, in his paper "A Method for the Construction of Minimum-Redundancy Codes", published in the year 1952 (while he was a Ph.D. student at MIT). Huffman's algorithm provided the first solution to the problem of constructing minimum-redundancy codes (even if equation (2.15) is not satisfied). The Huffman algorithm is also of importance because it has provided a foundation upon which other data compression techniques have built and a benchmark to which they may be compared.

Huffman algorithm consists of step by step reduction of the original source followed by a code construction, starting with the final reduced source and working backwards to the original source. The number of steps involved is given by λ, where

$$\lambda = (m - D)/(D - 1) \tag{2.24}$$

where '*m*' is the number of source symbols and D is the number of symbols in the code alphabet. Obviously 'λ' should be an integer. If it is not, **"dummy symbols"** with zero probabilities are added to '*m*', the source symbols, so that 'λ' becomes an integer. The step by step Huffman procedure is as follows:

1. The Source Symbols are arranged in the decreasing order of their probabilities.
2. Value of 'λ' is computed. If it is not an integer for the given values of '*m*' and 'D', suitable number of dummy symbols with zero probability is added to make 'λ' an integer. For binary codes 'λ' will always be an integer and hence this step is not required.
3. The last 'D' symbols are now grouped in to a **'single composite symbol'** whose probability of occurrence is equal to the sum of the probabilities of the 'D' symbols from which it is obtained. This results in a reduced source (S_1) with number of symbols equal to one less than the number symbols taken at step '1'.
4. Steps '1' and '3' are repeated on the resulting set of symbols until in the final step exactly 'D' symbols are left with.
5. The last 'D' composite source symbols of the final reduced source, are now assigned with 'D' different code symbols (i.e 0,1, 2, ….., D – 1).
6. Now we work backwards to the previous reduced sources and the decomposition of one code word each time is done in order to form new code words.
7. This procedure is continued till code words are assigned to all the source symbols. Codes of the dummy symbols, if any, are discarded.
8. Alternatively, following the steps carefully, a code tree diagram can be drawn and codes can be read directly.

Analysis of Huffman Algorithm

Going through the above step by step procedure of the Huffman Code we observe the following: In step '3' we have combined the last 'D' source symbols into a composite symbol. Now we are left with 'm – D + 1' symbols. In the next step we again combine the last 'D' symbols and the second reduced source now will have $(m – D + 1) – D + 1 = (m – 2D + 2)$ symbols. Continuing this way the 'k^{th}' reduced source will have $(m – kD + k) = [m – k(D – 1)]$ symbols. Accordingly if 'λ' steps are required and for the final reduced source to have exactly 'D' symbols, we must have $D = m – \lambda (D – 1)$. That is $\lambda = (D – m)/(D – 1)$ and hence equation (2.24) is proved.

Considering the working backwards procedure, let λ_k is the composite symbol obtained in the kth step, by combining the last D symbols of the $(k – 1)^{th}$ reduced source. Then the code assigned in λ_k will form the starting code sequence for the code words of its constituents in the $(k-1)^{th}$ reduction. The following two examples will illustrate the Huffman algorithm for the binary and ternary codes:

2.10.1. Huffman Coding with Binary Alphabet

***Example* 2.17:** Let the source symbols be **S = {a, b, c, d, e, f, g}** with probabilities of occurrences P = {0.05, 0.2, 0.1, 0.05, 0.3, 0.2, 0.1} and the code symbols **D = {0, 1}**. The code table is given below: (First the symbols are arranged in the decreasing order of probabilities).

Table 2.14: Code Table for (Huffman Coding with Binary Alphabet) Example (2.17)

Code	Symbols	p_i	Reduced Sources				
11	E	0.3	0.3	0.3	0.3	0.4	0.6 (1)
01	B	0.2	0.2	0.2	0.3	0.3(1)]0.6	0.4 (0)
00	F	0.2	0.2	0.2	0.2(1)]0.4	0.3(0) *l*	
100	C	0.1	0.1	0.2(1)]0.3	0.2(0) *k*		
1011	G	0.1	0.1(1)]0.2	0.1(0) *j*			
10101	A	0.05(1)]0.1	0.1(0) *i*				
10100	D	0.05(0) h					

The code table is formed following the steps of the Huffman coding algorithm explained. The flower bracket at the bottom of the columns for probability shows the combining of the two symbols with the least probability of occurrences and assigning a new name for the composite symbol and the probability. Using step 5 of the algorithm, last two composite source symbols ('*l*' and '*k*') of the final reduced source is assigned the two code symbols '1' and '0' respectively. Now using steps 6 and 7 of the algorithm the different composite symbols are assigned codes, till all the source symbols are assigned with code symbols. The final code word representing different source symbols are given in column 1 of the code Table (2.14) which is the Huffman code for the example (2.17). The assigning of the code symbols can be done freely at every step and there is no restriction. i.e'*l*' could have been assigned '0' instead of '1'and '*k*' could have been assigned with '1' instead of '0'. This is the case at every step. Thus for the example illustrated there can be a total of $2\times2\times2\times2\times2\times2 = 64$ possible instantaneous code patterns. For example the complement of the codes in the column 1 of the table forms another assignment. (However a set procedure can be followed to avoid any confusion at any stage). Moreover the length 'l_i'of the code word for symbol 's_i' will be constant for all possible codes.

The code tree diagram for the example can be drawn using the final step of the algorithm and decomposing the composite symbols towards the original source symbols. The code tree for this example is shown in Fig. (2.7).

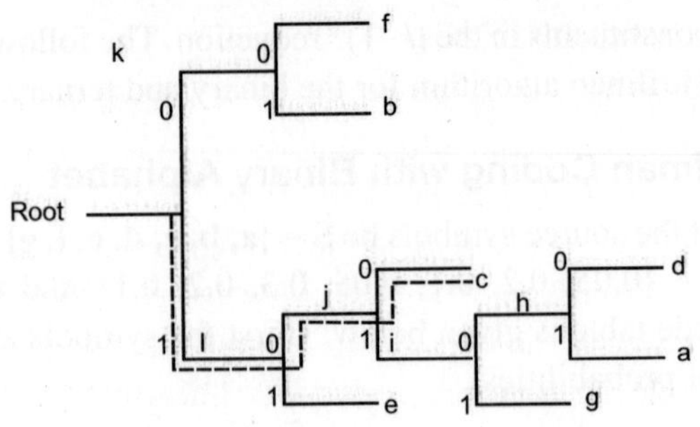

Fig. 2.7. Code Tree Diagram for Example (2.17)
(Huffman Coding for Binary Alphabets)

The code word for each source symbol S_i is obtained easily by traversing the binary tree from its root to the leaf node corresponding to the symbol S_i. For example the dashed line in the code tree represents the code word ('100') for the source symbol 'c'. It can be observed that the codes are originating from the same source (root) and diverge out into different branches and ending to a unique leaf, thus ensuring prefix property to the resultant code.

The average length of this Huffman code is given by

$$L = \sum_{i=1}^{7} p_i l_i = (0.05)\times2\times5 + (0.2)\times2\times2 + (0.1)\times1\times3 + (0.3)\times1\times2 + (0.1)\times1\times4$$

$$= \mathbf{2.6\ binits/symbol}$$

The Entropy of the source is given by

$$H(M) = \sum_{i=1}^{7} p_i \log(1/p_i) \text{ bits/message symbol}$$

$$= [(0.05)\log(1/0.05)]\times2 + [(0.2)\log(1/0.2)]\times2 + [(0.1)\log(1/0.1)]\times2 + (0.3)\log(1/0.3)$$

$$= 0.43223 + 0.92886 + 0.66445 + 0.52114 = 2.54668$$

$$= \mathbf{2.5467\ bits/message\ symbol.}$$

Coding efficiency is given by

$$\eta_c = \{H(M)/L\} = (2.5467/2.6) = 0.9795;$$

i.e., $\eta_c =$ **97.95 %**

Redundancy is $R_{\eta_c} = 1 - \eta_c = 1 - 0.9795 = 0.0205$;

i.e., $R_{\eta_c} =$ **2.05 %**

2.10.2. Compression Ratio

A parameter which gives a feel of the extent to which the coding scheme has compressed the source symbols is the Compression ratio. The compression ratio of a coding scheme is defined as the ratio of the code length in a fixed length scheme to the average code length of the coding scheme concerned. For the example (2.17) the 7 symbols of the given source can be represented by 3 bit code in a fixed length code representation. The average code length of the coding scheme in the example is 2.6 bit/source symbol. Therefore **Compression ratio** of the Huffman binary code in this example is 3/2.6 = **1.15**.

2.10.3. Variance of the Code Word Length

Another parameter of the coding scheme is the Variance of the code word length, which should be as low as possible for a code. The Variance of the code word length is defined as the average value of the difference between the actual code length and the average code length *i.e.*

$$\textbf{Variance of the code word length} = \sum_{i=1}^{7} p_i(l_i - L).$$

The variance of the code word length for the example (2.17) is

$$= \sum_{i=1}^{7} p_i(l_i - \mathrm{L}) = (0.05) \times 2 \times (5 - 2.6) + (0.2) \times 2 \times (2 - 2.6) + (0.1) \times 1$$

$$\times (3 - 2.6) + (0.3)\times 1\times(2 - 2.6) + (0.1)\times 1\times(4 - 2.6)$$

$$= 0.24 - 0.24 + 0.04 - 0.18 + 0.14 = 0$$

The variance here is equal to zero. This can be easily verified. The number of bits required to represent 7 symbols = $\log_2(7) = 2.5437$; and the average length of the code word by the coding scheme is 2.6 which is almost equal to the calculated value.

In general the variance of the code length should be as low as possible. One way to achieve this is to move the composite symbol as high as possible while arranging the symbols in descending order of probabilities (step 1 of the Huffman algorithm) in the reduced source symbol set. For example in Table (2.14) 'h' can be kept just after '*f*' under source $S_1$and '*i* ' can be kept just after '*e*' under source 'S_2' etc.,

2.10.4. Huffman Coding with Ternary Alphabet

***Example* 2.18:** Let the source symbols be **S = { a, b, c, d, e, f }**with probabilities of occurrences **P = {1/3, 1/4, 1/8, 1/8, 1/12, 1/12}** and for ternary code the code symbols D = {0, 1, 2}.

Consider equation (2.24) which gives the number of reduction steps required (λ), that is

$$\lambda = (m - D)/(D - 1)$$

For ternary code D = 3 and hence the value of 'λ' for ternary code is given by λ = $(m - 3)/2$. Hence, 'λ' will take integer values only for the odd values of 'm', and > 5. In the given example (2.18), the number of source symbols, $m = 6$. Hence, one dummy symbol, 'g' with zero probability is to be added to the set of source symbols.

Now the code table for the Ternary code is constructed using steps 1 to 7 of the Huffman algorithm, by combining last 3 symbols at a time and using the code alphabet D = {0, 1, 2}. This is shown in Table (2.15) below. Column 1 gives the code words assigned to the symbols.

Table 2.15: Code Table for (Huffman Coding with Ternary Alphabet) Example (2.18)

Codes	Source (S)		Reduced Sources		
1	A	1/3	A	1/3	I 5/12 (0)
2	B	1/4	B	1/4	A 1/3 (1)
01	C	1/8	H	1/6(0)]5/12	B 1/4 (2)
02	D	1/8	C	1/8 (1)	
000	E	1/12 (0)]1/6	D	1/8 (2)] *i*	
001	F	1/12 (1)			
002	G	0 (2)] h			

symbol 'G' → Dummy ; Code '002' → Discarded.

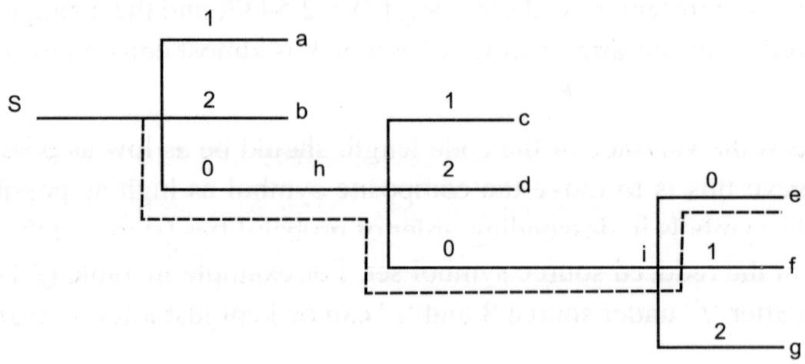

Fig. 2.8. Code Tree Diagram for (Huffman Coding for Ternary Alphabets) Example (2.18)

The code tree for the ternary code is also drawn as it is drawn for the binary code and is shown in Fig. (2.8). The dashed line in the code tree gives the code for the symbol 'e'. i.e., '000'.

The average length of this Huffman Ternary code is given by

$$L = \sum_{i=1}^{7} p_i l_i = 1/3\times1 + 1/4\times1 + 1/8\times2 + 1/8\times2 + 1/12\times3 + 1/12\times3$$

$= 19/12$ Trinits/symbol

$=$ **1.583 Trinits/symbol**

The Entropy of the source in binary bits/symbol is given by

$$H(M) = \sum_{i=1}^{7} p_i \log(1/p_i) \text{ bits/message symbol}$$

$= (1/3)\log(3) + (1/4)\log(4) + (1/8)\log(8) \times 2 + (1/12)\log(12) \times 2$

$= [6 \times \log(3) + 19]/12.$

$=$ **2.3758 bits/message symbol.**

The entropy in Ternary units/symbol is obtained using equation (2.10) with D = 3. That is

$$\mathbf{H_3(M) = \{H(M)/\log_2(3)\}}$$

$=$ **1.499 Trinits/source symbol.**

Coding efficiency of the Huffman Ternary code is given by

$$\eta_c = \{H_D(M)/L\} = (1.499/1.583) = 0.9469;$$

i.e., $\eta_c =$ **94.69 %**

Redundancy is $R_{\eta_c} = 1 - \eta_c = 1 - 0.9469 = 0.0531;$

i.e., $R_{\eta_c} = 5.31\%$

Compression ratio of the Huffman Ternary code in this example is 3/1.583 = **1.8951**

2.10.5. Huffman Coding with Quaternary Alphabet

***Example* 2.19:** Let us consider the same source of example (2.18) and the code symbols are D ={0, 1, 2 , 3}. Then the value of λ becomes

$$\lambda = (m - D)/(D - 1) = (6 - 4)/(4 - 1) = (2/3).$$

Hence we need a dummy symbol to make λ an integer and in this case $\lambda = 1$. The code table is drawn using Huffman coding algorithm and is given in Table (2.16).

Table 2.16: Code Table for (Huffman Coding with Quaternary Alphabet) Example (2.19)

Code	Symbol	p_i	Reduced Source
0	A	1/3	1/3 **(0)**
1	B	1/4	1/4 **(1)**
3	C	1/8	7/24 **(2)**
2 0	D	1/8 **(20)**	1/8 **(3)**
2 1	E	1/12 **(21)** 7/34	
2 2	F	1/12 **(22)**	
2 3	G	0 **(23)** h	

symbol '*G*' → Dummy ; Code '23' → Discarded.

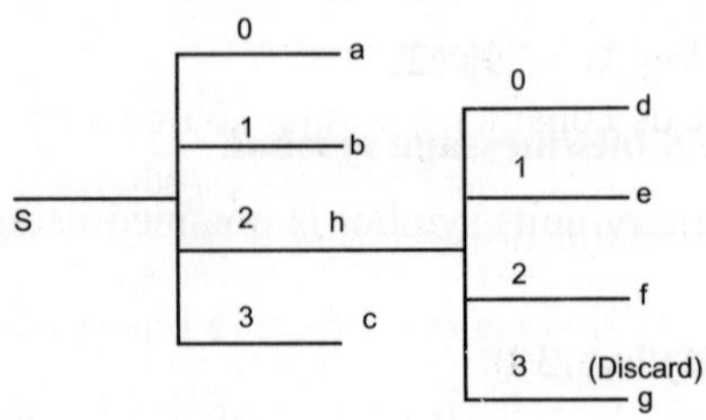

Fig. 2.9. Code Tree Diagram for (Huffman Coding for Quaternary Alphabets) Example (2.19)

The average length of this Huffman quaternary code is given by

$$L = \sum_{i=1}^{7} p_i l_i = 1/3 \times 1 + 1/4 \times 1 + 1/8 \times 1 + 1/8 \times 2 + 1/12 \times 2$$

$$+ 1/12 \times 2$$

$$= (31/24)$$

$$= \mathbf{1.2917\ Quadits/symbol}$$

The Entropy in quaternary units/symbol is obtained using equation (2.10) with D = 4. That is

$$H_4(M) = \{H(M)/\log_2(4)\}$$

$$= (2.3758/2) = 1.1879 \text{ quadits/source symbol.}$$

Coding efficiency of the Huffman Quaternary code is given by

$$\eta_c = \{H_4(M)/L\} = (1.1879/1.2917) = 0.91964;$$

i.e., $\eta_c = \mathbf{91.964\ \%}$

Redundancy is $\mathbf{R}_{\eta_c} = 1 - \eta_c = 1 - 0.91964 = 0.08036;$

i.e., $R_{\eta_c} = \mathbf{8.036\%}$

Compression ratio of the Huffman Quaternary code in this example is (3 /1.2917) = **2.3225.**

Considering the above examples it can be observed that the coding efficiency gradually decreases and the compression ratio increases as the value of D is increased. This is obvious as $\log(1/p_i)$ are indeed very close to integer values than $\log_3(1/p_i)$ or $\log_4(1/p_i)$, which guarantees minimum redundancy. Hence it is possible, by principle, to achieve a minimum redundancy coding by appropriate selection of 'D'.

2.10.6. Huffman Coding of an Extended Source

In section 1.4 we have seen that the entropy of the nth extension of a memoryless source is 'n' times the entropy of the basic memoryless source. Also by increasing the value of 'n' the coding efficiency can be increased to the maximum extend and the redundancy can be decreased to the maximum extend. In this section we will see the effects of Huffman coding on the coding efficiency and redundancy of extended sources through the following example (2.20).

***Example* 2.20:** Let us consider a simple source with alphabets $\dot{S} = \{s_1, s_2\}$ with probabilities P = {1/4, 3/4}. The best binary code for this source can be $s_1 \rightarrow 0$ and $s_2 \rightarrow 1$. The entropy of this primary source is

H(M) = (1/ 4) log 4 + (3/4) log (4/3) = 0.815 bits/symbol,

and Coding efficiency η_c = {H(M)/L} = (0.815/1) = 0.815;

The probability of binary '0' and '1' is given by $p(0) = p(s_1) = 1/4$ and $p(1) = p(s_2) = 3/4$.

Now considering the third extension of this source S, the alphabets will be S^3 = {000, 001, 010, 011, 100, 101, 110, 111} with P = {1/64, 3/64, 3/64, 9/64, 3/64, 9/64, 9/64, 27/64}.

Then the entropy of this source will be $H(M^3) = 3H(M) = 0.815 \times 3 = 2.445$ trinits/symbol. Now let us encode this source using Huffman algorithm. The codes for the new set of symbols are shown in Table (2.17).

Table 2.17: Code Table for (Huffman Coding for Extended Source) Example (2.20)

Codes	S_i	p_i	Reduced Sources					
1	x_8	27/64	27/64	27/64	27/64	27/64	27/64	37/64 (**0**)
001	x_7	9/64	9/64	9/64	10/64	18/64	19/64 (**0**)	27/64 (**1**)
010	x_6	9/64	9/64	9/64	9/64	10/64 (**0**)	18/64 (**1**)	
011	x_4	9/64	9/64	9/64	9/64 (**0**)	9/64 (**1**)		
00000	x_5	3/64	4/64	6/64 (**0**)	9/64 (**1**)			
00001	x_3	3/64	3/64 (**0**)	4/64 (**1**)				
00010	x_2	3/64 (**0**)	3/64 (**1**)					
00011	x_1	1/64 (**1**)						

The average length of the Huffman code for the extended source is given by

$$L = \sum_{i=1}^{8} p_i l_i = 1/64 \times 5 + 3/64 \times 5 \times 3 + 9/64 \times 3 \times 3 + 27/64 \times 1$$

$= (158/64)$

$=$ **2.47 trinits/symbol**

The Entropy of the extended source $H(M^3) = 3H(M) = 3\times 0.815$

$= 2.445$ trinits/symbol.

Coding efficiency of the Huffman code for extended source is given by

$\eta_c = \{H(M^3)/L\} = (2.445/2.47) = 0.98987;$

i.e., $\eta_c =$ **98.987%**

Redundancy is $R_{\eta_c} = 1 - \eta_c = 1 - 0.98987 = 0.01013;$

i.e., $R_{\eta_c} =$ **1.01013%**

The efficiency improves by (98.987% – 81.5%) = 17.487%

Thus we can see that there is a considerable and remarkable improvement over the coding efficiency without extension. Further extension of the source gives still better efficiency (as an example for S^4, of the above source, the efficiency is 99.1%).

2.10.7. Properties of Huffman Codes

In this section we will summarize some of the important properties of the Huffman code.

(a) Huffman code takes fixed length source symbols and produces variable length code words.

(b) The average Huffman code length (L) is related to the entropy of the source by $H(M) \leq L < H(M) + 1$. (The equality is achieved when all the symbol probabilities are inverse powers of 2).

(c) The code construction process has a complexity of the order of $O(m\log_2 m)$. With presorting of the input symbol probabilities, code construction schemes with complexity of the order of O(m) can be achieved. (Ref: **IEEE Transactions on ASSP – 38, pp. 1619–1626, Sept., 1990**).

(d) Huffman codes satisfy the prefix condition that is they are uniquely decodable and no code word is a prefix of any other code word.

(e) The complement of a binary Huffman code is also a valid Huffman code.

2.10.8. Decoding of Huffman Codes

There are two different well known decoding methods of Huffman codes. One the **Bit serial method** and the other the **Look-up table method,** which is explained below.

2.10.8.1. Bit-Serial decoding of Huffman codes

We assume that the binary coding tree is available to the decoder. In practice, Huffman coding tree can be reconstructed from the symbol-to-code word mapping table that is known to both the encoder and decoder. Following are the three steps involved in the bit serial decoding method of Huffman code:

***Step* 1:** Read the input compressed stream bit-by-bit and traverse the tree until a leaf node is reached.

***Step* 2:** As each bit in the stream is used, it is discarded. When the leaf node reached, the Huffman decoder outputs the symbol at the leaf node. This completes the decoding for that symbol.

***Step* 3:** Repeat step 1 and 2 until all the input is decoded.

Since the code word is variable in length, the decoding bit rate is not the same for all symbols. That is the decoding is characterized by fixed input bit rate and variable output symbol rate.

2.10.8.2. Lookup table based decoding of Huffman codes

The lookup table is constructed at the decoder from the symbol-to-code word mapping table. If the longest code word in the table is $\mathcal{L}$ bits, then a $2^{\mathcal{L}}$ entry lookup table is needed. (In practical systems the value of $\mathcal{L}$ used is 16 to 20).

Look-up table construction:

- Let D_i be the code word that corresponds to symbol s_i. Assume D_i has l_i bits. We form an $\mathcal{L}$ bit address in which the first l_i bits are D_i and the remaining $(\mathcal{L} - l_i)$ bits take on all possible combinations of '0' and '1'. Thus, for the symbol s_i; there will be $2^{(\mathcal{L} - l_i)}$ addresses.
- At each entry we form the two-tuple (s_i, l_i).

Decoding processes:

1. From the compressed input bit stream, we read in '$\mathcal{L}$' bits into a buffer.
2. We use the $\mathcal{L}$ bit word in the buffer as an address into the Look-up table and obtain the corresponding symbol, say 's_k'. Let the code word length be 'l_k'. We have now decoded one symbol.
3. We discard the first 'l_k' bits from the buffer and we append to the buffer with the next 'l_k' bits from the input, so that the buffer has again '$\mathcal{L}$' bits.
4. Repeat steps 2 and 3 until all of the symbols have been decoded.

(**Note:** Over the period of time many schemes for the Huffman coding was proposed and one such is the **"Memory Efficient and High-Speed Search Huffman Coding"** by R. Hashemian in **IEEE transactions on Communication, Oct., 1995, pp. 2576 – 2581)**

PROBLEMS

2.1. The following table gives several possible codes for a source with specified probabilities.

(*a*) Identify the uniquely decodable codes

(*b*) Which are instantaneous codes?

(*c*) Determine the average lengths of all uniquely decodable codes.

(*d*) Draw the code tree for all the codes.

Source Symbols	p_i	Code A	Code B	Code C	Code D	Code E
s_1	0.5	000	0	1	0	0
s_2	0.25	001	10	10	100	10
s_3	0.125	010	110	100	101	110
s_4	0.0625	011	1110	1000	110	1110
s_5	0.0625	100	1111	10000	111	1011

2.2. Find the smallest number of symbols (*r*) for devising a code with prefix property such that $w = \{2, 4, 5, 5\}$, where '*w*' is the set of number of words with word lengths 1, 2, 3, ...

2.3. Consider a zero memory binary source having $p(0) = 0.1$ and $p(1) = 0.9$

(*a*) Find the entropy, H(M).

(*b*) Find the average length of the compact code, 'L', for the source when $X = \{0, 1\}$.

(*c*) Find L_n / L for $n = 2, 3, 4$ and $n \to \infty$ when s^n is encoded into a compact binary code.

(*d*) Find the efficiency of the four codes.

2.4. Consider a 3-alphabet source with probabilities $p_1 = 0.4$, $p_2 = 0.3$ and $p_3 = 0.3$.

(*a*) Using Shannon's binary encoding algorithm design a code for the source.

(*b*) Calculate the coding efficiency for the code designed.

(*c*) Consider the 2nd order extended source S^2 with 9 alphabets. Design a new code for these alphabets using the same algorithm as in (a) and calculate the improvement in coding efficiency.

2.5. A source produces 6 symbols with probabilities p_1 to p_6 such that $p_1 \geq p_2 \geq p_3 \geq p_4 \geq p_5 \geq p_6$. Assuming $p_5 = p_6 = (1/12)$; $p_3 = p_4 = (2/12)$; $p_1 = p_2 = (3/12)$;

(*a*) Construct the Shannon—Fano binary code for the above messages.

(*b*) Determine the efficiency of the code so formed.

(*c*) Construct a Ternary code using the same procedure and find its efficiency and redundancy.

2.6. For a BSC, the transition probabilities are P (0/1) = P (1/0) = 1/3. The channel input consists of four symbols, $s_1 = 001$; $s_2 = 010$; $s_3 = 100$ and $s_4 = 111$, with probabilities 0.5, 0.25, 0.125 and 0.125 respectively.

(*a*) Show that the source probabilities of '0' and '1' are 7/12 and 5/12 respectively.

(*b*) Calculate the efficiency and redundancy of the code.

(*c*) Show that the capacity of the channel is 0.082 bits/binits.

2.7.(*a*) Find the capacity of a BSC with channel transition probability $p = 0.75$.

(*b*) The probabilities of the symbol produced by an information source are 1/4, 1/8, 1/4, 1/4 and 1/8. A practical coding scheme uses an average code length of 2 bits/message. Calculate the average efficiency of the scheme.

2.8. A discrete memoryless source has an alphabet of five symbols, s_0 to s_5 with probabilities 0.55, 0.15, 0.15, 0.10 and 0.05 respectively.

(*a*) Compute two different Huffman codes for this source.

(*b*) For each of the two codes, find

- The average code-word length
- The variance of the average code-word length over the ensemble of source symbols.

2.9. A discrete memory less source has an alphabet of seven symbols, s_0 to s_7 with probabilities 0.25, 0.25, 0.125, 0.125, 0.125, 0.0625, 0.0625 respectively.

(*a*) Compute the Huffman code for this source, moving the composite symbol as high as possible.

(*b*) Show that the efficiency of the computed source code has an efficiency of 100%. Explain the reason for the same.

2.10. A source produces two symbols X and Y with probabilities 15/16 and 1/16 respectively. Find a coding scheme to get a coding efficiency of at least 70%.

CHAPTER 3 COMMUNICATION CHANNELS

3.0 Introduction

It is very important to have an in depth knowledge of the medium through which we transfer the information from a source to the destination where the information is used. The design of transmission and receiving system depends on the characteristics of the medium. Each medium will have their own properties and affect the signal/information passing through them in different ways. In electrical or electronic communication theory the medium through which the information from a source is transferred or transmitted to a receiver or destination is referred to as a "***channel***". Typical transmission channels include telephone lines, high frequency, radio links, telemetry links, microwave links, satellite links etc. In these channels, the information is effected by various kind of noise disturbances. For example, on a telephone line, the disturbance may be due to the switching impulse noise, thermal noise, cross-talk from other lines (channels), ignition noise from vehicles, or lightning. The disturbances may be due to the amplitude and frequency response variations in the channel, variations in the characteristics of the channel or due to non-linearity in the channel. In military communication, the intentional jamming by hostile force also affects the communication. All these disturbances introduce error in data transmission and limit the maximum rate at which the information can be transferred over the channel. We have mentioned previously that for efficient transfer of information, it has to be changed from one form to another form before transmission and re-transformed back to the original form before delivered to the final destination. As far as the original information source is concerned all the units (encoders at the transmission side and decoders at the receiver side) which transforms the information from one form to another and back, can be considered as part of the channel. The imperfections in the design and operation of these units will also introduce, error present in the data transmission (typical example can be the catastrophic convolutional encoders). Depending on the type of (analog, digital or discrete) signals information the channel or part of channel are referred by different terminologies. In order to understand the concept of channel let us consider the block diagram of a typical data transmission system illustrated in Fig. 3.1.

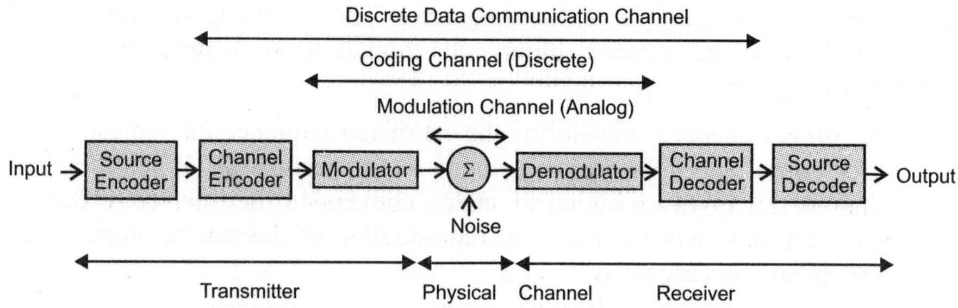

Fig. 3.1. Typical Data Communication System

A practical communication system will basically have a transmitter, a physical channel and a receiver. In a typical data transmission system the input can be from an information source which may be a person or digital system (like a digital computer). This source output which is to be transmitted to the destination can be either a continuous signal or a sequence of discrete symbols. This input is fed to a '***source encoder***'. In the case of a continuous signal the source encoder block first performs an analog to digital conversion, followed by the source encoding; in the second case it performs source encoding directly on the source symbol which is either binary or discrete. The source encoder output is a sequence of discrete or binary symbols. As we have seen in the previous chapter, the source encoder is ideally designed so that the number of bits per unit time required to represent the source output symbol is minimized and the source output can be reconstructed at the receiver without any ambiguity.

The source encoder output is fed to the '***channel encoder***', which transforms the code word into another code word belonging to different set of code words. This set of code words is also a sequence of binary or discrete symbols. The purpose of a channel encoder is to combat the noisy environment in which the code words must be transmitted. The processing of the information at the channel encoder (and channel decoder) provides the communication system the capability of detecting and correcting the channel induced errors. The subject of channel coding is dealt in chapters **4 through 7** of this book.

Discrete symbols are not suitable for transmission over physical channel. So the '***Modulator***' unit transforms each of the channel encoder output symbols into a waveform of appropriate period which is suitable for transmission over the physical channel. This waveform enters the channel and is corrupted by noise.

The '***demodulator***' processes each received waveform during the waveform period and produces an output that may be discrete. The sequence of demodulator outputs corresponding to the channel encoder output is called the received sequence. The '***channel decoder***' transforms the received sequence into a binary sequence called the estimated sequence. The channel decoding is based on the channel encoding algorithm and the noise characteristics of the channel so as to minimize the probability of decoding error. The estimated sequence will be a replica of the source encoder output, even though the noise might have caused some decoding error.

The '***source decoder***' transforms the estimated sequence into an estimate of the source output and delivers the estimate to the destination. When the source is continuous this involves digital to analog conversion. In a perfectly designed system the estimate will be a faithful reproduction of the source output except when the channel is very noisy.

We have already mentioned that as far as the original information source is concerned all the units (encoders at the transmission side and decoders at the receiver side) of the communication system can be considered as part of the channel. The physical channel between the modulator output and demodulator input handles analog electrical waveforms and hence is referred to as a '***Continuous or Modulation Channel***'. Examples of continuous channels are voice band and wideband telephone systems, high frequency radio systems and satellite links. The portion between the output of the channel encoder and the input of the channel decoder handles binary or discrete sequences of coded symbols. Hence this portion is referred to as a discrete ***'coding channel'***. The portion between the output of the source encoder and the input of the source decoder handles binary or discrete sequences of coded symbols and hence is referred to as '***Data Communication Channel***' or "***Discrete Communication Channel***". One of the important characteristics or the parameters of a data communication system is the "***Channel Capacity***". The channel capacity of a channel represents the maximum rate at which the data or information can be transmitted through the channel, with a tolerable probability of error. The following sections deal with the characteristics of the discrete communication channel and the concept of channel capacity.

3.1 Discrete Communication Channel

The coding channel between the channel encoder output and channel decoder input handles discrete sequences of coded symbols. When the channel encoder output code words pass through this channel they get corrupted mainly due to the noise in the physical channel and gets transformed to another set of code words. A common form of noise disturbance present in any communication system is *additive white Gaussian noise* (AWGN) which gets algebraically added to the signal passing through the channel. That is if the input waveform to the channel is $s(t)$ and the noise is $n(t)$ then the received waveform is $r(t) = s(t) + n(t)$.

Let us assume that the channel encoder output, that is the input to the coding channel, consists of a set of alphabet having 'M' discrete symbols. In this case we call the modulator unit as an 'M-ary' input modulator. Due to the channel induced error the output symbol of the channel may be different from the input symbol. Let the output of the coding channel be another set of alphabet having

'Q' discrete symbols (Q may not be equal to M). Then the demodulator is called a 'Q-ary' output demodulator. The probability of an input symbol being transformed into some other symbol depends on the noise conditions in the channel. If the demodulator (many times called as ***waveform detector***) output in a given interval depends only on the transmitted signal waveform in that interval, and not on any previous transmission, the channel is said to be memoryless. In this case, the combination of an M-ary input modulator, the physical channel and a Q-ary output demodulator can be modeled as a ***Discrete Memoryless Channel*** (DMC).

3.1.1. Transition Probability Diagram of Discrete Communication Channels

A discrete memoryless channel can be completely described by a set of input symbols, output symbols and a set of symbol transition probabilities $P(j/i)$ for $0 \leq i \leq (M-1)$ and $0 \leq j \leq (Q-1)$. Here 'i' represents a modulator input symbol and 'j' represents demodulator output symbol and $P(j/i)$ is the conditional probability of receiving the symbol 'j' given that 'i' was transmitted. The channel is represented by a '***transition probability diagram***'.

***Example* 3.1:** For example let us consider communication system in which

1. Binary modulation is used (M = 2),
2. The amplitude distribution of noise is symmetric,
3. The demodulator output is quantized to Q = 2 levels.

In this case a simple and practically important channel model, called the ***Binary Symmetric Channel*** (BSC) results. The transition probability diagram of BSC is shown in Fig. 3.2(a). We can see from the figure that the transition probability 'p' completely describes the channel. The transition probability 'p' can be calculated knowing the signal waveform used at the modulator to represent the code word, the probability distribution of noise in the channel and the output quantization threshold level of the demodulator.

The '***transition probability diagram***' is also called as '***channel diagram***' or '***noise diagram***'.

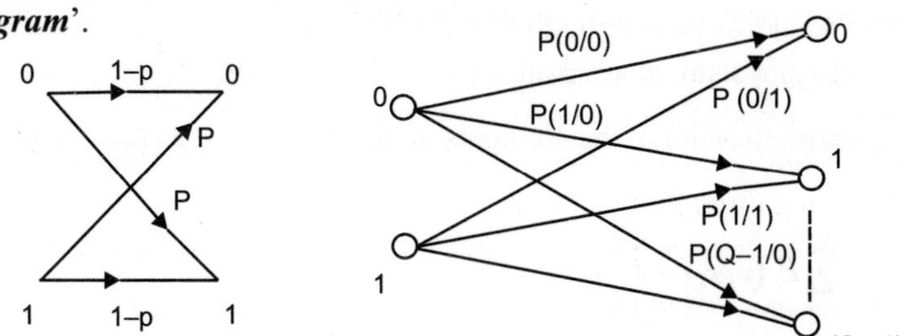

(a) Binary Symmetric Channel (b) Binary-input, Q-ary-output Discrete Memoryless Channel

Fig. 3.2. Transition Probability Diagram

Figure 3.2(b) illustrates the transition probability diagram for a binary input, Q-ary output discrete memoryless channel. Similarly the transition probability diagram can be drawn for an M-ary input, Q-ary output discrete memoryless channel. The transition probabilities can be calculated from the knowledge of the waveforms used in the modulator, the probability distribution of the noise, and the output quantization thresholds of the demodulator in the same way as it was done for the binary symmetric channel. It is important to mention here that models for channels with memory are difficult to construct and coding for these channels is normally done on ad-hoc basis.

3.1.2. Channel Matrix Representation of Discrete Communication Channel

A discrete communication channel can also be represented by a set of M-ary input symbols, Q-ary output symbols and a set of conditional probabilities $P(y_j / x_i)$ in the form of a matrix called ***Channel Matrix*** or ***Noise Matrix***. As any one of the input symbol may be received as any of the Q output symbols, there are M × Q conditional transition probabilities. Therefore, the Joint probability matrix will have a size of M × Q with input symbols represented row-wise and output symbols column-wise.

A typical Channel Matrix is given below:

$$P(y_j/x_i) = \begin{pmatrix} P(y_1/x_1) & P(y_2/x_1) & P(y_3/x_1) & \cdots\cdots & P(y_Q/x_1) \\ P(y_1/x_2) & P(y_2/x_2) & P(y_3/x_2) & \cdots\cdots & P(y_Q/x_2) \\ \cdot & \cdot & \cdot & \cdots\cdots & \cdot \\ \cdot & \cdot & \cdot & \cdots\cdots & \\ P(y_1/x_M) & P(y_2/x_M) & P(y_3/x_M) & \cdots\cdots & P(y_Q/x_M) \end{pmatrix} \quad (3.1)$$

It is obvious that when symbol x_1 is transmitted, due to noise, it may be received as y_1 or y_2 or……y_Q. Hence we have

$$P(y_1/x_1) + P(y_2/x_1) + P(y_3/x_1) + \ldots\ldots\ldots + P(y_Q/ x_1) = 1 \quad (3.2)$$

Therefore the sum of the elements in the first row of a channel matrix is unity. That is

$$\sum_{j=1}^{Q} P\,(y_j/x_1) = 1 \quad (3.3)$$

The equation (3.3) holds good for all 'M' values of the input symbol x_i.

Therefore we have $\sum_{j=1}^{Q} P\,(y_j / x_i) = 1$ (3.4)

From equation (3.4) it is clear that "*the sum of all the elements in any row of the channel matrix is equal to unity*". The symbols x_1, x_2, x_3, ……., x_M are the possible input symbols. Therefore, as per probability theory, the sum of the probabilities of their occurrences is equal to unity. Hence,

$$\sum_{i=1}^{M} P_i\,(x_i) = 1 \tag{3.5}$$

Therefore knowing the channel matrix elements, $P(y_j / x_i)$ and the probabilities of occurrence of the input symbols, $P(x_i)$, the probabilities of the output symbols, $P(y_j)$ can be written as:

$$\left.\begin{array}{llllll} P(y_1)=P(y_1/x_1)P(x_1) & + & P(y_1/x_2)P(x_2) & + & \ldots\ldots\ldots & + \; P(y_1/x_M)P(x_M) \\ P(y_2)=P(y_2/x_1)P(x_1) & + & P(y_2/x_2)P(x_2) & + & \ldots\ldots\ldots & + \; P(y_2/x_M)P(x_M) \\ :\;=\;: & : & : & : & : & : \\ :\;=\;: & : & : & : & : & : \\ P(y_Q)=P(y_Q/x_1)P(x_1) & + & P(y_Q/x_2)P(x_2) & + & \ldots\ldots\ldots & + \; P(y_Q/x_M)P(x_M) \end{array}\right\} \tag{3.6}$$

Knowing the channel matrix elements, $P(y_j / x_i)$, the probabilities of occurrence of the input symbols, $P(x_i)$, and the probabilities of the output symbols, $P(y_j)$, the "input conditional probabilities", $P(x_i / y_j)$, can be found out using Baye's Rule as:

$$P(x_i / y_j) = [P(y_j / x_i)P(x_i)] / P(y_j) \tag{3.7}$$

3.1.3. Joint Probability Matrix of Discrete Communication Channel

A discrete communication channel can also be represented by a joint probability matrix. The joint probability between any input symbol 'x_i' and any output symbol 'y_j' is given by

$$P(x_i, y_j) = P(x_i \cap y_j) = P(y_j / x_i)\, P(x_i) = P(x_i / y_j)\, P(y_j) \tag{3.8}$$

By multiplying the first row of the channel matrix of equation (3.1) by $P(x_1)$, second row by $P(x_2)$,………………., and M^{th} row by $P(x_M)$, we get

$$P(y_j / x_i)P(x_i) = \begin{pmatrix} P(y_1/x_1)P(x_1) & P(y_2/x_1)P(x_1) & \ldots\ldots\ldots & P(y_Q/x_1)P(x_1) \\ P(y_1/x_2)P(x_2) & P(y_2/x_2)P(x_2) & \ldots\ldots\ldots & P(y_Q/x_2)P(x_2) \\ . \quad . & . & \ldots\ldots\ldots & . \\ . \quad . & . & \ldots\ldots\ldots & . \\ . \quad . & . & \ldots\ldots\ldots & . \\ P(y_1/x_M)P(x_M) & P(y_2/x_M)P(x_M) & \ldots\ldots\ldots & P(y_Q/x_M)P(x_M) \end{pmatrix} \tag{3.9}$$

Using equation (3.8), equation (3.9) becomes

$$P(x_i, y_j) = \begin{pmatrix} P(x_1, y_1) & P(x_1, y_2) & \ldots\ldots\ldots & P(x_1, y_Q) \\ P(x_2, y_1) & P(x_2, y_2) & \ldots\ldots\ldots & P(x_2, y_Q) \\ . & . & . & \ldots\ldots\ldots \\ . & . & . & \ldots\ldots\ldots \\ . & . & . & \ldots\ldots\ldots \\ P(x_M, y_1) & P(x_M, y_2) & \ldots\ldots\ldots & P(x_M, y_Q) \end{pmatrix} \tag{3.10}$$

The matrix of equation (3.10) is called the "***Joint Probability Matrix***" (JPM) and the elements of this matrix are the joint probabilities between the input and output symbols. The JPM has very useful properties which are given below:

Property **1**: From equation (3.6) we have

$$P(y_1) = P(y_1/x_1)P(x_1) + P(y_1/x_2)P(x_2) + \ldots\ldots + P(y_1/x_M)\, P(x_M) \tag{3.11}$$

Using equation (3.8), equation (3.11) can be modified as

$$P(y_1) = P(x_1, y_1) + P(x_2, y_1) + \ldots\ldots\ldots\ldots\ldots\ldots\ldots + P(x_M, y_1) \tag{3.12}$$

Analyzing equation (3.12) and the JPM of equation (3.10) we see that the right hand side of equation (3.12) is the sum of the elements of the first column of the JPM. Hence, the probability of the first output symbol 'y_1' is equal to the sum of all elements of the first column of JPM. Therefore "***the probability of the output symbols can be obtained by adding the elements of JPM column wise***" that is

$$P(y_2) = P(x_1, y_2) + P(x_2, y_2) + \ldots\ldots\ldots\ldots\ldots\ldots + P(x_M, y_2) \tag{3.13}$$

$$\vdots \quad = \quad \vdots \qquad\qquad \vdots \qquad\qquad \vdots$$

$$\vdots \quad = \quad \vdots \qquad\qquad \vdots \qquad\qquad \vdots$$

$$P(y_Q) = P(x_1, y_Q) + P(x_2, y_Q) + \ldots\ldots\ldots\ldots\ldots\ldots + P(x_M, y_Q) \tag{3.14}$$

This property can be expressed as

$$P(y_j) = \sum_{i=1}^{M} P_i\ (x_i, y_j) \tag{3.15}$$

Property **2:** Knowing the probabilities of the output symbols, $P(y_j)$, and the conditional probabilities between the input and output symbols, we can write the expression for the probability of the input symbols as

$$P(x_1) = P(x_1/y_1)P(y_1) + P(x_1/y_2)P(y_2) + \;.........+ P(x_1/y_Q)\,P(y_Q)$$

Therefore $P(x_1) = P(x_1, y_1) + P(x_1, y_2) + \;........\; + P(x_1, y_Q)$ (3.16)

Similarly $P(x_2) = P(x_2, y_1) + P(x_2, y_2) + \;......\; + P(x_2, y_Q)$ (3.17)

$$\vdots \quad = \quad \vdots \qquad\qquad \vdots \qquad\qquad\qquad \vdots$$

$$\vdots \quad = \quad \vdots \qquad\qquad \vdots \qquad\qquad\qquad \vdots$$

$$P(x_M) = P(x_M, y_1) + P(x_M, y_2) + \;.............\; + P(x_M, y_Q) \tag{3.18}$$

It can be seen that the right hand sides of the above equations (3.15) to (3.18) are present in the successive 'rows' of the JPM. Hence the second property of the JPM can be stated as "***the probability of input symbols can be obtained by adding the elements of JPM row wise***".

This property can be expressed as:

$$P(x_i) = \sum_{j=1}^{Q} P\;(x_i, y_j) \tag{3.19}$$

***Property* 3:** Using equation (3.19) we can write the sum of the probabilities of all the input symbols as

$$\Sigma P(x_i) = \sum_{i=1}^{M} \left\{ \sum_{j=1}^{Q} P(x_i, y_j) \right\} \tag{3.20}$$

The right hand side of the above equation (3.20) is the sum of all the elements of JPM and as equation (3.5) the left hand side of equation (3.20) is equal to unity. Therefore the third property of JPM can be stated as "***the sum of all the elements of JPM is equal to unity***".

$$i.e., \quad \sum_{i=1}^{M} \; \sum_{j=1}^{Q} P\,(x_i, y_j) = 1 \tag{3.21}$$

The following example (3.2) illustrates the use of the JPM properties to find out certain parameters of the system, nature of the input and output symbols as well as certain conditional probabilities.

***Example* 3.2:** A discrete communication system uses a modulator with 3 inputs and a demodulator with 3 outputs. The system JPM with some marginal probabilities of these symbols is given below:

	y_1	y_2	y_3
x_1	1/12	#	5/36
x_2	5/36	1/9	5/36
x_3	#	1/6	#
$P(y_j)$	1/3	14/36	#

(*a*) Find the missing probabilities (#) in the matrix.

(*b*) Find $P(y_3/x_1)$ and $P(x_1/y_3)$.

(*c*) Are the symbols x_1 and y_1 statistically independent ? Give reasons for your answer.

Solution:

(*a*) From property1 of JPM, adding equations (3.12) to (3.14) we get

$$\sum_{j=1}^{a} P(x_i, y_j) = \sum_{i=1}^{M}\left\{\sum_{j=1}^{Q} P(x_i, y_j)\right\} = \text{the sum of all the elements of JPM}$$

$$\sum_{j=1}^{3} P(y_j) = 1$$

i.e., $P(y_1) + P(y_2) + P(y_3) = 1$

$1/3 + 14/36 + P(y_3) = 1$, i.e., $P(y_3) = 1 - (1/3 + 14/36)$

$\mathbf{P(y_3) = 5/18}$.

From property1 of JPM

$P(y_1) = P(x_1, y_1) + P(x_2, y_1) + P(x_3, y_1)$ (from first column of JPM)

i.e., $1/3 = 1/12 + 5/36 + P(x_3, y_1)$

i.e., $P(x_3, y_1) = 1/3 - (1/12 + 5/36) = 1/9$

Similarly

$P(y_2) = P(x_1, y_2) + P(x_2, y_2) + P(x_3, y_2)$ (from second column of JPM)

$14/36 = P(x_1, y_2) + 1/9 + 1/6$ or $P(x_1, y_2) = (14/36) - (1/9 + 1/6)$

$\mathbf{P(x_1, y_2) = 1/9}$

and $P(y_3) = P(x_1, y_3) + P(x_2, y_3) + P(x_3, y_3)$ (from third column of JPM)

$5/18 = 5/36 + 5/36 + P(x_3, y_3)$ or $P(x_3, y_3) = 5/18 - (5/36 + 5/36)$

$\mathbf{P(x_3, y_3) = 0}$

(*b*) From the definition of conditional probability and joint probability [equation (3.8)] we have

$$P(y_3/x_1) = [P(x_1 \cap y_3) / P(x_1)] = P(x_1, y_3)/P(x_1) \quad (3.22)$$

From property 2 of JPM

$$P(x_1) = P(x_1, y_1) + P(x_1, y_2) + P(x_1, y_3) \text{ (from first row of JPM)}$$

$$= 1/12 + 1/9 + 5/36 = 1/3$$

and $P(x_1, y_3) = 5/36$

Therefore equation (3.22) becomes $P(y_3 / x_1) = [(5/36) / (1/3)]$

i.e $P(y_3 / x_1) = 5/12$.

Similarly from equation (3.8) $P(x_1 / y_3) = P(x_1, y_3) / P(y_3) = [(5/36) / (5/18)]$
$= 1/2$.

(*c*) For x_1 and y_1 to be statistically independent they should satisfy

$$P(x_1 \cap y_1) = P(x_1, y_1) = P(x_1) P(y_1)$$

(i.e., the joint probability of two symbols should be equal to the product of their marginal probabilities)

However, $P(x_1) P(y_1) = (1/3)(1/3) = 1/9$

and $P(x_1, y_1) = 1/12$, that is $P(x_1, y_1) \neq P(x_1) P(y_1)$

Therefore x_1 and y_1 are not statistically independent. It is obvious because y_1 is the received symbol when x_1 is transmitted (of course with some probability) and hence y_1 is dependent on x_1.

3.2 Entropy Functions

From the discussions in the previous sections it is clear that for a discrete communication system there are five probability schemes of importance. They are the marginal probabilities of the input and output symbols namely P(*x*) and P(*y*) respectively; the joint probability between *x* and *y* namely P(*x*, *y*); the conditional probabilities P(*y*/*x*) and P(*x*/*y*). Based on these probabilities, five different entropy functions can be described for a discrete communication system.

3.2.1. Priori Entropy

The average information per input symbol, transmitted from the source, that is the entropy of the information source is defined as **"priori entropy"**, denoted by H(X) and given by

$$H(X) = \sum_{i=1}^{M} P_i(x_i) \log[1/P(x_i)] \text{ bits/message symbol} \quad (3.23)$$

3.2.2. Entropy of Receiver

The average information received per symbol at the receiver is defined as the ***"entropy of the receiver"*** or ***"entropy of output symbols"***, denoted by H(Y) and is given by

$$H(Y) = \sum_{j=1}^{Q} P(y_j) \log[1/P(y_j)] \text{ bits/message symbol} \quad (3.24)$$

3.2.3. Entropy of Discrete Communication System

The average information per pair of transmitted and received symbols is defined as the average uncertainty or the ***"entropy of the communication system"*** as a whole or ***"Joint Entropy"*** and is given by

$$H(X,Y) = \sum_{i=1}^{M}\sum_{j=1}^{Q} \{P(x_i, y_j)\log(1/P(x_i,y_j)\} \text{ bits/message symbol} \quad (3.25)$$

3.2.4. Posteriori (Conditional) Entropy

The entropy of input symbols, after transmission and reception of a particular output symbol 'y_j' is defined as ***"posteriori or conditional entropy"*** given by

$$H(X/y_j) = \sum_{i=1}^{M} P(x_i/y_j) \log[1/P(x_i/y_j)] \text{ bits/message symbol} \quad (3.26)$$

As '*j*' varies from 1 to Q there are Q such conditional entropies.

3.2.5. Average Conditional Entropy (Equivocation)

The average value of the conditional entropies when all the output symbols are taken into account is defined as the ***"Average Conditional Entropy or Equivocation"*** and is denoted by **H(X / Y)**. This entropy function gives a measure of information about a source when it is known that 'Y' is received. The Equivocation is obtained by averaging the equation (3.26) for all the values of 'y'.

$$H(X/Y) = E\{H(X/y_j)\}$$

$$= \sum_{j=1}^{Q} P(y_j)\, H(X/y_j)$$

Using equations (3.26) and (3.8) the above expression can be simplified as

$$H(X/Y) = \sum_{i=1}^{M}\sum_{j=1}^{Q} \{P(x_i y_j)\log[1/P(x_i/y_j)]\} \text{ bits/message symbol} \quad (3.27)$$

The equivocation is a measure of the amount of uncertainty when symbols are transmitted over the channel, hence represents the amount of information lost, with respect to any of the output symbol, as the symbols travel through the channel; and this loss is due to the noise and the error they induce.

Similar to H(X/Y), we can derive an expression to H(Y/X), the equivocation which gives a measure of information about the receiver. Using equation (3.27) we can get

$$H(Y/X) = \sum_{i=1}^{M}\sum_{j=1}^{Q}\{P(x_i,y_j)\log(1/P(y_j/x_i))\} \text{ bits/message symbol} \qquad (3.28)$$

3.3 Mutual Information

Observing equation (3.23) it is clear that, on an average H(X) bit of information is required to represent one input symbol. Whereas when we are in a position to monitor the output symbol corresponding to that input, only H(X/Y) bit of information is required to represent that input symbol. That is, on an average, monitoring a single output symbol provides us [H(X)–H(X/Y)] bits of information. This difference is known as the "Mutual Information" or "Transinformation" of the channel and is given by

$$I(X, Y) = [H(X) - H(X/Y)] \text{ bits/message symbol} \qquad (3.29)$$

Another way of interpreting Mutual information is as follows:

In section 3.2.5 we have mentioned that the average conditional entropy, H(X/Y), called equivocation is a measure of the amount of uncertainty when symbols are transmitted over the channel, hence, represents the amount of information lost while transmission through the channel. The amount of information transmitted, over the channel, from a source is given by the priori entropy H(X). Hence, the balance of information received at the receiver with respect to an observed output symbol is the mutual information, I(X,Y), given by equation (3.29). We can obtain the expression for the mutual information in terms of probabilities as follows:

The initial uncertainty (or self information), of the source symbol 'x_i' [as per equation (1.1)] is $\log[1/p(x_i)]$, and the final uncertainty about the same source symbol 'x_i' after receiving symbol 'y_j' is $\log[1/p(x_i/y_j)]$, which is the information received through the channel. We call the difference between the two as the mutual information. That is

$$I(x_i, y_j) = \log[1/p(x_i)] - \log[1/p(x_i/y_j)]$$

$$= \log[p(x_i/y_j)/p(x_i)], \text{ but } p(x_i/y_j) = p(x_i, y_j)/p(y_j)$$

Therefore, $\mathbf{I(x_i, y_j) = \log[p(x_i, y_j)/p(x_i)\,p(y_j)]}$ (3.30)

From equation (3.30) we can write $I(x_i) = I(x_i, x_i) = \log[p(x_i/x_i)/p(x_i)] = \log[1/p(x_i)]$, which is the expression for the self information with which we started the discussion on information of a symbol.

Also as $p(x_i, y_j) / p(x_i) = p(y_j / x_i)$ we can write

$$I(x_i, y_j) = \log [p(y_j / x_i) / p(y_j)]$$

$$= \log [1/ p(y_j)] - \log [1/ p (y_j / x_i)]$$

$$I(x_i, y_j) = I(y_j) - I(y_j / x_i)$$

$$I(x_i, y_j) = I(y_j, x_i)$$

$$I(X, Y) = I(Y, X) \tag{3.31}$$

Thus the mutual information is symmetrical with respect to its arguments and hence the name.

From this symmetric property we can write the expression for mutual information using the entropy of the output symbol H(Y), as:

$$I(Y, X) = [H(Y) - H(Y / X)] \text{ bits/message symbol} \tag{3.32}$$

From equation (3.31) we can write

$$H(X) - H(X / Y) = H(Y) - H(Y / X)$$

i.e., $H(X) + H (Y / X) = H(Y) + H(X / Y);$

but $H(X) +H(Y / X) = H(X, Y)$ (3.33)

and $H(Y) +H (X / Y) = H(Y, X)$

Therefore, $H(X, Y) = H(Y, X)$

From equation (3.32) we can write $H(Y/X) = H(Y) - I(Y, X)$

i.e., $H(Y / X) = H(Y) - I(X, Y)$, and using this relation in equation (3.33) we get,

$$H(X) +H(Y) - I(X, Y) = H(X, Y)$$

or $$I(X, Y) = H(X) + H(Y) - H(X, Y) \tag{3.34}$$

The relation between the different entropy functions are shown in Fig. 3.3 below. The entropies H(X) and H(Y) are represented by the two circles shown. The intersection area of the two circles represents the mutual information I(X, Y). The hatched area on the left circle represents the average conditional entropy (equivocation) H(X/Y) and that on the right circle represents H(Y/X) and they are as per equations (3.29) and (3.32). From this figure it is also clear that the joint entropy H(X, Y) = H(X) + H(Y) – I(X, Y) which satisfies equation (3.34). It can also be observed that

$$H(X,Y) = H(X) + H(Y/X)$$

$$= H(Y) + H(X/Y.)$$

From Fig. 3.3 it is clear that **H(X) ≥ H(X / Y)** and **H(Y) ≥ H(Y / X)**. Therefore it is clear that **"even though for a particular received symbol 'y_j', H(X) – H(X / y_j) may be negative, when all the valid output symbols are considered, the average mutual information is always non negative".** i.e., **I(X, Y) ≥ 0**.

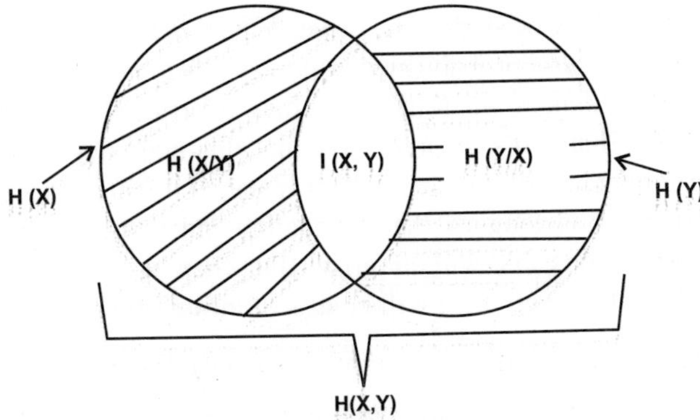

Fig. 3.3. Entropies Relation Diagram.

3.3.1. Properties of Mutual Information

The properties of the mutual information with respect to the various entropies are summarized below:

1. The mutual information of a channel is symmetric, i.e., **I(X, Y) = I(Y, X)**
2. The mutual information is always non negative, i.e., **I(X, Y) ≥ 0**.
3. The mutual information of a channel can be expressed in terms of the entropy of the channel output as **I(X, Y) = [H(Y) – H(Y / X)]**
4. The mutual information and the joint entropy of the channel are related as **I(X, Y) = H(X) + H(Y) – H(X, Y)**

3.4 Channel Capacity: Shannon's Theorem

The term communication channel encompasses all the features and component parts of the transmission system which introduce noise or limit the bandwidth. When we are studying about a channel, one will be very eager to know, how much of information we can transmit through a given channel or what is the rate R at which we can transmit and receive error free information. Shannon's (second) theorem which is fundamental to the theory of communication, says that it is possible, at least in principle, to conceive a method by which a communication system can transmit information with an arbitrarily small probability of error provided that the information rate R is less than or equal to a rate C called the ***channel capacity***. The technique used to achieve this limit is called coding. The formal statement of the theorem is given in two ways, a positive statement and negative statement.

Positive statement of Shannon's channel capacity theorem

"Given a source that generates information at a rate R, and a channel with capacity C, then if $R \leq C$, then there exists a coding technique which ensures transmission over the channel with as small a probability of error as possible even in the presence of channel noise".

This statement may be surprising, because there is category of noise, found in practical channels, whose probability density extends to infinity (Gaussian noise). So there may be instances when the noise may override the signal and result in errors. However, Shannon's theorem says that this need not cause a symbol to be in error. There are similar examples, for instance if we consider an amplitude modulation system, as long as the modulation index $m \leq 1$, transmission of modulated signal is possible without error. But when $m > 1$, there will be errors due to over modulation and there will be loss of information.

Negative statement of Shannon's channel capacity theorem:

"Given a source that generates information at a rate R, if $R > C$ then the probability of decoding error is close to unity".

That is when $R > C$ reliable transmission of information is not possible without error. No coding technique, however complex it may, will ensure reliable communication, as increased complexity increases the probability of error.

3.4.1. Capacity of Discrete Memoryless Channel

The capacity of a discrete memoryless channel is defined as the maximum possible rate of information transmission over the channel.

That is the channel capacity C is defined as

$$C = \text{Max } \{R_t\}$$

where 'R_t' is the average rate of information transmission. The value for 'R_t' can be obtained from the entropy functions of the discrete communication system.

For a discrete memoryless channel accepting symbols at a rate of 'r_s' message symbols per second, the average rate at which information is fed into the channel is given by

$$R_{in} = H(X)\, r_s \text{ bits/sec}$$

At the receiver we will not be able to guarantee the unambiguous reconstruction of the symbols because of the errors caused by the channel noise. Thus some amount of information is lost in the channel, which is the average conditional entropy (equivocation) H(X/Y) given by equation (3.27). The net amount of information transferred is the mutual information I(X, Y), given by equation (3.29) as:

$$I(X, Y) = [H(X) - H(X/Y)] \text{ bits/message symbol}$$

Then the average rate of information transmission 'R_t' is given by

$$R_t = I(X, Y)\, r_s \text{ bits/sec}$$

or $$R_t = \{[H(X) - H(X / Y)]\, r_s\} \text{ bits/sec} \quad (3.35)$$

As $I(X, Y) = I(Y, X) = [H(Y) - H(Y/ X)]$ we can also express R_t as

$$R_t = \{ [H(Y) - H(Y/X)]\, r_s \} \text{ bits/sec}$$

Therefore, the channel capacity of a discrete memoryless channel can be written as:

$$C = \text{Max } \{[H(X) - H(X/Y)]\, r_s\} \text{ bits/sec} \quad (3.36)$$

3.5 Channel Efficiency and Redundancy

The **"Channel Efficiency"** is given by

$$\eta_{ch} = [R_t / C] \times 100\ \%$$

Substituting the values of R_t and C from equations (3.35) and (3.36) we have

$$\eta_{ch} = \{[H(X) - H(X/Y)]\, r_s\} / \text{Max } \{[H(X) - H(X/Y)]\, r_s\} \times 100\%$$

or $$\eta_{ch} = I(X, Y) / \text{Max } [I(X,Y)] \times 100\% \quad (3.37)$$

The **"Channel Redundancy"** is given by

$$R_{\eta_c} = 1 - \eta_{ch} \quad (3.38)$$

3.6 Different Communication Channels

In section 3.1.1 under example (3.1) we have mentioned simple and practically important discrete memoryless channel models, like ***Binary Symmetric Channel*** (BSC) and a **binary input, Q-ary output channel**. There are several other special channel models which are of great interest in the field of communication systems. In this section, we will find out the channel capacity of these special channels. The important channel models are as given below:

(*a*) Symmetric/Uniform Channels
(*b*) Binary Symmetric Channels (BSC)
(*c*) Binary Erasure Channels (BEC)
(*d*) Deterministic Channel
(*e*) Noiseless Channel
(*f*) Cascaded Channel

3.6.1. Symmetric/Uniform Channel

"A channel is set to be symmetric or uniform channel, if it affects all the symbols passing through it, in a manner such that the set of transition probabilities of an input symbol transformed into the possible output symbols is the same for all the input symbols except for the order".

This definition can be given in terms of the channel matrix as, "a channel is said to be symmetric or uniform channel, if the second and subsequent rows of the channel matrix contain the same elements as that of first row, but in a different order". That is the elements of the second and subsequent rows are exactly the same as those of the first row except for their position. This is illustrated by the following matrix of equation (3.39)

$$P(Y/X)=\begin{pmatrix} p_1 & p_2 & p_3 & \cdots\cdots & p_{n-1}p_n \\ p_2 & p_3 & p_4 & \cdots\cdots & p_n\ p_1 \\ p_{n-1} & p_n & p_1\cdots\cdots & p_{n-3} & p_{n-2} \\ \cdot & \cdot & \cdot & \cdots\cdots & \cdot \\ \cdot & \cdot & \cdot & \cdots\cdots & \cdot \\ p_n & p_{n-1} & p_{n-2}\cdots\cdots & p_2 & p_1 \end{pmatrix} \tag{3.39}$$

The important property of the conditional probability matrix of the channel is that, the sum of the elements in any row should be equal to 1. Accordingly for the channel matrix in equation (3.39) we have

$$\sum_{j=1}^{n} P_j = 1 \tag{3.40}$$

The conditional entropy H(Y/X) for this channel can be computed from

$$H(Y/X) = \sum_{i=1}^{m}\sum_{j=1}^{n}\{p(x_i, y_j)\log(1/p(y_j/x_i))\}$$

$$H(Y/X) = \sum_{i=1}^{m} P(x_i)\sum_{j=1}^{n}\{p(y_j/x_i)\ \log(1/p\ (y_j/x_i))\} \tag{3.41}$$

For the channel under consideration we have

$$\sum_{j=1}^{n}\{p(y_j/x_i)\log(1/p(y_j/x_i))\} = \sum_{j=1}^{n}\{p_j \log\ (1/p_j) = h, \text{ a constant}$$

and $\sum_{i=1}^{m} P\ (x_i)$ = sum of the probabilities of all source symbols = 1.

Therefore, the conditional entropy of the channel, **H(Y/X) = *h*,** a constant.

Thus the conditional entropy of a symmetric or uniform channel can be computed from the elements of any row of the channel matrix, that is

$$H(Y/X) = h = \sum_{j=1}^{n}\{p_j \log(1/p_j)\}$$

Accordingly the mutual information of a symmetric or uniform channel is given by

$$I(X,Y) = H(Y) - H(Y/X)$$

i.e., $I(X,Y) = H(Y) - h$

Hence the Capacity of this channel $C = \text{Max}\ \{I(X, Y)\}$

$$= \text{Max}\ \{H(Y) - h\}$$

$$= \text{Max}\ \{H(Y)\} - h$$

Since H(Y) will be maximum iff all the received symbols are equiprobable and as there are '*n*' output symbols, we have $H(Y)_{Max} = \log(n)$. Hence **the capacity of a symmetric or uniform channel** is given by

$$\mathbf{C = \log(n) - \mathit{h}} \qquad (3.42)$$

Since, I(X,Y) = H(X) – H(X/Y) we can so that the channel capacity is also given by

$$C = \text{Max}\ \{H(X)\} - h, \text{ were } h' \text{ is a constant.}$$

Since H(X) will be maximum iff all the source symbols are equiprobable and as there are '*m*' input symbols, we have $H(X)_{Max} = \log(m)$. Hence **the capacity of a symmetric or uniform channel** is also given by

$$\mathbf{C = \log(\mathit{m}) - \mathit{h'}} \qquad (3.43)$$

Thus the computation of the channel capacity of a symmetric or uniform channel is simple and straight forward. The following example (3.3) illustrates the computation of the channel capacity of a symmetric channel.

***Example* 3.3:** Let us consider the channel represented by the channel diagram given in Fig. 3.4.

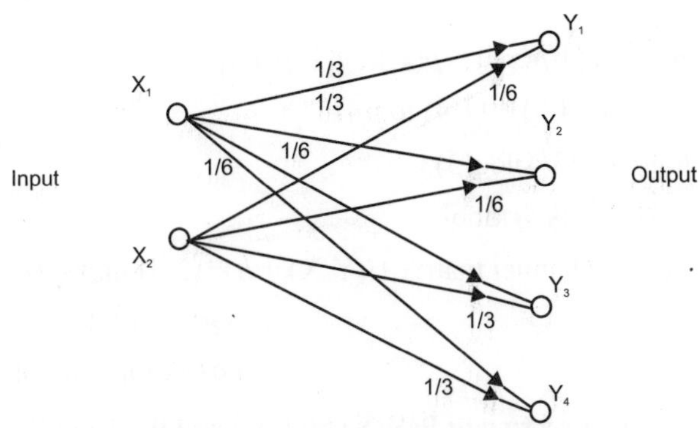

Fig. 3.4. A Symmetric Channel

The channel matrix can be formed from the channel diagram as

$$P(Y/X) = \begin{pmatrix} 1/3 & 1/3 & 1/6 & 1/6 \\ 1/6 & 1/6 & 1/3 & 1/3 \end{pmatrix}$$

It is seen that the second row of the matrix is a permutation of the first row (written in the reverse order) and hence the channel illustrated is a uniform (symmetric) channel. Then the conditional entropy is computed from any one of the rows as;

$$H(Y/X) = 2\,(1/3)\log(3) + 2(1/6)\log(6) = 1.9183 \text{ bits/symbol.}$$

i.e., $h = 1.9183$ bits/symbol

Hence the capacity of the channel $C = \log(n) - h = \log(4) - 1.9183$

$$= 2 - 1.9183$$

i.e., $C = 0.0817$ bits/symbol

***Example* 3.4:** The channel matrix of a binary channel is given as below:

$$\begin{matrix} & 0 & 1 \\ 0 & 2/3 & 1/3 \\ 1 & 1/3 & 2/3 \end{matrix}$$

(*a*) The input symbol (X) transmission probabilities are $p(x_1) = 3/4$ and $p(x_2) = 1/4$, find the priori H(X), receiver entropy H(Y), joint entropy H(X, Y), conditional channel entropy H(Y/X), and the mutual information I(X,Y).

(*b*) Find the channel capacity, efficiency and redundancy of the channel.

(*c*) What are the source probabilities that correspond to the channel capacity?

Solution: Given: $p(x_1) = (3/4)$ and $p(x_2) = (1/4)$; $\quad P(Y/X) = \begin{bmatrix} 2/3 & 1/3 \\ 1/3 & 2/3 \end{bmatrix}$

(*a*) $H(X) = p(x_1)\log[1/p(x_1)] + p(x_2)\log[1/p(x_2)]$

$$= (3/4)\log(4/3) + (1/4)\log(4)$$

$$= \log(4) - (3/4)\log(3)$$

$$= 0.8113 \text{ bits/symbol}$$

From the given channel matrix $H(Y/X) = h = (2/3)\log(3/2) + (1/3)\log(3)$

$$= \log(3) - (2/3)$$

$$= 0.9183 \text{ bit/symbol}$$

Multiplying the first row of P(Y/X) by $p(x_1)$ and the second row by $p(x_2)$, we get the joint probability P(X, Y) as:

$$P(X, Y) = \begin{pmatrix} (2/3)\times(3/4) & (1/3)\times(3/4) \\ (1/3)\,(1/4) & (2/3)\times(1/4) \end{pmatrix} = \begin{pmatrix} 1/2 & 1/4 \\ 1/12 & 1/6 \end{pmatrix}$$

Adding the elements of the matrix P(X, Y), column wise, we get

$p(y_1) = (7/12);\ p(y_2) = (5/12)$

Dividing the elements of the first column of P(X, Y) by $p(y_1)$ and those of column two by $p(y_2)$ we get

$$P(X/Y) = \begin{pmatrix} 6/7 & 3/5 \\ 1/7 & 2/5 \end{pmatrix}$$

$H(Y) = p(y_1)\log[1/p(y_1)] + p(y_2)\log[1/p(y_2)]$

$= (7/12)\log(12/7) + (5/12)\log(12/5) = 0.9799$ bits/symbol

From P(X,Y) we have $H(X,Y) = (1/2)\log 2 + (1/4)\log 4 + (1/12)\log 12 + (1/6)\log 6$

$= 1.7296$ bits/symbol.

Using the elements of P(X,Y) and P(X/Y) we have

$H(X/Y) = (1/2)\log(7/6) + (1/4)\log(5/3) + (1/12)\log 7 + (1/6)\log(5/2)$

$= 0.7497$ bits/symbol.

Using the elements of P(X,Y) and P(Y/X) we have

$H(Y/X) = (1/2)\log(3/2) + (1/4)\log(3) + (1/12)\log 3 + (1/6)\log(3/2)$

$= (2/3)\log(3/2) + (1/3)\log 3 = \log(3) - (2/3) = 0.9183$ (as before)

$= h$

$I(X,Y) = H(X) - H(X/Y) = 0.8113 - 0.7497 = 0.0616$ bits/symbol.

$= H(Y) - h = 0.0616$ bits/symbol.

(*b*) The channel capacity $C = \log(n) - h = \log 2 - h = 1 - h$

$= 0.0817$ bits/symbol.

If the input symbol rate $r_s = 1000$ symbols/sec, then the channel capacity in bits/sec is given by $C = (r_s) \times (0.0817) = 1000 \times 0.0817 = 81.7$ bits/sec.

The channel efficiency $= \eta = [I(X,Y)/C] = 0.7536$ or 75.36 %

The channel redundancy $= 1 - \eta = 0.2464$ or 24.64 %.

(*c*) Let $p(x_1) = p$ and $p(x_2) = q = (1 - p)$, the JPM can be written as

$$P(X,Y) = \begin{pmatrix} (2/3)p & (1/3)p \\ (1/3)(1-p) & (2/3)(1-p) \end{pmatrix}$$

Adding the elements of the above matrix P(X.Y), column wise, we get

$p(y_1) = (1/3)(1+p);\ p(y_2) = (1/3)(2-p)$

For $H(Y) = H(Y)_{Max}$, $p(y_1)$ should be equal to $p(y_2)$, hence $(1 + p) = (2 - p)$ or $p = 1/2$.

Therefore, the source probabilities corresponding to the channel capacity are

$$p(x_1) = p(x_2) = 1/2.$$

3.6.2. Binary Symmetric Channels (BSC)

BSC is the simple, practically important, most common and widely used Discrete Memoryless Channel (DMC) model. The channel diagram and channel matrix are given in Fig. 3.5. below: Line of Sight (LOS) communication channels and Satellite communication channels are examples of BSC.

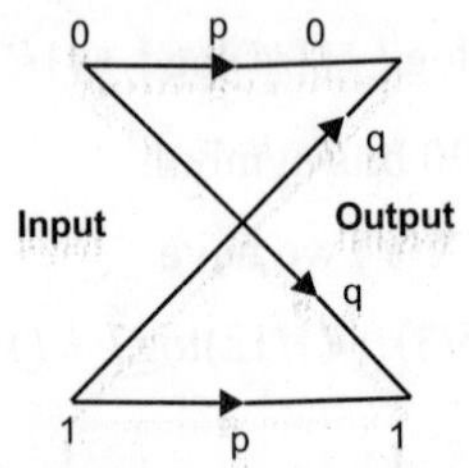

$$P(Y/X) = \begin{array}{c} \\ 0 \\ 1 \end{array}\begin{array}{c} \begin{array}{cc} 0 & 1 \end{array} \\ \begin{pmatrix} p & q \\ q & p \end{pmatrix} \end{array}$$

Fig. 3.5. Channel Diagram and Channel Matrix of a BSC.

Here 'p' and 'q' are the transition probabilities such that "$p + q = 1$". Let the input symbol probabilities are $P(X = 0) = \beta$, and $P(X = 1) = 1 - \beta$. In a BSC the source symbol 'x_1' is coded as '0' and 'x_2' is coded as '1'. Also we see that the P(Y/X), channel matrix, is symmetric and hence the channel is symmetric channel.

For a symmetric channel we have $H(Y/X) = h = \sum_{j=1}^{n} \{p_j \log(1/p_j)\}$

$$\text{Therefore for the BSC } H(Y/X) = h = p\log(1/p) + q\log(1/q) \tag{3.44}$$

$$H(Y) = \sum_{j=1}^{2} \{p(y_j)\log(1/p(y_j))\}$$

$$H(Y) = P(y_1)\log(1/P(y_1))\} + P(y_2)\log(1/P(y_2))\} \tag{3.45}$$

Using equation (3.6)

$$P(y_1) = P(y_1 / x_1)P(x_1) + P(y_1 / x_2)P(x_2) = q\beta + p(1-\beta)$$

$$P(y_2) = P(y_2 / x_1)P(x_1) + P(y_2 / x_2)P(x_2) = p\beta + q(1-\beta)$$

Substituting the values of $P(y_1)$ and $P(y_2)$ in equation (3.45) we get

$$H(Y) = [q\beta + p(1-\beta)] \log\{1/[q\beta + p(1-\beta)]\} + [p\beta + q(1-\beta)] \times \log\{1/[p\beta + q(1-\beta)]\} \tag{3.46}$$

Therefore, the mutual information of a BSC is given by equation (3.32)

$$I(Y,X) = I(X,Y) = [H(Y) - H(Y/X)]$$

Therefore, using equation (3.46) and (3.44)

$$I(X,Y) = \{[q\beta + p(1-\beta)] \log\{1/[q\beta+p(1-\beta)]\}+[p\beta +q(1-\beta)] \log\{1/[p\beta + q(1-\beta)]\} - [p\log(1/p) + q\log(1/q)]\} \text{ bits/message symbol} \quad (3.47)$$

From the above equation it is clear that if 'p' and 'q' are interchanged in the channel matrix the mutual information of the channel remains unaltered.

Since BSC is a symmetric channel, the channel capacity with the input symbol rate $r_s = 1$ symbol/sec is obtained from equation (3.42) as;

$$C = \{\log(n) - h\} \text{ bits/sec}$$

$$= \log(2) - h = 1 - h$$

Then using the value of 'h' from equation (3.44), we have,

$$\mathbf{C = 1- [p\log(1/p)+q\log(1/q)]} \text{ bits/sec} \quad (3.48)$$

Suppose $\beta = (1- \beta) = 1/2$ then from equation (3.46) we have

$$H(Y) = (1/2)\ (p + q)\log[1/(1/2)(p + q)] + (1/2)\ (p + q)\log[1/(1/2)\ (p + q)]$$

$$= (1/2) \log2 + (1/2) \log2, \text{ since } (p + q) = 1$$

Therefore, H(Y) = 1

Also the mutual information I(X,Y) becomes

$$I(X,Y) = [H(Y) - H(Y/X)] = 1 - h$$

$$\textbf{Therefore, } \mathbf{I(X,Y) = 1 - [p\log(1/p) + q\log(1/q)]} \quad (3.49)$$

Thus when the input symbols are equiprobable, the mutual information I(X,Y) maximizes and becomes equal to the channel capacity C.

3.6.3. Binary Erasure Channels (BEC)

Another binary discrete channel which is one of the most important channels in digital communication is the Binary Erasure Channel. The noise conditions in the channel are such that a symbol 'y' is received with which the receiver (decoder) is not in a position to make a decision whether a binary '1' was transmitted or a binary '0' was transmitted. In other words, the information is erased in the channel and hence the name ***Binary Erasure Channel***. Such erasure channel has an input alphabet of size 'q' and an output alphabet of size 'q+1', and the **extra symbol is called an erasure**. In order to have a reliable communication through such channels, digital communication systems use what is known as Automatic Repeat Request (ARQ) strategy. In such systems whenever the symbol 'y' is received an immediate request is sent through a reverse (reliable) channel, for retransmission of the transmitted symbol till a correct symbol is received. However this scheme requires a reverse channel. The channel transition probability diagram of a BEC is shown in Fig. 3.6a. The BEC is a symmetric channel and the channel matrix can be derived from the channel diagram as shown in Fig. 3.6b.

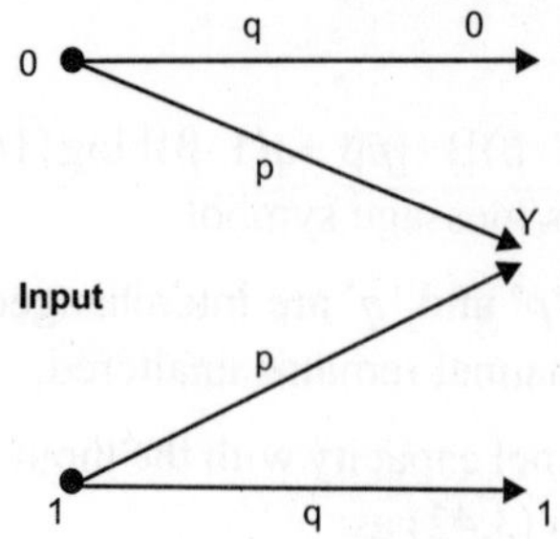

$$P(Y/X) = \begin{array}{c} \\ 0 \\ 1 \end{array}\begin{array}{c} \begin{array}{ccc} 0 & y & 1 \end{array} \\ \begin{pmatrix} q & p & 0 \\ 0 & p & q \end{pmatrix} \end{array}$$

(a) Channel Diagram (b) Channel Matrix

Fig. 3.6. Binary Erasure Channel

Here again 'p' and 'q' are the transition probabilities, that is the probability of '0' received as 'y' and '1' received as 'y' is 'p'; the probability of '0' received as '0' and '1' received as '1' is 'q'. Also "$p + q = 1$". Let the input symbol probabilities are $P(X = 0) = \beta$, and $P(X = 1) = (1 - \beta)$. In a BEC the source symbol 'x_1' is coded as '0' and 'x_2' is coded as '1'. Also we see that P(Y/X), the channel matrix is symmetric and hence the channel is symmetric channel.

For a symmetric channel we have $H(Y/X) = h = \sum_{j=1}^{n} \{p_j \log(1/p_j)\}$

Therefore, for the BEC $\mathbf{H(Y/X) = h = \mathit{p}\log(1/\mathit{p}) + \mathit{q}\log(1/\mathit{q})}$ (3.50)

From equation (3.23) $\mathbf{H(X) = \beta\log(1/\beta) + (1-\beta)\log[1/(1-\beta)]}$ (3.51)

We know from the equation (3.8) that $P(x_i, y_j) = P(y_j / x_i)P(x_i)$. Therefore, by multiplying the first row of the channel matrix of BEC by $P(x_1) = \beta$ and the second row by $P(x_2) = (1-\beta)$, we get the JPM of BEC as,

$$P(X,Y) = P(x_i, y_j) = \begin{array}{c} \\ x_1 \\ x_2 \end{array}\begin{array}{c} \begin{array}{ccc} y_1 & y & y_2 \end{array} \\ \begin{pmatrix} q\beta & p\beta & 0 \\ 0 & p(1-\beta) & q(1-\beta) \end{pmatrix} \end{array} \quad (3.52a)$$

From property 1 of JPM we have

$P(y_1) = q\beta + 0 = q\beta$; $P(y) = p\beta + p(1-\beta) = p$; $P(y_2) = 0 + p(1-\beta) = p(1-\beta)$

From equation (3.8) we also have that $P(x_i / y_j) = P(x_i, y_j) / P(y_j)$

The input conditional probability matrix P(X/Y) can then be constructed as below

$$P(X/Y) = P(x_i / y_j) = \begin{pmatrix} q(1-\beta)/q(1-\beta) & p\beta/p & 0 \\ 0 & p(1-\beta)/p & q(1-\beta)/q(1-\beta) \end{pmatrix}$$

$$P(X/Y) = P(x_i / y_j) = \begin{pmatrix} 1 & \beta & 0 \\ 0 & (1-\beta & 1 \end{pmatrix} \quad (3.52b)$$

From equation (3.27) we have

$$H(X/Y) = \sum_{i=1}^{2}\sum_{j=1}^{3}\{P(x_i, y_j)\log(1/P(x_i/y_i))\}$$

$$= q\beta\log(1) + p\beta\log(1/\beta) + p(1-\beta)\log[1/(1-\beta)] + q(1-\beta)\log(1)$$

$$\mathbf{H(X/Y) = p\ \{\beta\log(1/\beta) + (1-\beta)\log[1/(1-\beta)]\}} \qquad (3.53)$$

Therefore, the mutual information of a BEC is given by

$$I(X, Y) = H(X) - H(X/Y)$$

$$= \beta\log(1/\beta) + (1-\beta)\log[1/(1-\beta)] - p\ \{\beta\log(1/\beta) + (1-\beta)\log[1/(1-\beta)]\}$$

$$= (1-p)\{\beta\log(1/\beta) + (1-\beta)\log[1/(1-\beta)]\}$$

Therefore, $I(X, Y) = q\ H(X)$ (3.54)

Therefore, the channel capacity of BEC is given by

$$C = \text{Max}\ [\ I(X, Y)\ ] = \text{Max}\ [\ q\ H(X)\ ]$$

$$= q\ H(X)_{\text{Max}}$$

$$= q\log(2) = q.\ [\text{since } H(X)_{\text{Max}} = \log(Q) = \log(2) = 1]$$

Therefore, the capacity of a BEC is given by $\mathbf{C = q}$ (3.55)

It is to be noted here that even though BEC is a symmetric channel, the formula $I(X, Y) = H(Y) - H(Y/X)$, cannot be used since H(Y) involves 'y' which is rejected at the receiver.

3.6.4. Deterministic Channel

If a particular received symbol is always due to the transmission of the same set of (one or more) input symbols, the channel is called as a Deterministic Channel. The channel matrix which satisfies this condition will have the following two properties:

- Each row will have one and only one non zero element (equal to '1').
- The sum of all the elements in every row is equal to unity.

In order to understand the concept let us consider the following example (3.5)

***Example* 3.5:** Fig. 3.7., below shows the channel matrix and the corresponding channel diagram of a channel.

$$P(Y/X) \text{ or } P(y_j/x_i) = \begin{pmatrix} 1 & 0 & 0 \\ 1 & 0 & 0 \\ 0 & 1 & 0 \\ 0 & 1 & 0 \\ 0 & 1 & 0 \\ 0 & 0 & 1 \end{pmatrix}$$

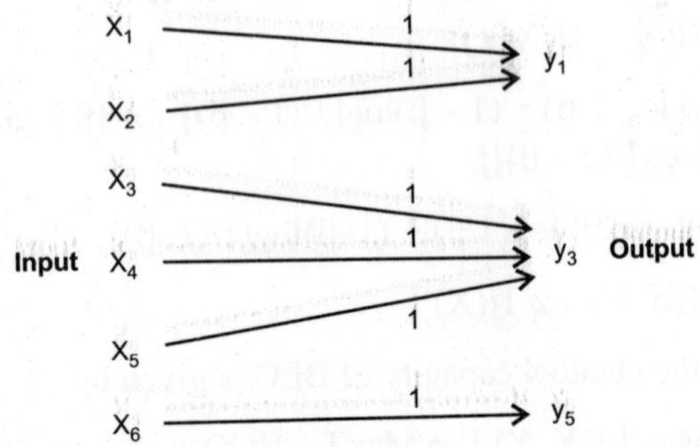

Fig. 3.7. Channel Matrix and Channel Diagram of Deterministic Channel.

The channel matrix above is symmetric and has the property for a given 'i' and 'j', $P(y_j / x_i) = 1$ and all other entries are zero. Hence, given 'x_i', probability of receiving it as 'y_j' is one. Hence, it is clear that we can determine which symbol is to be received at the encoder stage itself. That is for a given input we can say which is going to be the output. However by observing the output no decision can be made about the transmitted symbol.

Using equation (3.28) we can find out the equivocation as:

$$H(Y/X) = \sum_{i=1}^{6}\sum_{j=1}^{3}\{P(x_i, y_j)\log(1/P(y_j/x_i))\}$$

From equation (3.8) we have $P(x_i, y_j) = P(y_j / x_i)P(x_i)$

$$\text{Therefore } H(Y/X) = \sum_{i=1}^{6}\sum_{j=1}^{3} P(x_i)\{P(y_j / x_i)\log(1/P(y_j / x_i))\}$$

However, The terms inside the flower bracket is of the form either $\{0\times\log(1/0)\}$ or $\{1\times \log(1/1)\}$ which is always zero.

Therefore, $H(Y/X) = 0$

Therefore, the mutual information is given by

$$I(X, Y) = H(Y) - H(Y/X) = H(Y) - 0$$

$$\text{Therefore, } I(X, Y) = H(Y) \tag{3.56}$$

Therefore, the channel capacity of a deterministic channel is given by

$$C = \text{Max}\,[I(X, Y)] = H(Y)_{\text{Max}} = \log\,(\text{number of output symbols}) = \log 3$$
$$= 1.585 \text{ bits/symbol}$$

Therefore, in general, for 'Q' output symbols, the capacity of a deterministic channel is given by

$$\mathbf{C = log(Q)\ bits/message\ symbol} \qquad (3.57)$$

3.6.5. Noiseless Channel

It is possible to design a system in which a received symbol is due to the transmission of a particular source symbol only. At the same time one or more received symbols may be due to the transmission of the same source symbol. In other words a particular source symbol may be transformed into the same set of (one or more) output symbols always. The channel matrix which satisfies these conditions will have the following two properties:

- Each column will have one and only one non zero element.
- The sum of all the elements in any row is equal to unity.

In order to understand the concept, let us consider the following example (3.6) and prove that the channel matrix of above type represents a noiseless channel.

***Example* 3.6:** Fig. 3.8., shows the channel matrix and the corresponding channel diagram of a channel.

As it is seen from the channel diagram, x_1 is having connection only to y_1 and y_2, x_2 to y_3, y_4, y_5 and x_3 to y_6 and not to any other output symbols.

$$P(Y/X) = P(y_j/x_i) = \begin{pmatrix} 1/2 & 1/2 & 0 & 0 & 0 & 0 \\ 0 & 0 & 1/4 & 1/4 & 1/2 & 0 \\ 0 & 0 & 0 & 0 & 0 & 1 \end{pmatrix}$$

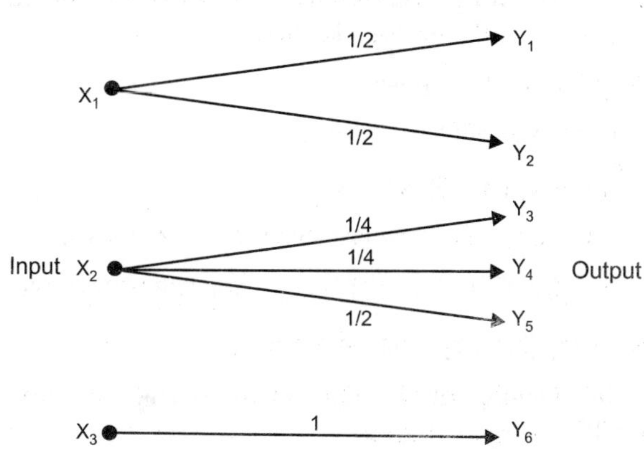

Fig. 3.8. Channel Matrix and Diagram of a Noiseless Channel.

Therefore, the conditional probabilities can be written as;

$P(x_1/y_1) = 1,\ P(x_2/y_1) = 0,\ P(x_3/y_1) = 0$

$P(x_1/y_2) = 1,\ P(x_2/y_2) = 0,\ P(x_3/y_2) = 0$

$P(x_1/y_3) = 0,\ P(x_2/y_3) = 1,\ P(x_3/y_3) = 0$

$P(x_1/y_4) = 0,\ P(x_2/y_4) = 1,\ P(x_3/y_4) = 0$

$P(x_1/y_5) = 0,\ P(x_2/y_5) = 1,\ P(x_3/y_5) = 0$

$P(x_1/y_6) = 0,\ P(x_2/y_6) = 0,\ P(x_3/y_6) = 1$

Therefore, the conditional probability matrix can be written as;

$$P(x_i/y_j) = \begin{pmatrix} 1 & 1 & 0 & 0 & 0 & 0 \\ 0 & 0 & 1 & 1 & 1 & 0 \\ 0 & 0 & 0 & 0 & 0 & 1 \end{pmatrix} \tag{3.58}$$

So, it is clear that $P(x_i / y_j)$ is either '0' or '1' for all 'i' and 'j'. Using equation (3.27) the average conditional entropy (equivocation) can be written as;

$$H(X/Y) = \sum_{i=1}^{3}\sum_{j=1}^{6} \{P(x_i, y_j) \log(1/P(x_i/y_j))\}$$

$$= \sum_{i=1}^{3}\sum_{j=1}^{6} \{P(y_j)\}\{P(x_i/y_j) \log(1/P(x_i/y_j))\}$$

The terms inside the flower bracket is of the form either $\{0 \log(1/0)\}$ or $\{1 \log(1/1)\}$ which is always zero.

Therefore, $\mathbf{H(X/Y) = 0}$

That is the equivocation which represents the amount of information lost in the channel due to noise is zero and hence the channel represented by the matrix considered here is also known as ***Noiseless Channel;*** and the channel matrix of example (3.6) represents a Noiseless channel.

The mutual information is given by

$$I(X, Y) = H(X) - H(X/Y) = H(X) - 0$$

$$\text{Therefore, } I(X, Y) = H(X) \tag{3.59}$$

The channel capacity of a noiseless channel is given by

$$C = \text{Max}\ [I(X, Y)] = H(X)_{\text{Max}} = \log\ (\text{number of input symbols})$$

$$= \log 3 = 1.585 \text{ bits/symbol}$$

Therefore, in general, for 'r' input symbols, the capacity of a Noiseless channel is given by

$$\mathbf{C = \log(\mathit{r})} \text{ bits/message symbol} \tag{3.60}$$

We can notice in this case that, one output symbol uniquely specifies the input symbol whereas for a given input symbol, we cannot make any decision about the received symbol. This is exactly the complement of the deterministic channel case. Hence, sometimes this channel is also called as a deterministic channel.

***Example* 3.7:** Let us consider a special case of a square matrix with all entries except the ones on the principal diagonal are zero, as the channel matrix. In this case $P(x_i / y_i) = 1$ and $P(x_i / y_j) = 0$ for $i \neq j$. That is P(Y/X) is an Identity Matrix of order 'n'. For this case we can verify that

$$P(X/Y) = P(Y/X) \text{ and } P(x_i, y_j) = P(x_i) = P(y_j).$$

For such a channel, it follows that H(X/Y) = H(Y/X) = 0 and I(X, Y) = H(X) = H(Y) = H(X,Y). Therefore, there is a one to one correspondence between the input and output symbols. No information will be lost in such a channel. Hence, such a channel is called as "Noiseless channel". ***A unique decision exists with regard to the transmitted or received symbols from both (transmitter and receiver) end***; which is not the case in example (3.5) and (3.6). ***This uniqueness property is vital in calling the channel a "Noiseless Channel***". Such a channel is also referred to as an ***"Ideal Channel"***. If all the symbols occur with equal probability the channel capacity of such a channel is given by

$$C = I(X,Y)_{Max} = H(X)_{Max} = H(Y)_{Max} = H(X,Y) = \log(n) \text{ bits/symbol.}$$

Thus, a Noiseless Channel is symmetric and deterministic with respect to both matrix P(Y/X) and P(X/Y).

3.6.6. Cascaded Channel

In many cases, it is required for the information to travel through more than one channel, one after the other. For example, from home or office in a building the information may be connected first by a cable to the nearby switching centre and then through a Microwave link to the destination. That is the two channels are in cascade to transfer the information. Let us consider two channels connected in series (cascade) as shown in Fig. 3.9. The source X emits m-symbol alphabet. The output, Y(*a* 'q' symbol alphabet), of channel1is fed as the input of channel 2. The final output, Z, of channel 2 is a 'r' symbol alphabet. A symbol 'x_i' transmitted from X will appear as 'y_j' at the input of channel 2. This in turn will appear as 'z_k' at the output of channel 2.

Fig. 3.9. Cascade of Two Channels.

The output symbol 'z_k' depends on the source symbol 'x_i' only through 'y_j'. The probability of getting 'z_k' depends only on the probability of 'y_j', whatever may the way it was produced. The property of the cascaded channel can be given by the equation (3.61).

$$P(z_k / y_j, x_i) = P(z_k / y_j) \text{ for all } i, j, k \tag{3.61}$$

Similarly using Baye's probability theorem, we have

$$P(x_i / y_j, z_k) = [P(x_i, y_j, z_k)] / P(y_j, z_k)$$

$$= [P(y_j)P(x_i / y_j)P(z_k / y_j, x_i)] / [P(y_j)P(z_k / y_j)]$$

$$\text{i.e., } P(x_i / y_j, z_k) = P(x_i / y_j) \tag{3.62}$$

The equation (3.61) can also be taken as the definition for the cascaded channels. Also equations (3.61) and (3.62) is valid for the cascaded configuration of figure(3.9) only.

When symbols pass through the channel 1 from X to Y, there will be loss of information due to noise in channel 1 and the mutual information available at Y is I(X, Y) = H(X) – H(X/Y). When this resultant information is passed through channel 2, there will be further loss of information and the final mutual information available at Z will be I(X, Z) = H(Z) – H(Z/X) .

$$\text{It is clear that } I(X, Z) < I(X, Y) \tag{3.63}$$

However if channel 2 is noiseless then I(X, Z) = I(X, Y) (or in general $I(X, Z) \leq I(X, Y)$).

The following example (3.8) illustrates the concept of two BSCs in cascade.

***Example* 3.8:** Let us assume that both the channels have the same channel diagram of Fig. 3.5. The channel diagrams can be cascaded as shown in Fig 3.10.

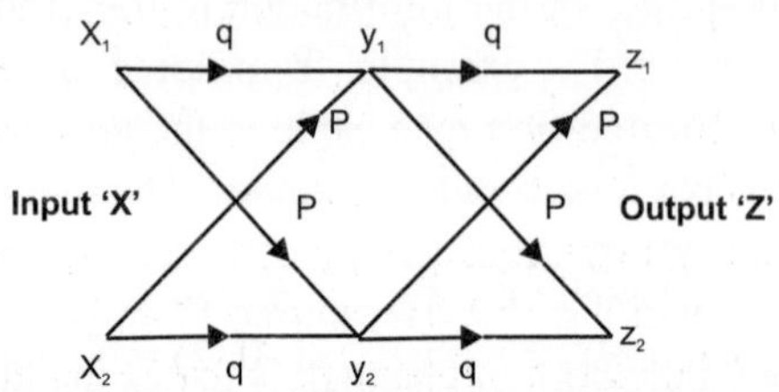

Fig. 3.10. Channel Diagram of Cascaded BSCs

From the channel diagram of the two BSCs, we can write as

$$P(z_1 / x_1) = q^2 + p^2 = (p + q)^2 - 2pq = 1 - 2pq = P(z_2 / x_2)$$

$$\text{Let} \quad 2pq = u$$

$$\text{Then } P(z_1 / x_1) = P(z_2 / x_2) = 1 - u = \overline{u}$$

$$\text{Similarly } P(z_1 / x_2) = pq + pq = 2pq = u = P(z_2 / x_1)$$

$$\text{That is } P(z_1 / x_2) = P(z_2 / x_1) = u$$

Therefore, the combined channel matrix with X as input (transmitter) and Z as out (receiver) is given by

$$P(Z/X) = \begin{matrix} & z_1 & z_2 \\ x_1 & \bar{u} & u \\ x_2 & u & \bar{u} \end{matrix} \quad (3.64)$$

Comparing the channel matrix of the BSC (Fig. 3.5.) and that of the cascaded channel (equation 3.64) it is clear that both are same except that the transition probability (probability of error) of the cascaded channel is 'u' instead of 'p'. Therefore, it can be concluded that when we cascade two BSCs, the resultant combined channel is also a BSC with transition probability equal to 'u'. Also the channel diagram of the cascaded channel can be redrawn as given in Fig. 3.11.

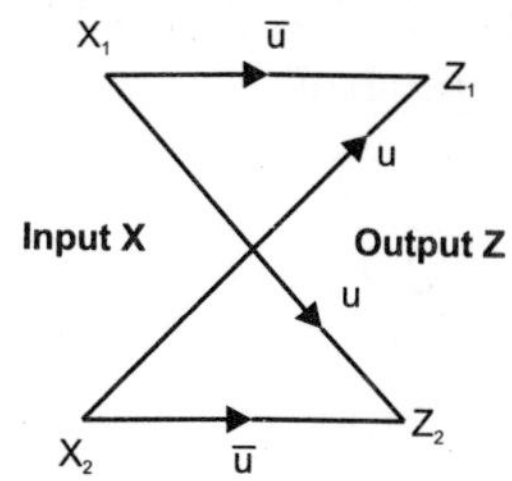

Fig. 3.11. Channel Diagram of Cascaded BSCs.

The channel capacity of individual BSCs, is given by equation (3.48) as

$$C_{XY} = C_{YZ} = 1 - [p\log(1/p) + q\log(1/q)] \quad (3.65)$$

Similarly, the channel capacity of the cascaded channel which is also a BSC of Fig. 3.10. is given by

$$C_{YZ} = 1 - [\bar{u}\log(1/\bar{u}) + u\log(1/u)] \quad (3.66)$$

If $p = 0.2$, then $u = 2pq = 2\,(0.2)\,(1\text{–}0.\,2) = 0.32$ and $\bar{u} = 1 - u = 0.68$

Therefore, $C_{XY} = C_{YZ} = 1 - [(0.2)\log(1/(0.2)) + (0.8)\log(1/(0.8))]$

$= 0.278$ bits/symbol

$C_{XZ} = 1 - [(0.68)\log(1/(0.68)) + (0.32)\log(1/(0.32))]$

$= 0.0956$ bits/symbol

Therefore, $C_{XZ} < C_{XY}$

3.6.7. Channels with Independent Input-output Characteristics

Let us consider a channel with the following JPM,

$$P(X,Y)=\begin{pmatrix} p_1 & p_1 & p_1 & \cdots\cdots & p_1 & p_1 \\ p_2 & p_2 & p_2 & \cdots\cdots & p_2 & p_2 \\ p_3 & p_3 & p_3 & \cdots\cdots & p_3 & p_3 \\ \cdot & \cdot & & & \cdot & \cdot \\ \cdot & \cdot & & & \cdot & \cdot \\ p_m & p_m & p_m & \cdots\cdots & p_m & p_m \end{pmatrix} \quad (3.67)$$

With $\sum_{i=1}^{M} P_{i_i} = (1/n)$. That is $P(y_j)$ = sum of elements in first column of $P(X, Y) = (1/n)$, for $j = 1, 2, 3, \ldots\ldots\ldots, n$.

This means that there is no correlation between 'x_i' and 'y_j' and an input 'x_i' may be received as any one of 'y_j' with equal probability. In other words the input – output statistics are independent.

As we have $P(x_i, y_j) = p_i$

$$= n.p_i \,.\, \sum_{i=1}^{M} p_i$$

$$= P(x_i).\, P(y_j)$$

Therefore, $P(x_i / y_i) = P(x_i) = np_i$

$P(y_j / x_i) = P(y_j) = (1/n)$ and we have

$$H(X, Y) = n.\ \sum_{i=1}^{M} P_i \ \log(1/p_i);$$

$$H(X) = \sum_{i=1}^{m} (np_i)\log[1/(np_i)]$$

i.e., $H(X) = n\sum (p_i) \log[1/p_i] + \log[1/n]$;

$$H(Y) = \sum_{j=1}^{n} P\,(y_j) \log\,[1/P(y_j)] = \log(n),$$

$$\left.\begin{aligned} &H(X/Y)=H(X),\\ &H(Y/X)=H(Y)\text{ and}\\ &I(X,Y)=0 \end{aligned}\right\} \quad (3.68)$$

Since, the mutual information I(X, Y) = 0, such a channel conveys no information at all. Thus a channel with independent input-output characteristics is similar to a purely resistive network with heavy and largest internal loss. This channel is in contrast to a noiseless channel which resembles a lossless network.

The characteristics of the channels seen so far can be summarized as follows:

- A channel with channel matrix having only one non-zero element in each column is termed as "**Lossless or Noiseless Channel**"; in this case **H(X/Y)=0** and **I(X, Y) = H(X)**, the Source Entropy.
- A channel with channel matrix having only one non-zero element in each row (which is necessarily a one) is termed as "**Deterministic Channel**"; in this case **H(Y/X) = 0** and **I(X, Y) = H(Y)**.
- A channel with channel matrix having only one non-zero element in each column and in each row (i.e., a diagonal matrix) is termed as an "**Ideal Channel**"; in this case **H(X/Y) = H(Y/X) = 0** and **I(X, Y) = H(X) = H(Y)**. An ideal channel is both lossless and deterministic.
- A channel with channel matrix having identical columns except for permutations and identical rows except for the permutations is termed as a "**Uniform Channel**". When the channel matrix is square, then every row and every column are simply the permutations of the first row.
- When each row and column of the channel matrix are made up of all non-zero elements and each equal to (1/n) and consequently both the rows and columns are linearly dependent, then the channel will have independent input – output characteristics.
- The capacity of a cascaded channel will be less than that of the individual channels which are cascaded; If two BSCs are cascaded the resultant channel will also be a BSC.

3.6.8. Muroga's Algorithm for Channel Capacity Estimation

The channel matrix of a binary channel can be written as;

$$P(y_j/x_i) = \begin{array}{c} \\ x_1 \\ x_2 \end{array} \begin{array}{c} \begin{array}{cc} y_1 & y_2 \end{array} \\ \begin{pmatrix} p_{11} & p_{12} \\ p_{21} & p_{22} \end{pmatrix} \end{array} \tag{3.69}$$

The algorithm, suggested by Dr. S. Muroga for the computation of the channel capacity of a binary channel, estimates two quantities Q_1 and Q_2 from the channel matrix. The procedure is as follows:

$$\begin{pmatrix} p_{11} & p_{12} \\ p_{21} & p_{22} \end{pmatrix} \begin{pmatrix} Q_1 \\ Q_2 \end{pmatrix} = \begin{pmatrix} p_{11}\log(p_{11}) & p_{12}\log(p_{12}) \\ p_{21}\log(p_{21}) & p_{22}\log(p_{22}) \end{pmatrix} \tag{3.70}$$

i.e.,

$$\left.\begin{aligned} p_{11}Q_1 + p_{12}Q_2 &= p_{11}\log(p_{11}) + p_{12}\log(p_{12}) \\ p_{21}Q_1 + p_{22}Q_2 &= p_{21}\log(p_{21}) + p_{22}\log(p_{22}) \end{aligned}\right\} \quad (3.71)$$

The above two equations being simultaneous equations, can be solved for Q_1 and Q_2. Then the channel capacity is computed based on the following expression:

$$C = \log\,[2^{Q_1} + 2^{Q_2}] \text{ bits/message symbol} \quad (3.72)$$

The above result can be generalized for a channel matrix of size "$m \times m$" as follows:

$$P(y_j/x_i) = \begin{pmatrix} p_{11} & p_{12} & \cdots\cdots & p_{1m} \\ p_{21} & p_{22} & \cdots\cdots & p_{2m} \\ \vdots & \vdots & \cdots\cdots & \vdots \\ p_{m1} & p_{m2} & \cdots\cdots & p_{mm} \end{pmatrix} \quad (3.73)$$

The quantities $Q_1, Q_2, \ldots\ldots\ldots\ldots, Q_m$ can be computed as follows:

$$\begin{pmatrix} p_{11} & p_{12} & \cdots & p_{1m} \\ p_{21} & p_{22} & \cdots & p_{2m} \\ \vdots & \vdots & \cdots & \vdots \\ pm_1 & p_{m22} & \cdots & p_{mm} \end{pmatrix} \begin{pmatrix} Q_1 \\ Q_2 \\ \vdots \\ Q_m \end{pmatrix} = \begin{pmatrix} p_{11}\log p_{11} + & p_{12}\log p_{12} + & \cdots & + p_{1m}\log p_{1m} \\ p_{21}\log p_{21} + & p_{22}\log p_{22} + & \cdots & p_{2m}\log p_{2m} \\ \vdots & \vdots & \vdots & \vdots \\ p_{m1}\log p_{m1} + & p_{m2}\log p_{m2} + & \cdots & p_{mm}\log p_{mm} \end{pmatrix} \quad (3.74)$$

Then the channel capacity 'C' is given by

$$C = \log\,[2^{Q_1} + 2^{Q_2} + \ldots\ldots\ldots + 2^{Q_m}] \text{ bits / message symbol} \quad (3.75)$$

3.6.9. Discrete Channels with Memory

So far we were discussing on the aspects of discrete memoryless channels used in digital communication. In these memoryless channels we have seen that the demodulator (detector) output at the receiver in a given interval depends only on the symbols transmitted in that interval and output is independent of the symbols transmitted in the previous intervals. Also the noise in such channels affects the transmitted symbols independently. That is the errors in different intervals are independent of one another and the transmission errors occur randomly. For this reason the ***memoryless channels are also known as random error channels***. This is not the case in most of the channels used in practical communication systems. The symbol received in an interval depends not only on the symbol transmitted during that interval but on the symbols transmitted during previous intervals also. A fading radio channel used in mobile communication is a good example for this case, since the multipath reflections destroys the independence from interval to interval. Landline telephone channels which are affected by switching transients is another example. The noise in these channels is not independent from transmission to transmission and the transmission errors occur in bursts. These channels are said to have memory. That is, ***if the demodulator (detector) output in a given interval depends on the transmitted symbol in previous intervals, as well as the transmitted symbol in the present interval, the channels is said to be a discrete channel with memory***. It is difficult to model a perfect discrete channel with memory because of the complex nature of the noise in such channels. A simplified model of a channel with memory is shown in Fig. 3.12(a).

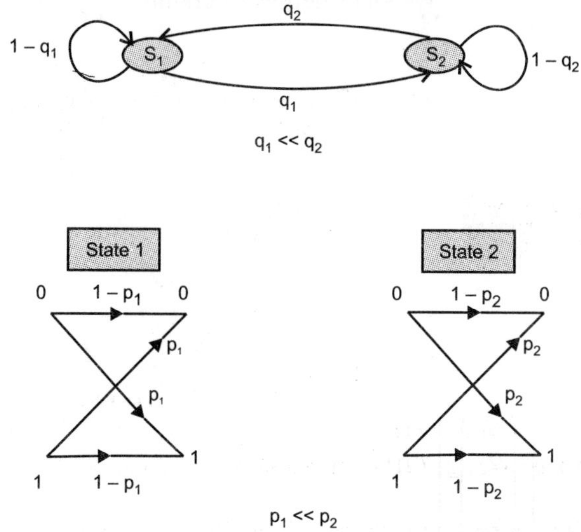

Fig. 3.12(a). Simplified Model of a Channel with Memory

The model consists of two states, ***"good state"***, in which transmission errors occur infrequently, $p_1 \approx 0$, and a ***"bad state"***, in which transmission errors are highly probable, $p_2 \approx 0.5$. In each state the channel is modeled as a BSC. The channel is in good state most of the time, but on occasion goes to the bad state due to the change in channel characteristics (a deep fade caused by multipath transmission, switching transient in a telephone line, due to an ignition noise from an automobile or jamming signal). As a result, transmission errors occur in clusters or bursts because of the high transition probability in the bad state. For this reason channels with memory are called **burst error channels**. Examples of burst error channels are radio channels, where the error bursts are caused by signal fading due to multipath transmission, wire and cable transmission, which is affected by impulsive switching noise and crosstalk.

Some channels contain a combination of both random and burst errors. These are called compound channels. Example of a compound channel is a mobile communication channel. A communication system installed in a mobile, when the mobile moves in open terrain, is subjected to only random errors (due to thermal noise), as there is no multipath reflections; when the mobile moves into an area where there are large number of buildings, towers with iron structure which can reflect EM waves and result in multipath reflection, is subjected to burst errors.

Thus the system encounters a compound channel as it moves from one area to another.

Another popular model which represents a channel with memory is the ***Gilbert Model***. Here the channel is modeled as a discrete memoryless BSC, where the error probability is a varying function of time. The changes in error probability can be represented by a Markov process. The details of the Gilbert model is shown in Fig. 3.12(b) below.

State	1	2	3
P_e	0.5	10^{-2}	10^{-6}

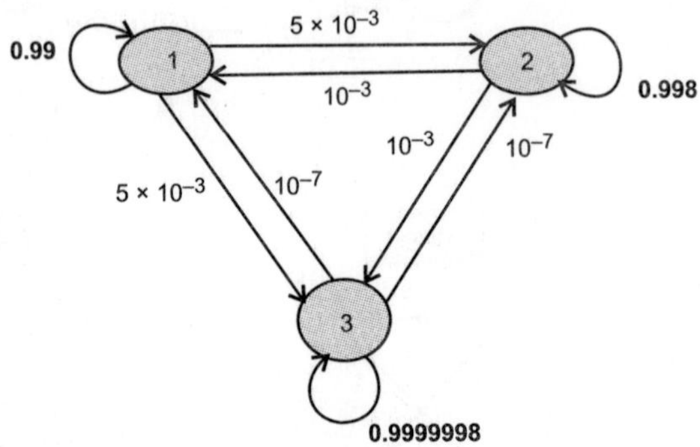

Fig. 3.12 (b). A 3-State Gilbert Model for Communication Channels

The channel assumes three states and the transition from one state to another are modeled by a discrete stationary Markov process. When the channel is in state 2, the bit error probability during a symbol (bit) interval is 10^{-2} and the channel stays in this state during the succeeding symbol interval with a probability 0.998. Occasionally the channel may go to state1 wherein the error probability (P_e) is 0.5. Since the system stays in this state with probability 0.99, errors tend to occur in cluster or bursts. State 3 represents a low symbol error rate and errors in this state are produced by Gaussian noise and errors very rarely occur in bursts. The channel capacity of the channel represented by the Gilbert model can be computed for each state of the channel using BSC model of the channel corresponding to each of the three states.

3.7 Continuous Channels

We have mentioned in section 3.0 that discrete symbols are not suitable for transmission over physical channel; so the modulator unit transforms each of the channel encoder output symbols into a waveform (analog), of appropriate period, which is suitable for transmission over the physical channel. Most of the communication channels (radio) are suitable for transmission of continuous analog signals. This waveform enters the channel and is corrupted by additive noise. We assume this additive noise as Gaussian; this is because the noise that results from summed effects of many sources produce thermal noise and radiation noise which are generally Gaussian in nature. The modulator and demodulator pairs are designed to handle the Gaussian noise. The physical channel between the modulator output and demodulator input handles analog electrical waveforms and hence is referred to as a ***'Continuous or Modulation Channel'***. Examples of continuous channels are voice band and wideband telephone systems, high frequency radio systems and satellite links.

We shall define a continuous channel as one whose input is a sample point from a continuous sample space and the output is a sample point belonging to either the same sample space or to some other sample space. Further we shall define ***"zero memory continuous channels"*** as the one in which the channel output statistically depends on the corresponding channel input. Here in the study of continuous channels we will concentrate only on continuous channels without memory. The analog ***'Continuous Channel'*** [of Fig. (3.1)] can be modeled as shown in Fig. 3.13. below:

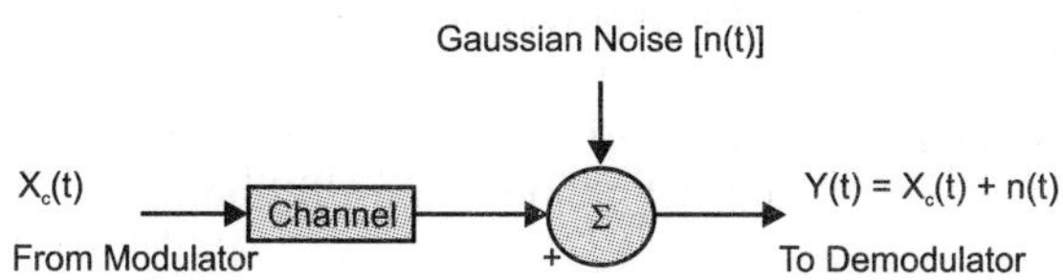

Fig. 3.13. Continuous Channel Model

The input to the channel, $\mathbf{X_c(t)}$, consisting of the waveforms from the modulator, is a random process. The bandwidth, of the channel and the input waveform $\mathbf{X_c(t)}$, is assumed as **B** Hz. The additive noise added to the waveform is a zero mean, band limited, white Gaussian Noise, **n(t)**. The modulator can be considered as a continuous message source.

3.7.1. Entropy of Continuous Channels – Differential Entropy of Continuous Ensembles

In this section we will compute the entropy of the continuous channel model defined above. For the discrete channel or the discrete message the entropy of the message source is

$$\mathbf{H(M)} = \sum_{i=1}^{m} \{P(S_i)\log[1/P(S_i)]\} \tag{3.76}$$

In the case of a continuous message source the above equation (3.76) can be extended, by considering the analog or continuous message as a sum of infinite number closely spaced discrete messages. For the continuous channel model considered in Fig. 3.13, let us assume that the input to the channel is a continuous random variable (CRV) given by

$$X = i\Delta x,$$

i.e., $X = 0, \pm\Delta x, \pm 2\Delta x, \pm 3\Delta x, \ldots\ldots\ldots\ldots, \pm i\,\Delta x, \ldots\ldots\ldots$ and $\Delta x \to 0$.

By definition, the continuous random variable X takes a value in the interval $(x_i, x_i + i\Delta x)$ with probability $f(x_i)\Delta x$, [i.e., $P\{x \le X \le x + dx\} = f(x)dx$]; where $f(x)$ is the probability density function (pdf) of a random variable. In the limit, when $\Delta x \to 0$, the error in the approximation would tend to become zero. Hence, by using equation (3.76), the entropy of a continuous random variable, X, can be given by

$$H(X) = \lim_{\Delta x \to 0} \{\Sigma f(x_i)\, \Delta x.\, \log[1/(f(x_i)\Delta x]\}$$

$$= \lim_{\Delta x \to 0} \{\Sigma f(x_i)\, \Delta x.\, \log[1/(f(x_i)] - \Sigma f(x_i)\, \Delta x.\, \log(\Delta x)\}$$

When $\Delta x \to 0$, $f(x_i).\ \Delta x \to f(x)dx$ and the summation can be replaced by integration. Hence we have

$$H(X) = \int_{-\infty}^{+\infty} f(x)\log[1/f(x)]dx - \lim_{\Delta x \to 0} \{\log\Delta x \int_{-\infty}^{+\infty} f(x)dx\}$$

Since, $f(x)$ is a pdf, $\int_{-\infty}^{+\infty} f(x)dx = 1$.

Therefore, $H(X) = \int_{-\infty}^{+\infty} f(x)\log[1/f(x)]dx - \lim_{\Delta x \to 0} \{\log\Delta x\}$ (3.77)

However, as Δx tends zero, $\log(\Delta x)$ tends to ∞. That is the entropy of a continuous random variable or source is infinitely large. This is valid, because a continuous random variable can take any value in the interval $(-\infty,\infty)$. This situation can be avoided by going for a relative measure of the entropy instead of the absolute value. That is we can take H(X) as **"differential entropy"**, by taking the term "**– log Δx**" as a reference. The information conveyed by this way will be meaningful as all the entropy measure is given with respect to the same common reference. Hence the **differential entropy of a continuous message can be defined as;**

$$H(X) \triangleq \int_{-\infty}^{+\infty} f(x)\log[1/f(x)]dx \text{ bits/sample} \tag{3.78}$$

It should be always remembered that the above definition is only a relative measure and not an absolute value. The following example (3.9) will illustrate this.

***Example* 3.9:** Let X is a uniform random variable (URV) over the interval (0, 2). Hence, we have,

$$f(x) = 1/2 \text{ for } 0 \le x \le 2$$
$$= 0 \text{ elsewhere.}$$

Then using equation (3.78) we have $H(X) = \int_0^2 \left(\frac{1}{2}\right)\log(2)dx = 1$ bits/sample.

Suppose X is the input to an amplifier with gain 4. Then the output of the amplifier would be

$$Y = 4X \text{ and } f(y) = 1/(2x) = 1/8 \text{ for } 0 \le y \le 8.$$

Therefore, the entropy, H(Y), of the amplifier output becomes

$$H(Y) = \int_0^8 \left(\frac{1}{8}\right) \log(8) dy = 3 \text{ bits/sample.}$$

i.e., the entropy of the output is thrice that of the input. However since Y is determined by X uniquely and as an (ideal) amplifier does not alter the information except for amplification the entropy, H(Y), should have been 4 bits/sample. This discrepancy is because we have not taken the reference level into account. The reference level of X, using equation (3.77), is given by

$$R_x = \lim_{\Delta x \to 0} \{-\log \Delta x\},$$

Similarly the reference level of Y is given by

$$R_y = \lim_{\Delta y \to 0} \{-\log \Delta y\},$$

$$\text{Therefore } R_x - R_y = \lim_{\Delta x \to 0} \{-\log \Delta x\} - \lim_{\Delta y \to 0} \{-\log \Delta y\}$$

$$= \lim_{\Delta x,\, \Delta y \to 0} \{\log \Delta y - \log \Delta x\}$$

$$= \lim_{\Delta x, \Delta y \to 0} \{\log(\Delta y / \Delta x)\}$$

$$= \log[\lim_{\Delta x, \Delta y \to 0} \{\Delta y / \Delta x\}]$$

$$= \log (dy / dx) \text{ [by definition of differentiation]}$$

$$R_x - R_y = \log(4) = 2 \text{ bits/sample}$$

Therefore, (absolute entropy of X-absolute entropy of Y)

$$= [R_x + H(X)] - [R_y + H(Y)]$$

$$= (R_x - R_y) + [H(X) - H(Y)]$$

$$= 2 + (1 - 3)$$

$$= 0$$

That is the absolute values of both X and Y are equal, there is no loss of information and they carry equal uncertainties.

3.7.2. Maximization of Entropy

Practical communication systems have many limitations and resource constraints. For example, the system in mobile vehicles, airborne systems, on board satellite systems etc., (battery) power is always a limitation. As a result the transmitted power of the communication system is limited. The aim of any communication system will be to maximize the amount of information transfer or transmission; that is to maximize the entropy function under the existing constraints. There are five general constraints and are as listed below:

1. **Normalizing constraint:** which follows from the basic property of any probability density function that is

   $$\int_{-\infty}^{+\infty} f(x)dx = 1.$$

2. **Peak signal limitation:** when the signal is limited to $\pm$ M, then,

   $$\int_{-M}^{+M} f(x)dx = 1.$$

3. **Average signal limitation:**

   $$\int_{-\infty}^{+\infty} xf(x)dx = \text{'}m\text{'} = E\{X\} = \text{a constant.}$$

4. **Average power limitation:**

   $$\int_{-\infty}^{+\infty} x^2 f(x)dx = \text{'}m_2\text{'} = E\{X^2\} = \text{the 2}^{\text{nd}}\text{ moment} = \text{a constant.}$$

5. **Average power limitation with unidirectional distribution (causal systems which produce response only after the input is fed):**

   $$\int_{0}^{+\infty} x^2 f(x)dx = \text{a constant.}$$

The objective now will be to arrive at a probability density function, $f(x)$ that will maximize the entropy function H(X) given by the equation (3.78). This maximization is a problem of constrained optimization and is a particular case of the Isoperimetric problem of the calculus of variations. Euler—Lagrange's method of constrained maximization can be used to maximize the entropy function H(X). Let us try to maximize the following integral

$$I = \int_a^b \phi(x, f)dx \tag{3.79}$$

with the following constraints:

$$\begin{aligned} \int_a^b \phi_1(x, f)dx &= \lambda_1 \\ \int_a^b \phi_2(x, f)dx &= \lambda_2 \\ \vdots \qquad & \quad \vdots \\ \int_a^b \phi_r(x, f)dx &= \lambda_r \end{aligned} \tag{3.80}$$

where $\lambda_1, \lambda_2, \lambda_3, \ldots\ldots, \lambda_r$ are pre-assigned constants. Then the function $f(x)$ that satisfies all the above constraints and makes 'I' a maximum is calculated by solving the equation,

$$\frac{\delta\phi}{\delta f}+\alpha_1\frac{\delta\phi_1}{\delta f}+\alpha_2\frac{\delta\phi_2}{\delta f}+\ldots\ldots\ldots\ldots\ldots\ldots+\alpha_r\frac{\delta\phi_r}{\delta f}=0 \tag{3.81}$$

The coefficients $\alpha_1, \alpha_2, \ldots, \alpha_r$ in the above equation, called **"Lagrangian Multipliers"** or **"Lagrangians"** are determined substituting the value of $f(x)$ in equations (3.80) successively. Now let us consider the constraints (2) to (5) mentioned above and determine the corresponding pdf, $f(x)$.

Maximization of Entropy under Peak Signal Limitation

Many practical systems have peak signal limitation. The communication system which uses AM, FM and Pulse code modulation transmitters has always the peak signal limitation. Let the signal peak, a random variable, is limited to ± M. Then using the equation (3.78), the entropy function is given by

$$\mathrm{H(X)}=\int_{-\mathrm{M}}^{+\mathrm{M}} f(x)\log\left[\frac{1}{f(x)}\right]dx \tag{3.82}$$

The only constraint on $f(x)$ is the normalizing condition based on the property of a pdf, that is

$$\int_{-\mathrm{M}}^{+\mathrm{M}} f(x)dx=1 \tag{3.83}$$

Comparing equations (3.79) & (3.82); and (3.80) & (3.83) we have

$\phi(x, f) = -f(x)\log[f(x)]$,

$\phi_1(x, f) = f(x)$ with $\lambda_1 = 1$

In equation (3.81) we will have only the first two terms, given by

$$\frac{\delta\phi}{\delta f}+\alpha_1\frac{\delta\phi_1}{\delta f}=0.$$

$\therefore$ by substituting the values of $\phi(x, f)$ and $\phi_1(x, f)$ we have

$$\frac{\delta}{\delta f}[-f(x)\log f(x)]+\alpha_1\frac{\delta}{\delta f}[f(x)]=0$$

$$\frac{\delta}{\delta f}\left[-f(x)\frac{\ln[f(x)]}{\ln(2)}\right]+\alpha_1=0$$

$$-\frac{1}{\ln(2)}\left[f(x)\left(\frac{1}{f(x)}\right)+\ln[f(x)]\right]+\alpha_1=0,\left[\text{that is}\frac{d}{dx}(\log x)=\frac{1}{x}\right]$$

$$\therefore -\frac{1}{\ln(2)}\left[1+\ln[f(x)]\right]=-\alpha_1$$

$$\therefore f(x) = e^{[\alpha_1 \ln(2) - 1]} \qquad (3.84)$$

Therefore, substituting equation (3.84) in (3.83) we have

$$\int_{-M}^{M} e^{[\alpha_1 \ln(2)-1]}dx = 1$$

$$\therefore\ e^{[\alpha_1 \ln(2) - 1]}[x]_{-M}^{+M} = 1$$

$$\therefore\quad e^{[\alpha_1 \ln(2) - 1]} = \frac{1}{2M}$$

$$\therefore\quad \text{The pdf, } f(x) = e^{[\alpha_1 \ln(2) - 1]} = \frac{1}{2M} \text{ for } (-M \le x \le +M)$$

$= 0$ elsewhere , i.e., $f(x)$ is uniformly distributed in the interval $(-M, +M)$

Therefore, the entropy is maximum, under peak signal constraint, if the signal is uniformly distributed over the signal interval $(-M, +M)$.

The **maximum entropy in terms of peak signal level** can be obtained by substituting the value of $f(x) = (1/2M)$ in equation (3.82). That is

$$H(X)_{max} = \int_{-M}^{+M} \frac{1}{2M} \log[2M]dx$$

$$= \frac{\log(2M)}{2M}[x]_{-M}^{+M}$$

$$\mathbf{H(X)_{max} = \log(2M)} \text{ bits/sample}$$

When 'M' is the peak signal voltage the normalized peak signal power P_m is given by M^2. Then ***the maximum entropy in terms of the peak signal power can be given by***

$$H(X)_{max} = \log(2M)$$

$$= \log[2\sqrt{P_m}\,] = \log(\sqrt{(4P_m)}\,)$$

i.e., $$H(X)_{max} = (1/2)\log(4P_m) \text{ bits/sample}$$

When the signal is band limited to 'B'Hz and is sampled at Nyquist rate of $r_s = 2B$ samples/sec, ***the maximum entropy in terms of the bandwidth and peak signal power*** is given by

$$H(X)_{max} = (2B)\{(1/2)\log(4P_m)\} \text{ bits/sec, or}$$

$$\mathbf{H(X)_{max} = (B)\log(4P_m) \text{ bits/sec}}$$

For uniformly distributed signal over the interval (– M, + M) the mean (average) value is given by

$$\mu = [(+M) + (-M)] = 0$$

and the variance is given by $\sigma^2 = [\,M - (-M)\,]^2/12 = M^2/3$

Since mean = 0, the variance represents the average power P_{av} and hence we have

$$P_{av} = M^2/3 = P_m / 3; \text{ and } P_m = 3P_{av}$$

Then ***the maximum entropy in terms of average power P_{av}*** is given by

$$H(X)_{max} = (B)\log(4P_m) \text{ bits/sec}$$

$$\mathbf{H(X)_{max} = (B)\log(12P_{av}) \text{ bits/sec}}$$

Maximization of Entropy under Average Signal Limitation

The average signal constraint is observed when the continuous signal which is a random variable has a unidirectional distribution with a specified average value. This situation is found in systems using Pulse Amplitude Modulation, Pulse Position Modulation and Amplitude modulation with average carrier amplitude limitation.

Since the signal X has unidirectional distribution and a specified average value, say λ, the Normalizing constraint equation and Average signal constraint equations can be modified as

$$\int_0^\infty f(x)dx = 1 \tag{3.85}$$

and $$\int_0^\infty xf(x)dx = \lambda \tag{3.86}$$

Then the entropy is given by the modified equation

$$H(X) = \int_0^\infty f(x)\log\left[\frac{1}{f(x)}\right]dx\,.$$

or $$H(X) = -\int_0^\infty f(x)\log[f(x)]dx \tag{3.87}$$

Then comparing equations (3.79) and (3.87) we have

$\phi(x,f) = -f(x)\log[f(x)]$,

Comparing equations (3.80) and (3.85) we have $\phi_1(x,f) = f(x)$

Comparing equation (3.80) and (3.86) we have

$\phi_2(x,f) = xf(x)$

$\therefore$ In equation (3.81) we will have only the first three terms, that is

$$\frac{\delta\phi}{\delta f} + \alpha_1 \frac{\delta\phi_1}{\delta f} + \alpha_2 \frac{\delta\phi_2}{\delta f} = 0. \tag{3.88}$$

Then using above values of ϕ, ϕ_1, ϕ_2 we have

$$\frac{\delta}{\delta f}[-f(x)\log[f(x)]] + \alpha_1 \frac{\delta}{\delta f}[f(x)] + \alpha_2 \frac{\delta}{\delta f}[xf(x)] = 0$$

$$\therefore \frac{\delta}{\delta f}\left[-f(x)\frac{\ln[f(x)]}{\ln(2)}\right] + \alpha_1 + \alpha_2 x = 0$$

i.e., $-\dfrac{1}{\ln(2)}[1 + \ln[f(x)]] = -\alpha_1 - \alpha_2 x$

$\therefore \ln[f(x)] = -[1 - \alpha_1 \ln(2)] + [\alpha_2 \ln(2)]x$

$\therefore \quad f(x) = e^{-[1-\alpha_1 \ln(2)+[\alpha_2 \ln(2)]x}$

$$\therefore f(x) = e^{-[1-\alpha_1 \ln(2)]} e^{-ax} \tag{3.89}$$

where $a = -\alpha_2 \ln(2)$ is greater than zero when α_2 is negative. Then using this value $f(x)$ in equation (3.85), we get

$$\int_0^\infty e^{-[1-\alpha_1 \ln(2)} . e^{-ax} dx = 1$$

$$\therefore e^{-[1-\alpha_1 \ln(2)]}[e^{-ax}/(-a)]_0^\infty = 1$$

$$\therefore \quad e^{-[1-\alpha_1 \ln(2)]}[1/a)] = 1$$

or $\quad e^{-[1-\alpha_1 \ln(2)]} = a$

Therefore, equation (3.89) becomes $f(x) = ae^{-ax}$ (3.90)

Hence equation (3.86) becomes $\int_0^\infty x\, ae^{-ax} dx = \lambda$

$$\text{Therefore, } a\int_0^\infty x\, e^{-ax}\, dx = \lambda \tag{3.91}$$

As $\int_0^\infty x^n e^{-ax}\,dx = [(n!)/a^{n+1}]$; for $n = 1$ we have $\int_0^\infty x\; e^{-ax}\,dx = [1/a^2]$

Therefore equation (3.91) becomes $a[1/a^2] = \lambda$, or $a = [1/\lambda]$.

Using this value of 'a' in equation (3.90) we get the expression for the pdf as

$$f(x) = [1/\lambda]\; e^{-(1/\lambda)x} \text{ for } x > 0 \qquad (3.92)$$

Hence, the probability density function (pdf) of the continuous random variable which maximizes the entropy H(X) is having exponential distribution with the parameter $(1/\lambda)$ when the signal is limited to a specified average value.

The maximum value of the entropy can be obtained by using this value of $f(x)$ in equation (3.87) as

$$\mathrm{H(X)}_{\max} = -\int_0^\infty \left(\frac{1}{\lambda}\right) e^{-(x/\lambda)} \log\left[\frac{1}{\lambda}\; e^{-(x/\lambda)}\right]dx$$

$$= -\left(\frac{1}{\lambda}\right)\int_0^\infty e^{-(x/\lambda)} \left\{\log\left[\frac{1}{\lambda}\right] + \log\left[e^{-(x/\lambda)}\right]\right\}dx$$

$$= -\left(\frac{1}{\lambda}\right)\int_0^\infty e^{-(x/\lambda)} \log\left[\frac{1}{\lambda}\right]dx - \left(\frac{1}{\lambda}\right)\int_0^\infty e^{-(x/\lambda)}(-x/\lambda)\log(e)dx$$

$$= -\left(\frac{1}{\lambda}\right)\log\left(\frac{1}{\lambda}\right)\left[e^{-x/\lambda}/-\left(\frac{1}{\lambda}\right)\right]_0^\infty + [\log(e)/\lambda]\int_0^\infty e^{-(y)}(y\lambda)dy$$

where $[x/\lambda] = y$, $\therefore\ dx = (\lambda dy)$ and y also varies from 0 to ∞.

Therefore, the maximum entropy is

$$\mathrm{H(X)}_{\max} = \left\{-\left(\frac{1}{\lambda}\right)\log\left(\frac{1}{\lambda}\right)\right\}[\lambda] + \log(\mathrm{e})\,[1!/1^{(1+1)}]$$

(again using the relation $\int_0^\infty x^n e^{-ax}\,dx = [(n!)/a^{n+1}]$; for $n = 1$ and $a = 1$)

$$= \log(\lambda) + \log(e)$$

$$\mathbf{H(X)_{max} = \log(\lambda e)\ bits/sample}$$

Therefore, the entropy in a bandwidth of 'B' Hz sampled at Nyquist rate is

$$\mathbf{H(X)_{max} = (2B)\log(\lambda e)\ bits/sec.}$$

Maximization of Entropy under Average Power Limitation (Symmetrical Distribution)

We come across average power limitations when the channel noise is random with a specified variance, in systems dealing with audio frequency telephonic signals and with signals which has symmetrical distribution. Let us consider a signal represented by a continuous random variable, X having symmetrical distribution, zero mean and average power limitation specified as variance σ^2. Now, we will show that the entropy of X maximizes when it is a Gaussian distribution with zero mean and variance σ^2. Also we will find out the value of maximum entropy.

The constraint equations are

$$\int_{-\infty}^{\infty} f(x)dx = 1 \tag{3.93}$$

$$\text{and} \int_{-\infty}^{\infty} x^2 f(x)dx = \sigma^2 \tag{3.94}$$

$$\text{Also the entropy } H(X) = -\int_{-\infty}^{\infty} f(x)\log[f(x)]dx \tag{3.95}$$

Comparing equation (3.95) with (3.79) we have, $\phi(x,f) = -f(x)\log[f(x)]$,

Comparing equation (3.93) and (3.80) we have, $\phi_1(x,f) = f(x)$,

Comparing equation (3.94) and (3.80) we have, $\phi_2(x,f) = x^2 f(x)$,

$\therefore$ In equation (3.81) we will have only the first three terms, that is,

$$\frac{\delta\phi}{\delta f} + \alpha_1 \frac{\delta\phi_1}{\delta f} + \alpha_2 \frac{\delta\phi_2}{\delta f} = 0$$

Then using above values of ϕ, ϕ_1, ϕ_2 in the above equation we have,

$$\frac{\delta}{\delta f}[-f(x)\log[f(x)] + \alpha_1 \frac{\delta}{\delta f}[f(x)] + \alpha_2 \frac{\delta}{\delta f}[x^2 f(x)] = 0$$

$$\therefore \frac{\delta}{\delta f}\left[-f(x)\frac{\ln[f(x)]}{\ln(2)}\right] + \alpha_1 + \alpha_2 x^2 = 0$$

$$\text{i.e., } -\frac{1}{\ln(2)}[1 + \ln[f(x)]] = -\alpha_1 - \alpha_2 x^2$$

$$\therefore \ln[f(x)] = -[1-\alpha_1\ln(2)] + [\alpha_2\ln(2)]x^2$$

$$\therefore f(x) = e^{-[1-\alpha_1\ln(2)] + [\alpha_2\ln(2)]x^2}$$

$$\therefore f(x) = e^{-[1-\alpha_1\ln(2)]}e^{-ax^2} \tag{3.96}$$

where $a = -\alpha_2\ln(2)$ and $a > 0$. Then using this value $f(x)$ in equation (3.93) we get

$$\int_{-\infty}^{\infty} e^{-[1-\alpha_1 ln(2)]}e^{-ax^2}dx = 1$$

$$e^{-[1-\alpha_1 ln(2)]}\int_{-\infty}^{\infty} e^{-ax^2}dx = 1 \qquad (3.97)$$

Since, the term (e^{-ax^2}) in the above equation is an even function (i.e., 'x' replaced by '$-x$' will not alter the value of the term) equation (3.97) can be written as,

$$2\, e^{-[1-\alpha_1 ln(2)]}\int_{0}^{\infty} e^{-ax^2}dx = 1$$

Let, $ax^2 = y$, then $x = \sqrt{\left(\frac{y}{a}\right)}$ and $dx = \left(\frac{1}{\sqrt{a}}\right)\left[\frac{1}{2\sqrt{y}}\right] dy$,

that is $dx = \left[\frac{1}{2\sqrt{ay}}\right] dy$

And 'y' varies from 0 to ∞.

Therefore $2\, e^{-[1-\alpha_1 \ln(2)]}\int_{0}^{\infty} e^{-y}\left[\frac{1}{2\sqrt{ay}}\right] dy = 1$

i.e., $2\, e^{-[1-\alpha_1 \ln(2)]}\left(\frac{1}{2\sqrt{a}}\right)\int_{0}^{\infty} y^{-1/2}e^{-y}dy = 1$

$$2\, e^{-[1-\alpha_1 \ln(2)]}\left(\frac{1}{2\sqrt{a}}\right)\int_{0}^{\infty} y^{[(1/2)-1]}e^{-y}dy = 1$$

$$\frac{1}{\sqrt{a}}\, e^{-[1-\alpha_1 \ln(2)]}\int_{0}^{\infty} y^{[(1/2)-1]}e^{-y}dy = 1$$

But $\int_{0}^{\infty} y^{(p-1)}e^{-y}dy$ = the Gamma function = $\gamma(\text{p})$

$$\therefore \int_{0}^{\infty} y^{[(1/2)-1]}e^{-y}dy = \gamma\left(\frac{1}{2}\right) = \sqrt{\pi}$$

Substituting this value we get, $\frac{1}{\sqrt{a}}\, e^{-[1-\alpha_1 \ln(2)]}\sqrt{\pi} = 1$; or $e^{-[1-\alpha_1 \ln(2)]} = \sqrt{\frac{a}{\pi}}$

Substituting this value in (3.96) we get

$$f(x) = \sqrt{\frac{a}{\pi}}\, e^{-ax^2} \qquad (3.98)$$

Using this value of '$f(x)$' in equation (3.94) we have

$$\int_{-\infty}^{\infty} x^2\sqrt{\frac{a}{\pi}}\, e^{-ax^2}dx = \sigma^2$$

Again the above integral is an even function and hence can be written as

$$2\{\sqrt{\frac{a}{\pi}}\int_0^\infty x^2 e^{-ax^2}dx\} = \sigma^2$$

Again let $ax^2 = y$, then $dx = \left[\frac{1}{2\sqrt{ay}}\right]dy$

Therefore, $2\sqrt{\frac{\alpha}{\pi}}\int_0^\infty \frac{y}{a}e^{-y}\left[\frac{1}{2\sqrt{ay}}\right]dy = \sigma^2$

As $$\frac{1}{a\sqrt{\pi}}\int_0^\infty y^{[(3/2)-1]}e^{-y}dy = \sigma^2$$

$$\frac{1}{a\sqrt{\pi}}\gamma\left(\frac{3}{2}\right) = \sigma^2$$

Since $\gamma(p+1) = p\,\gamma(p)$, we have $\gamma\left(\frac{3}{2}\right) = \gamma\left(1+\frac{1}{2}\right) = (1/2)\gamma\left(\frac{1}{2}\right) = (1/2)\sqrt{\pi}$

Therefore $\frac{1}{a\sqrt{\pi}}\gamma\left(\frac{3}{2}\right) = \sigma^2$ becomes

$$\frac{1}{a\sqrt{\pi}}(1/2)\sqrt{\pi} = \sigma^2$$

i.e., $a = (1/2\,\sigma^2)$

Substituting this value of 'a' in equation (3.98) we have

$$f(x) = \sqrt{\frac{a}{\pi}}\,e^{-ax^2}$$

$$f(x) = \frac{1}{\sigma\sqrt{2\pi}}\exp\left[-\left(\frac{1}{2}\right)\left(\frac{x}{\sigma}\right)^2\right] \tag{3.99}$$

Equation (3.99) above is the pdf of Gaussian distribution with mean $\mu = 0$ and variance $= \sigma^2$.

Substituting the value of '$f(x)$' from equation (3.99) in equation (3.95) we get the expression for the maximum entropy.

$$H(X)_{max} = -\int_{-\infty}^{\infty} f(x)\log\left\{\frac{1}{\sigma\sqrt{2\pi}}\exp\left[-\left(\frac{1}{2}\right)\left(\frac{x}{\sigma}\right)^2\right]\right\}dx$$

$$= -\int_{-\infty}^{\infty} f(x)\log\left\{\frac{1}{\sigma\sqrt{2\pi}}\right\}dx - \int_{-\infty}^{\infty} f(x)\log\left\{\exp\left[-\left(\frac{1}{2}\right)\left(\frac{x}{\sigma}\right)^2\right]\right\}dx$$

$$H(X)_{max} = \log(\sqrt{2\pi}\,\sigma) \int_{-\infty}^{\infty} f(x)dx + [\log(e)/2\sigma^2] \int_{-\infty}^{\infty} x^2 f(x)dx$$

Further using the constraint equations (3.93) and (3.94) the expression for $H(X)_{max}$ can be simplified as

$$H(X)_{max} = \log(\sqrt{2\pi}\,\sigma)\,(1) + [\log(e)/2\sigma^2]\,(\sigma^2)$$

$$= \log(\sqrt{2\pi}\,\sigma) + \log(\sqrt{e})$$

$$= \log(\sqrt{2\pi e}\,\sigma)$$

$$H(X)_{max} = \left(\frac{1}{2}\right) \log(2\pi e\sigma^2) \text{ bits/sample.} \qquad (3.100)$$

Therefore the entropy in a bandwidth of 'B' Hz sampled at Nyquist rate is given by

$$H(X)_{max} = (2B)\left(\frac{1}{2}\right) \log(2\pi e\sigma^2) \text{ bits/sec}$$

$$H(X)_{max} = (B) \log(2\pi e\sigma^2) \text{ bits/sec.} \qquad (3.101)$$

If 'X' represents a Gaussian Noise with an average power $\sigma^2 = N$, then

$$H(X)_{max} = (B) \log(2\pi eN) \text{ bits/sec} \qquad (3.102)$$

Maximization of Entropy under Unidirectional Average Power Limitation

Many times the signal X may have unidirectional distribution with average power limitation (for example Amplitude Modulated systems have average power limitation). In such cases the constraint equations become

$$\int_{-\infty}^{\infty} f(x)dx = 1 \text{ and } \int_{-\infty}^{\infty} x^2 f(x)dx = P_0$$

Using the procedure followed in the previous case we can show that

$$f(x) = (2/\pi\, P_0)^{(1/2)} \exp[-(x^2 / P_0)], \text{ for } 0 \le x \le \infty$$

$$H(X)_{max} = (1/2) \log(\pi e P_0 / 2) \text{ bits/sample}$$

and $$H(X)_{max} = (B) \log(\pi e P_0 / 2) \text{ bits/sec.}$$

3.7.3. Mutual Information of Continuous Noisy Channels

The expressions for the entropy of continuous signals derived under different conditions, in the previous section, are differential and are relative to some reference. Also logarithmic properties were used in the derivation. All the entropy relations we have considered for the discrete signals also apply for the continuous signals too. Hence, the joint entropy and average value of all conditional entropies (Equivocation) are given respectively by

$$H(X, Y) = -\iint_{-\infty}^{\infty} f(x, y)\log[f(x, y)]\, dx\, dy$$

and $$H(X/Y) = -\iint_{-\infty}^{\infty} f(x, y) \log[f(x/y)]\, dx\, dy$$

where $f(x, y)$ is the joint pdf and $f(x/y)$ is the conditional pdf of X given that $Y = y$.

Then the mutual information is given by

$$I(X, Y) = H(X) - H(X / Y)$$

or $$I(X, Y) = H(Y) - H(Y / X)$$

and $$I(X, Y) = H(X) + H(Y) - H(X, Y)$$

Also the channel capacity is given by $C = \text{Max}\,[\, I(X, Y)\,]$

Similarly we can verify various relations among entropies for the continuous signal.

***Example* 3.10:** In the continuous channel model of figure (3.13) we have seen that as the transmitted signal (X) passes through the physical channel, noise (N) gets algebraically added to it. In this example let us find out the differential entropy of this combined signal (Y) fed to the receiver.

Let Y be a random variable defined by $Y = X + N$, where X and N are statistically independent.

The equivocation for continuous signal is given by

$$H(Y / X) = -\iint_{-\infty}^{\infty} f(x, y)\log[f(y / x)]\, dx\, dy$$

Since $f(x, y) = f(y/x)\, f(x)$ we have

$$H(Y / X) = -\iint_{-\infty}^{\infty} f(x) f(y/x)\log[f(y / x)]\, dx\, dy$$

Therefore $$H(Y / X) = -\int_{-\infty}^{\infty} f(x)dx \int_{-\infty}^{\infty} f(y/x) \log f(y/x) dy \qquad (3.103)$$

Given that $Y = X + N$, the conditional pdf $f(y/x)$ depends on $(y-x)$ and not on 'x' or 'y'. Hence $f(y/x)$ is identical to that of 'N' except for a translation of 'x', the sample value of the random variable 'X'. Let $f_N(n)$ be the pdf of the additive noise 'N'. Then we have

$$f(y/x) = f_N(y-x)$$

Then $$-\int_{-\infty}^{\infty} f(y/x) \log f(y/x) dy = -\int_{-\infty}^{\infty} f_N(y-x) \log f_N(y-x)\, dy$$

$$= -\int_{-\infty}^{\infty} f_N(n) \log f_N(n)\, dn \quad [\text{ i.e } y = x + n,\ dy = dn, \text{ as '}x\text{' a constant }]$$

$$= H(N) \tag{3.104}$$

Using equation (3.104) in (3.103) we have

$$H(Y/X) = -\int_{-\infty}^{\infty} f(x)dx\, H(N)$$

since $-\int_{-\infty}^{\infty} f(x)dx = 1$,

$$\therefore\ H(Y/X) = H(N) \tag{3.105}$$

That is the conditional differential entropy of Y, given X is the differential entropy of 'N'.

3.7.4. Rate of Transmission

The rate of transmission of continuous signals (in bits/sample) is given by

R_t = Mutual information I(X, Y)= H(Y) – H(Y / X)

Using equation (3.105) $\therefore$ $R_t = H(Y) - H(N)$

The Channel Capacity $C = (R_t)_{max} = [H(X) - H(N)]_{max} = [H(Y) - H(N)]_{max}$

(3.106)

3.7.5. Properties of Mutual Information of Continuous Channels

It can be shown that the mutual information for a continuous channel has the following properties:

(*a*) I (X, Y) = I (Y, X)

(*b*) I (X, Y) $\geq$ 0

(*c*) I (X, Y) = H(X) – H(X/Y)

(*d*) I (X, Y) = H(Y) – H(Y/X)

3.8 Capacity of Band Limited Channels with AWGN—(Shannon-Hartley Theorem)

Shannon-Hartley theorem is complementary to the Shannon's channel capacity theorem which we have discussed in section (3.4). This theorem applies to a channel in which the noise is additive white Gaussian. In practice most of the channel noise encountered are almost white (i.e., the power spectral density is uniform) and the amplitude distribution is Gaussian with zero mean and with certain variance. Hence, this noise is known as Additive White Gaussian Noise (AWGN) and the channel is called an AWGN channel. This theorem, although stated for the Gaussian channel, is of fundamental importance. First, the channels encountered in practical systems generally are, at least, approximately Gaussian. Second, the results obtained for a Gaussian channel often provide a lower bound on the performance of a system operating over a non-Gaussian channel. Thus, if we can design a system (encoder – decoder pair) for Gaussian channel with an error probability P_e, then with a non-Gaussian channel another encoder – decoder pair can be designed so that the P_e will be lesser. It is seen in practice that the channel capacity equations based on Shannon-Hartley theorem have been derived for a number of non-Gaussian channels.

Shannon-Hartley Theorem (or Law)

The capacity of a white, band limited Gaussian channel is

$$C = B \log_2(1 + S/N) \text{ bits/sec.} \tag{3.107}$$

where **B** is the channel bandwidth in Hz,

S is the signal power in watts.

N is the total noise power in watts, within the channel band width, and $N = \eta B$ with $(\eta/2)$ the (two sided) power spectral density.

Proof : Let us consider the channel model of Fig. 3.13. The received signal will be a combination of the transmitted signal 'X' plus noise. When the signal 'X' and the noise are independent, using equation (3.106), the channel capacity is given by

$$C = [\, H(Y) - H(N) \,]_{max} \tag{3.108}$$

Considering the noise to an AWGN with a noise power of N in a band width of 'B' Hz then from equation (3.102) we have

$$H(N)_{max} = (B) \log(2\pi e N) \text{ bits/sec} \tag{3.109}$$

Further, if the input signal is also limited to an average power S, over the same band width, and when the received signal is $Y = X + N$ with X and N being independent, then the received signal power is the variance of the received signal and is given by

$$(\sigma_Y)^2 = (\, S + N \,) \tag{3.110}$$

During the derivation of the equation (3.101) we have seen that for a given mean square value, the entropy is maximum if the signal is Gaussian and the maximum entropy is given by

$$H(Y)_{max} = (B) \log[\, 2\pi e(\sigma_Y)^2 \,] \text{ bits / sec.} \tag{3.111}$$

And using the value of $(\sigma_Y)^2$ from equation (3.110) in equation (3.111) we have

$$H(Y)max = (B) \log[\, 2\pi e(S + N) \,] \text{ bits/sec.} \tag{3.112}$$

Then using equations (3.109) and (3.112) in equation (3.108) we have

$$C = (B) \log[\, 2\pi e\, (S + N) \,] - (B) \log(2\pi e N)$$

$$= B \log \{ [2\pi e(S + N)] / [2\pi e N] \}$$

$$C = B \log_2(1 + S/N) \text{ bits / sec.} \qquad (3.113)$$

The importance of the above equation (3.113), the formula to compute the capacity of a channel, is that it is possible to transmit information, over a channel of band width 'B' Hz and in the presence of AWGN, at a rate of C bits/sec with an arbitrarily small probability of error if the signal (information) is transformed (in the modulator) to waveforms of Gaussian nature. In order to obtain an errorless transmission for $R \leq C$ it requires complicated communication systems and hence we may have to accept a performance less than optimum. One possible method of attaining the performance predicted by Shannon's theorem, in the presence of Gaussian noise, is to use orthogonal waveforms (codes) to represent the information symbols.

3.8.1. Bandwidth and 'S/N' Ratio Trade Off

From equation (3.113) it is clear that the channel capacity is proportional to the channel bandwidth 'B' and the signal to noise power ratio 'S/N'. Therefore these two parameters can be adjusted to obtain a required channel capacity. For example let (S/N) = 7 and B = 4 KHz. Then from equation (3.113), C = 12000 bits/sec. Suppose the bandwidth B available is only 3 KHz, then to have the same channel capacity the S/N is to be increased to 15. But the noise power N depends on the bandwidth B and is given by

$$N = \eta B \qquad (3.114)$$

where $(\eta/2)$ is the (two sided) power spectral density of the white Gaussian noise. Hence when bandwidth is reduced the noise power also reduces. Let us see the implication of this on the signal power.

$S_1/N_1 = 7$, $B_1 = 4$ KHz ;$S_2/N_2 = 7$, $B_2 = 3$ KHz. Then

$(S_1/N_1) / (S_2/N_2) = (7 / 15)$, then using equation (3.114)

$(S_1/\eta B_1)(B_2/\eta S_2) = 7/15$

$(S_1/S_2) = (7/15)(B_1/B_2) = (7/15)(4/3) = (28/45) = 0.62222$

$S_2 = (45/28)\, S_1 = (1.607)\, S_1$.

That is for a 25% reduction in bandwidth, the signal power S_1 is to be increased by about 60% to maintain the same channel capacity. The exact significance of this can be looked into by drawing the "trade-off" curve.

Equation (3.113) can be modified as

$(B/C) \log_2(1 + S/N) = 1$.

Using this relation, the variation of (B/C) with respect to (S/N) is shown in the Table (3.1) and Fig. (3.14):

Table 3.1: (B/C) Vs (S/N) Trade off table.

(S/N)	0.5	1	2	5	10	15	20	30
(B/C)	1.71	1.0	0.63	0.37	0.289	0.25	0.23	0.20

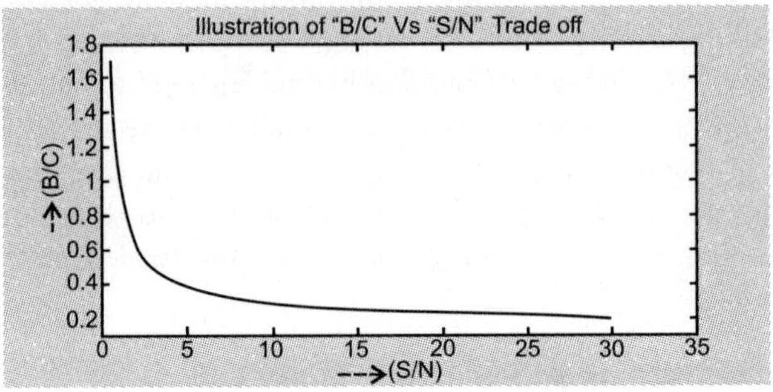

Fig. 3.14. "(B/C) Vs (S/N) Trade off Curve."

It is clear from the table (3.1) that we can maintain a required channel capacity (C) by increasing the bandwidth (B) when the S/N is small. It is also clear that there exists a threshold point say at (S/N) $\approx$ 10, beyond which any increase in (S/N) do not produce any significant change in (B/C). That is up to (S/N) $\approx$ 10, the exchange rate of bandwidth with (S/N) is advantageous and for (S/N) > 10 the deduction in B with increasing (S/N) is poor. The use of larger bandwidth for smaller (S/N) is generally known as "coding upwards" and the use of smaller bandwidth with larger (S/N) is called "coding downwards". FM, PM and PCM systems which use larger bandwidth for improvement in (S/N) are examples of "coding upwards". Multilevel discrete PAM can be considered as examples of coding downwards, where the bandwidth reduction depends on the signal power available.

Let us consider systems in which (S/N) >> 1. The channel capacity equation can be written as:

$$C = B_1 \log_2(1+(S_1/N_1)) = B_2 \log_2(1+(S_2/N_2))$$

or $$(1+(S_2/N_2)) = (1+(S_1/N_1))^{(B_1/B_2)}$$

$$(S_2/N_2) = (S_1/N_1)^{(B_1/B_2)} [\because (S/N) >> 1]$$

That is there is an exponential improvement in (S/N) ratio with bandwidth, for an ideal system where we can achieve very large signal to noise ratio. For conventional demodulation methods used in FM and PM, the S/N varies as the square of the transmission bandwidth, which obviously is inferior to the above ideal system case. However better performance, that approaches the Shannon bound in the limit as bandwidth is made infinite, can be obtained by using optimum demodulators (like phase lock loop).

3.8.2. Capacity of Channels with Infinite Bandwidth

From the Shannon-Hartley channel capacity formula of equation (3.113) we may decide that the noiseless channels for which (S/N) = ∞, has infinite capacity. However, that is not so. The channel capacity does not become infinite as the bandwidth is made infinite. This is because the noise power (N) also becomes infinite as the bandwidth B is made infinite, as N = ηB. Using this relation of equation (3.114), the channel capacity equation (3.113) can be modified as

$$C = B\log_2\left(1+\frac{S}{\eta B}\right)\text{bits/ sec}$$

$$= \frac{S}{\eta}\times\frac{\eta B}{S}\log_2\left(1+\frac{S}{\eta B}\right)$$

$$= \frac{S}{\eta}\times\frac{1}{x}\log_2(1+x) \qquad \text{where } x = \frac{S}{\eta B}$$

or

$$C = \frac{S}{\eta}\times\log_2(1+x)^{(1/x)}$$

Accordingly, when B → ∞, $x \to 0$, and we have

$$C_{max} = \lim_{B\to\infty} C = \frac{S}{\eta}\times\log_2\left(\lim_{x\to 0}(1+x)^{(1/x)}\right)$$

Since $\lim_{x\to 0}(1+x)^{(1/x)} = e$ (the naperian base) and $\log_2(e) = 1.44$, we have

$$C_{max} = \lim_{B\to\infty} C = \frac{S}{\eta}\times\log_2(e)$$

$$\mathbf{C_{max} = 1.44\left(\frac{S}{\eta}\right)\textbf{bits / sec}} \tag{3.115}$$

Equation (3.115) gives an upper limit on the channel capacity with increasing bandwidth.

3.8.3. Bandwidth Efficiency—Shannon's Limit

The aim of any communication system will be to transmit the information at the highest rate possible; and this rate can be the channel capacity. So an "ideal system" is defined as one that transmits information at a rate R equal to the channel capacity C. If the transmitted energy per bit is E_b, then the average transmitted power is given by

$$S = E_b C \tag{3.116}$$

Then using equations (3.114) and (3.116), the channel capacity equation (3.113) can be modified to get the channel capacity of an ideal system as

$$C = B\log_2(1 + (E_b/\eta)(C/B))$$

or $$(C/B) = \log_2 (1 + (E_b/\eta)(C/B)) \tag{3.117}$$

The ratio (C / B) is called **"Bandwidth Efficiency"** of the system and E_b/η is given by

$$E_b/\eta = [\, 2^{(C/B)} - 1 \,] / (C/B) \tag{3.118}$$

If C/B =1, then from the above equation (3.118) we have $E_b = \eta$, that is the signal power is equal to the noise power. Suppose $B = B_0$ for which S = N, then the bandwidth efficiency equation (3.117) can be modified as

$$(C/B_0) = (B/B_0) \log_2[1 + (B_0/B)] \tag{3.119}$$

and $$\lim_{B\to\infty} (C/B_0) = (1/B_0) \log_2 \lim_{B\to\infty} [\, 1 + (B_0/B) \,]^B$$

$$= (1/B_0) \log_2(e)^{B_0}$$

$$C_{max} / B_0 = \log(e)$$

$$C_{max} = B_0 \log(e)$$

Therefore $$\mathbf{C_{max} = 1.44\, B_0} \tag{3.120}$$

i.e., **"The maximum transmission rate for a given signal power S is 1.44 bits/sec/Hz in the bandwidth over which the signal power can be spread without its falling below the noise level".**

Now equation (3.116) can be written as

$$S = E_b C = \eta B_0$$

or $$E_b/\eta = (B_0 / C)$$

and as $B \to \infty$, we have

$$\lim_{B\to\infty} E_b/\eta = \lim_{B\to\infty} (B_0/C)$$

$$= B_0/C_{max}$$

$$= B_0/ [B_0 \log(e)]$$

$$= [1/\log(e)] = 0.6931$$

$$(E_b/\eta)_{min} = 0.6931 \tag{3.121}$$

or $(E_b/\eta)_{min}$ in dB $= 10 \log_{10}(0.6931)$

i.e., $$\mathbf{(E_b/\eta)_{min} = -1.5917\ dB} \tag{3.122}$$

The value (E_b/η)min = – 1.5917 dB is known as the "the Shannon's Limit" for transmission at maximum capacity C_{max} = 1.44 (S/η). When (E_b/η) < the Shannon's limit (0.6931), the communication fails.

We can plot a graph of Bandwidth efficiency $C/B = R_t / B$, as a function of the Energy per bit to Noise power spectral density ratio E_b/η. The graph is known as "Bandwidth Efficiency Diagram" and is shown in Fig. 3.15.

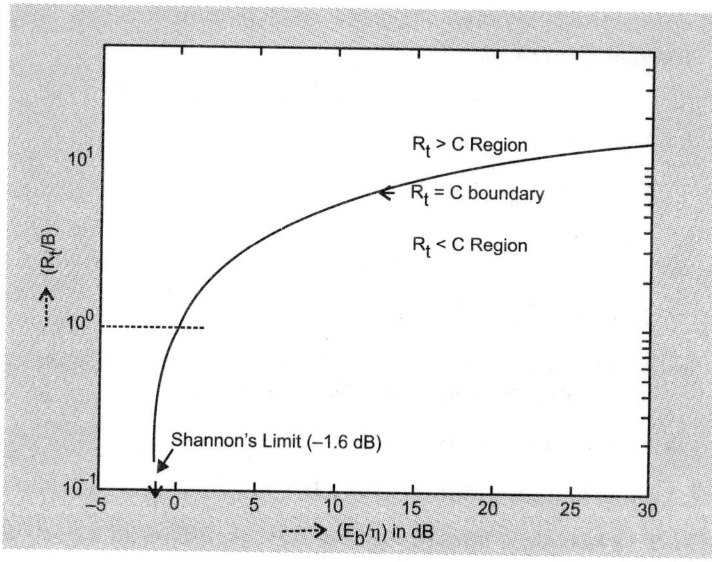

Fig. 3.15. "Bandwidth Efficiency Diagram"

The figure (3.15) shows the following:

- The capacity boundary is defined by the curve $R_t = C$.
- The region in which $R_t > C$ (above the curve) for which error free transmission is not possible.
- The region in which $R_t < C$ (below the curve) for which system parameter supports error free transmission.

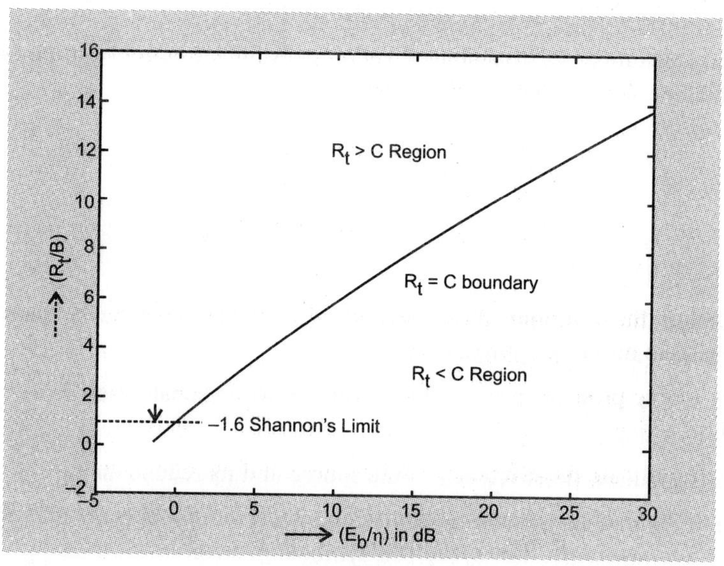

Fig. 3.16. "Bandwidth Efficiency Diagram"
(Axis in linear Scale)

3.8.4. Bandwidth-limited Regime and Power-limited Regime of AWGN Channel

Consider the capacity of a white, band limited Gaussian (AWGN) channel, given by equation (3.107) as

$$\mathbf{C = B \log_2 \left(1 + \frac{S}{N}\right) \text{bits / sec.}}$$

In the above expression, when the SNR is large (S/N >> 0 dB), the capacity

$$\mathbf{C \approx B \log_2 (S / N)}$$

That is the capacity is logarithmic in power and approximately linear in bandwidth. This is called the **bandwidth-limited regime.**

When the SNR is small (S/N<< 0 dB), the capacity

$$\mathbf{C \approx (S/\eta) \log_2 (e) \; [\text{ that is ' } N = \eta B \text{' }]}$$

That is the capacity is linear in power but insensitive to bandwidth. This is called the power-limited regime.

PROBLEMS

3.1. Consider the BSC shown in figure (3.5) with $p = 0.9$ and P(X=0) = P(X=1) = 1/ 2.

(*a*) Find the rate of transmission over this channel, for a symbol rate of r_s=1000 symbols/sec.

(*b*) What conclusions can be made from the results obtained when the probability of error, q, approaches 1/ 2 ?

3.2. Find the rate of transmission of 3 equi-probable messages over the following channel, for r_s = 1000 symbols/sec.

$$\begin{pmatrix} 1/2 & 1/2 & 0 & 0 & 0 & 0 \\ 0 & 0 & 1/2 & 1/2 & 0 & 0 \\ 0 & 0 & 0 & 0 & 1/2 & 1/2 \end{pmatrix}$$

3.3. Obtain the communication capacity of a noiseless channel transmitting 'n' discrete message symbols /sec.

3.4. A source produces two symbols X and Y with probabilities P(X) = 0.4 and P(Y) = 0.6

(*a*) Calculate the efficiency of the source and its redundancy.

(*b*) Calculate the system transmission rate, if the symbols are received on an average with 4 in every 100 symbols in error.

3.5. For the channel matrix

$$P(Y/X) = \begin{pmatrix} 0.6 & 0.2 & 0.2 \\ 0.5 & 0.2 & 0.3 \\ 0.0 & 0.6 & 0.4 \end{pmatrix}$$

(*a*) Compute the channel capacity and the corresponding source probabilities.

(*b*) If $P(x_1) = P(x_2) = P(x_3) = 1/3$, compute I(X, Y), channel efficiency and redundancy.

3.6. Consider the cascading of two channels having the following noise matrices.

$$P(B/A) = P(C/B) = \begin{pmatrix} 0.9 & 0.1 \\ 0.1 & 0.9 \end{pmatrix}$$

(*a*) Find the noise matrix of the cascaded channel.

(*b*) Find I(A, B) and I(A, C).

3.7. An one dimensional random variable has a pdf given by $f(x) = x$ for $0 \le x \le 4$.

(*a*) Determine the entropy of the source producing this variable.

(*b*) If this message is passed through a linear amplifier of gain 16 determine the entropy of the output.

3.8. Consider the cascade of three BSCs (each with an error probability of 0.1) as shown in figure below, with P(0)= ¼ and P(1) = ¾

Cascaded channels for Problem 3.8

(*a*) Calculate H(B), H(C) and H(D)

(*b*) Calculate I(A, C) and I(A, D). Is I(A, C) >I(A, D)?

3.9. A key-board (KB) is used to enter alphanumeric characters into a computer. The KB is connected through a voice grade telephone line having usable bandwidth of 3 KHz and an output S/N of 10 dB. Assume that the KB terminal has 128 characters and that the data sent from the KB consists of independent sequences of equi-probable characters.

(*a*) Find the capacity of the channel.

(*b*) Find the average information content per character.

(*c*) Find the maximum rate at which data can be sent from the KB to the computer without errors.

[**Ans:** (*a*) 10378.295 bits/sec (*b*) 7 bits/char (*c*) 1482.614 char/sec]

3.10. A line printer needs 8 bits of data per character and prints out 80 characters per line at a speed of 30 lines per minute. Is it possible to design a system for transmitting the output of a mini computer to the line printer over a voice grade telephone line with bandwidth of 4 KHz and S/N of 30 dB?

CHAPTER 4 ERROR CONTROL CODING

4.0 Introduction

Digital communication systems are now in common use and have swept away even such time honoured analog communication systems as are used now in radio, television and telephone. We have reached a stage where virtually anything we want to do in the field of communication is possible. The constraints are no longer technical, but economical, legal or political. Thus if one wants to transmit an entire "Encyclopedia" around the world in few seconds it can be done. The changing social needs and the development of new technologies are bringing the evolution of new communication capabilities, in the form of video text, electronic mail, high speed fax, teleconferencing, interactive CATV's etc., The demand has been accelerated by the emergence of large scale, high speed data networks for the exchange, processing and storage of digital information in the military, governmental, public and private spheres. All these are possible because of the developments in efficient and reliable digital data communication and storage systems. This requires an absolutely error free transmission. For example transmission within computer systems usually are intolerant of even very low error rates, because a single error can destroy an entire computer programme.

Conventional methods of achieving an error free transmission is (i) Increased power and (ii) Diversity systems.

An increase in power level invariably implies increase in size, weight and cost of transmitting equipment. Even if the added cost is acceptable, there are applications, such as space communication systems and satellite systems, where mobility is very important, and restrict limitations on size and weight of the equipment must be complied with. In yet other applications, particularly in very low regions of the radio spectrum, enormous amount of energy are required to radiate usable signals, and therefore transmitted signal power is always a dominant factor in the design of a transmission system.

Diversity systems, either Frequency or Space diversity systems, requires multiple transmission systems of the same information or multiple receiving systems for the same information. Hence the diversity systems are complex and cost ineffective which cannot be used in most of the applications.

Communication systems are affected by noise, interference from neighbouring systems and fading; storage systems by deterioration and failure of the recording medium; processing systems by faulty operations and other imperfections. The system designer, seeking acceptable performance, needs to protect information against errors caused by these imperfections. In applications where the conventional methods of error free transmission, mentioned above, are not suitable, the concept of channel coding comes as a cure for communication calamities. Channel coding, commonly known as error control coding is a cost effective way of achieving the required protection for the transmitted information particularly, where the channels are being pushed to the limits of their capacity. Error control coding can achieve reliable communication even when the received signal power is close to the thermal noise power; also when the electromagnetic spectrum becomes even more crowded with man made noise, they can provide reliable communication. In military applications error control codes are essential to protect the information against the intentional enemy interference (jamming). Error control codes are excellent way to reduce power needs, because messages received weakly at their destinations can be recovered correctly with the aid of the code. The phonograph records, tapes, CDs and television waveforms employ digital messages protected by error control codes. Scratches in a record or interference in a received signal will be completely suppressed by the codes as long as the errors are less serious than the designed capability of the code used. Digital data processing systems, control systems, digital switching systems, military communication and radar systems all use error control codes for error free data transfer. Thus the error control codes play a central role in all modern communication systems.

4.1 Principle of Channel Coding or Error Control Coding

The process of encoding the information for the purpose of combating the channel induced errors is called ***Channel Coding*** *or* ***Error Control Coding***.

"A message with content and clarity
Has gotten to be quite a rarity,
To combat the terror
Of serious error
Use bits of appropriate parity".

The above statement by Solomon Golomb is the basic principle of error control coding. The purpose of channel (error control) coding is to trade of information rate capacity with information error rejection, based on the Shannon theorem. This is achieved by adding a set of symbols, called the parity symbols, which do not carry information; but which are derived from the information symbols in such a way that they can detect and correct errors in the information carrying symbols, at the receiver.

The advantage of the systems using channel coding is quantified by the term ***Coding Gain,*** which is defined as the difference in E_b/N_0 required to attain a particular error rate without coding and with coding. E_b is the energy per bit of information transmitted and N_0 is the noise power spectral density.

4.2 Error Control Schemes

There are two main error control strategies or schemes used in practical communication systems. They are the Forward Error Correction (FEC) and the Automatic Repeat Request (ARQ) schemes, (sometimes also referred to as Backward Error Correction [BEC]).

4.2.1. FEC Scheme

When the transmission (or recording) is strictly in one direction, from the transmitter to the receiver, error control for such one way system must be accomplished using forward error correction. That is by employing error correcting codes that automatically correct errors detected at the receiver. Examples are (*i*) magnetic tape storage systems, in which information recorded on tape, may be replayed weeks or even months after it is recorded; (*ii*) Space communication systems, where the relatively simple encoding equipment can be placed aboard the space craft (like missiles, rockets, satellites etc.,), but the much more complex decoding procedure can be performed on earth; (*iii*) one way repeater stations and (*iv*) Real time secret systems which cannot retransmit the signals for security, use the FEC scheme. Most of the coded systems in use today employ some form of FEC, even if the channel is not strictly one way.

4.2.2. ARQ Scheme

In cases where a transmission system can be two-way; that is, information can be sent in both directions and the transmitter also can act as a receiver (Transceiver), and vice-versa. Examples of two-way systems are telephone channels, some satellite communication systems and duplex radio systems. Error control for two-way systems can be accomplished using error detection and retransmission, called Automatic Repeat Request (ARQ). In this scheme when errors are detected at the receiver, a request is sent for the transmitter (via a reverse channel) to repeat the message, and this continues until the message is received correctly.

4.2.3. Merits and Demerits of FEC and ARQ Schemes

- The major advantage of ARQ over FEC is that error detection requires much simpler decoding equipment or circuitries than does error correction.

- ARQ is adaptive, in the sense that information is retransmitted only when errors occur.
- But, when the channel error rate is high, re-transmission will be too frequent and the system throughput (i.e., the rate at which newly generated messages are correctly received) is lowered by ARQ. Whereas FEC gives a constant throughput rate.
- ARQ always requires a reliable reverse channel to transmit the request for retransmission; or at least a full duplex channel.

In channels with high error rate, a combination of FEC for the most frequent error patterns, together ARQ scheme for the less likely error patterns, is more efficient than ARQ alone. This hybrid ARQ (HARQ) error control strategy has the potential for improving throughput in two-way systems subject to a high error rate. In this book we will concentrate on the study of error control coding based on FEC only.

4.3 Types of Channel, Noise and Error

Before taking up the detailed study on channel coding, that is the design of the channel encoder and decoder pair of the "Typical Data Communication System" of Fig. 3.1, it is important to have the knowledge of the characteristics of the physical channel, the type of noise and the type of errors that are introduced as the data passes through the physical channel.

The important physical channels which are encountered in typical communication systems are Discrete Memoryless Channel, Binary Symmetric Channel, Binary Erasure Channel and Channel with Memory. We had a detailed study on characteristics of these channels in chapter 3.

The unwanted electrical disturbances or the error that occur in the, received information signal are defined as noise. The difference between the transmitted symbol and the received symbol is called the error. The common form of noise disturbances present in any communication system are Additive White Gaussian Noise (AWGN) and Impulse noise. The common types of error encountered are Random Errors, Burst Errors and Compound Errors.

4.3.1. White Noise

The noise whose power spectral density is uniform over the entire frequency range of interest is known as white noise. The term white is used in analogy with white light, which is a super position of all visible spectral components. Thus, it can be assumed that over the entire spectrum (including positive and negative frequencies) the power spectrum is given by $G_n(f) = n/2$, where 'n' is a constant as shown in Fig. 4.1.

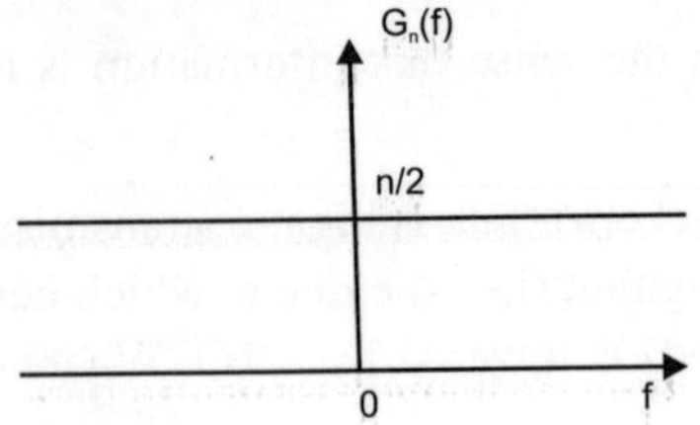

Fig. 4.1. White Noise Spectrum.

4.3.2. Gaussian Noise

The noise for which the observed samples values depend on the sum of large number of relatively independent samples (perturbations) is known as Gaussian Noise. This concept is based on the central limit theorem which indicates that the probability density of a sum of 'N' independent random variables tend to approach a Gaussian density as the number N increases. Such a process is met in many situations, such as thermal noise in resistors, galactic noise in radio astronomy and space communication, thermal and shot noise in transmitting and receiving equipment, thermal noise in the channel, radiation pickups by the receiving antennas etc. The bell shaped probability density function of the Gaussian process is given by

$$f(x) = \frac{1}{\sigma\sqrt{2\pi}} \exp\left\{-\frac{1}{2}\left[\frac{x-m}{\sigma}\right]^2\right\} \tag{4.1}$$

where 'm' is the mean value and 'σ' is the standard deviation.

4.3.3. Additive White Gaussian Noise (AWGN)

A white noise with uniform power spectral density and Gaussian amplitude distribution is called a White Gaussian Noise. When this noise gets algebraically added to the information signal is called an Additive White Gaussian Noise (AWGN). There are noises which multiplies the magnitude of the information signal.

4.3.4. Impulse Noise

An Impulse noise is characterized by long quiet intervals followed by high amplitude noise bursts. This type of noise results from many natural and man made reasons such as lightning and switching noise in telephone lines, jamming, fading, time dispersion etc. When noise bursts i.e., impulse noise occurs, it affects more than one symbol or bit, and there is usually a dependence of errors in successive transmitted symbols. Thus errors occur in bursts.

4.3.5. Random Errors

On memoryless channels, the noise affects each transmitted symbol independently. Hence transmission errors occur randomly in the received sequence, and such memoryless channels are called **Random error Channels**. Such independent occasional errors are called **Random Errors**.

Good examples of the random error channels are the deep-space communication and many satellite communication channels. Most Line-of-Sight (LOS) communication facilities are also affected primarily by random errors. Gaussian Noise produces random errors.

4.3.6. Burst Errors

On channels with memory, the noise is not independent from transmission to transmission and channel encounters impulsive noise. As a result, transmission errors occur in cluster or in bursts. As a result, channels with memory are called a **Burst Error Channels** and the errors in this type of channel with memory are called **Burst Errors**.

Examples of burst error channels are (*i*) wireless radio channels, where the error bursts are caused by signal fading due to multipath transmission, (*ii*) wire and cable transmission, which is affected by impulsive switching noise and cross talk, (*iii*) magnetic recording, which is subject to tape dropouts due to surface defects and dust particles or scratches in CDs. (*iv*) Frequency Hopping transmission in which, due to frequency selective fading one or more frequencies may be fully affected and all the data transmitted in those frequency goes together in error.

4.3.7. Compound Errors

Some channels cause a combination of both random and burst errors. Such combined errors are called **Compound Error** and the channels are called **"Compound Error Channels"**. Mobile radio channel is good example of a compound error channel. When a mobile moves in a plain terrain it encounters random errors due to thermal noise. The moment it enters urban areas, where there are high raised buildings, towers etc., which causes multipath reflections burst errors are encountered. Therefore, the channel coding technique used in a mobile communication system should cater for both random and burst errors.

4.4 Simplified Model of Coded System

In order to concentrate the study on the channel coding the block diagram of the general digital communication system can be simplified as shown in Fig. 4.2. The digital source consists of the information source and source encoder with output 'u'; the coding channel consists of the modulator, the channel and demodulator with input 'v' and output 'r'; digital sink represents the source decoder and destination with input '$\tilde{u}$'.

The major engineering problem is to design and implement the channel encoder/decoder pair such that (1) Information can be transmitted in a noisy environment as fast as possible, (2) Reliable reproduction of the information can be obtained at the receiver and (3) The cost of the system is within acceptable limits.

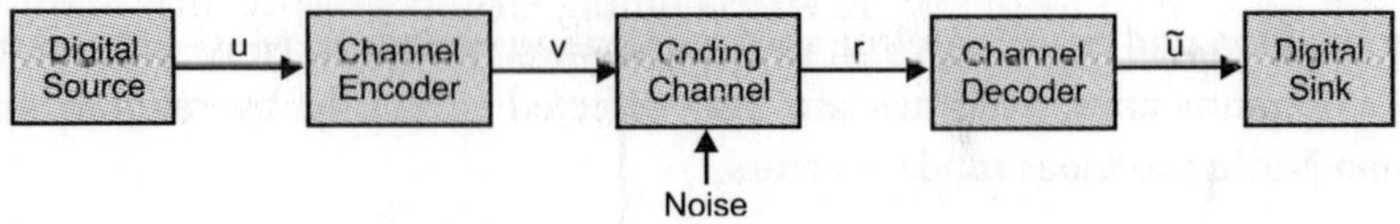

Fig. 4.2. Simplified Model of Coded System.

4.5 Types of Error Control Codes

There are two types of error control codes used in general. They are the Block codes and Convolutional codes.

Block Codes : For the purpose of encoding messages for the error protection, the long information sequence is broken into message blocks each consisting of, say, 'k' bits of information. Each message block is represented by a binary k-tuple given by

$u = (u_0, u_1, u_2, \ldots\ldots\ldots, u_{k-1})$ called a **Message**.

Next the encoder transforms each message 'u' independently into an n-tuple,

$v = (v_0, v_1, v_2, \ldots\ldots\ldots, v_{n-1})$ of discrete symbols called a **Code Word or Code Vector**. This is done by adding (n–k) redundant bits (normally called the parity bits) to the k-bit message, according to certain rules of coding. The code vector, 'v', an n-tuple is now transmitted through the noisy channel. At the receiver, the 'k' message bits are decoded from the received (n-tuple) code word (which may be erroneous) using suitable algorithms, such that the errors in the received code vector are detected and corrected. As the message block is a k-tuple, 2^k code words are possible at the output of the encoder corresponding to the 2^k different possible messages. This set of 2^k code words each of length 'n' is called an **(n , k) Block Code**. The ratio $\mathbf{R = (k / n)}$ is called the *code rate*, and is interpreted as the number of information bits entering the encoder per transmitted code word (symbol). Since the code word which is an n-tuple depends only on the k-bit input message, the **Encoder is memoryless** and can be implemented with a **"Combinational" Logic Circuit**.

Convolutional Codes: The convolutional code Encoder also accepts a k-bit message block 'u' and produces an encoded sequence (code word) 'v' or n-symbol block. However each encoded block depends not only on the corresponding k-bit message block at the same time unit, but also on 'm' previous message blocks. Hence the encoder has a memory of order 'm'. The set of code words produced by a k-input, n-output encoder of memory order 'm' is called a (n, k, m) **Convolutional Code**.

The ratio R = k/n is called the code rate. Since the encoder contains memory it must be implemented with a **"Sequential" Logic Circuit.** Typically 'k' and 'n' are small integers and more redundancy is added by increasing the memory order 'm' of the code while keeping 'k' and 'n' and hence the code rate 'R' fixed. An example of a binary convolutional encoder with $k = 1$ and $n = 2$ and $m = 3$ is shown in Fig. 4.3. below. How to use/choose the redundant bits or the memory to achieve reliable communication over a noisy channel is the major issue in the design of the encoder.

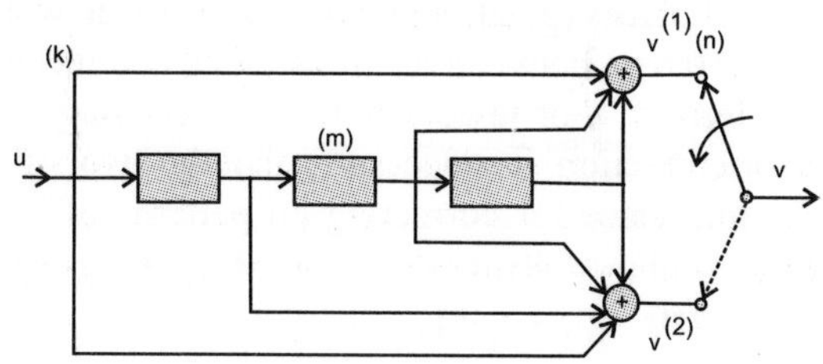

Fig. 4.3. Binary Convolutional Encoder with $n = 2$, $k = 1$ and $m = 3$.

4.6 Code Mapping Principle

The problem of encoding and decoding the message blocks can be thought of as a problem of mapping the source output to match the channel characteristics as shown in Fig. 4.4. above. The encoder generates 2^k distinct code words, $\{v_j\}$, which are corrupted by the noise in the channel. Then the received code words are $r_j = v_j + e_j$, where 'e_j' is the error vector. There are 2^n possible received vectors.

Fig. 4.4. Code Mapping Principle.

The decoder now observes $\{r_j\}$ and decides which particular code word was most likely transmitted. This amounts to optimum partitioning of the n-dimensional space enclosing 2^n vectors (each an n-tuple), into 2^k distinct spheres or subsets, each containing one code word. If now the received word $\{r_j\}$ falls within the sphere around v_j, that is r_j is one of the words clustered around v_j then the decoder decides that the transmitted code was v_j .This means that the correct decision is possible only when the received code r_j differs from v_j in few bit positions only. For this the code words themselves should differ by a minimum distance. This minimum distance is called the Hamming distance of the coding technique, which in turn depends on 't' the number of errors to be detected and corrected.

4.7 Hamming Distance

R.W. Hamming first developed the concept of Single Error Correcting codes, and the Hamming Distance (HD). The Hamming distance between two codes is defined as the number of places (positions) in which the code words differ. Then a single error results in a HD = 1 between the transmitted and received words. Thus, for detecting error pattern of 't' or fewer errors, it is necessary and sufficient that **HD ≥ (t + 1)**. For error correction, it is necessary that the errors should be detected and located as well, and hence for correcting all patterns of 't' or fewer errors, **HD ≥ (2t + 1)** and for combined error correction of 't_1' errors and error detection of 't_2' errors and $t_2 \geq t_1$, **HD ≥ ($t_1 + t_2$ +1)**.

4.8 Concept of Distance

To put the above in perspective consider that 3 bit code words are plotted at the vertices of a cube, as shown in the Fig. 4.5. below. It is seen that if the code words used are {000, 101,011,110} then H.D between the words is 2 and thus they may be used as code words with capability of detecting one error per word only. It is easily seen any error will locate the received words at the vertices which are not the code words, and thus be recognized as single errors. The codes with HD = 3 are only two i.e., {000, 111} or {101, 010} or {110, 001} or {011, 100}. Suppose the pair {000, 111} is used as code words and if a code word '000' is received as 100, or 010 or 001 then it is seen that these are nearer to '000' than to '111'. Hence the decision is made that the transmitted word is '000'. For 'n' bit code words we may think that the code words are placed at the vertices of an n-dimensional cube with the required HD (= d). Then the number of code words N with $d \geq (2t+1)$ is calculated as given below:

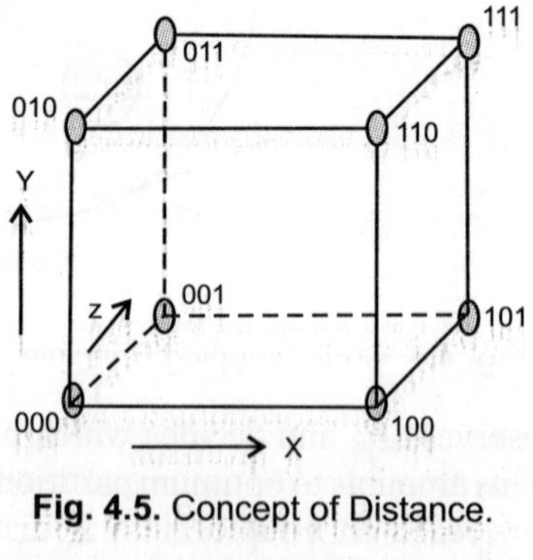

Fig. 4.5. Concept of Distance.

For 't' error correction, in code mapping principle, the n-tuples forming one sphere or subset around a code word should differ from it only in 't' or less places.

The numbers of n-tuples that differ in that way are

$$= {}^nC_0 + {}^nC_1 + {}^nC_2 + \ldots\ldots\ldots\ldots\ldots\ldots + {}^nC_t = \sum_{i=0}^{t} {}^nC_i \tag{4.2}$$

Therefore the total number of 'n-tuples' in all 2^k subsets is equal to $2^k \times \sum_{i=0}^{t} {}^nC_i$. But the total number of n-tuples in the vector space is 2^n.

Therefore $$2^n \geq 2^k \times \sum_{i=0}^{t} {}^nC_i\,.$$

Therefore the number of code words, N (n, d) with $d \geq (2t + 1)$ is

$$2^k \leq (2^n) / \sum_{i=0}^{t} ({}^nC_i)$$

i.e., $$N(n, d) \leq (2^n) / \sum_{i=0}^{t} ({}^nC_i)$$

and $$2^{n-k} \geq \sum_{i=0}^{t} {}^nC_i \tag{4.3}$$

Equation (4.3) gives the relation between the number of parity bits, length of message block (k) and the number of errors (t). This relation is called the **"Hamming Bound"**. For example for a single error correcting (n, k) code the requirements of parity check bits is given by $2^{n-k} \geq n + 1$.

4.9 Linear Block Codes

In block coding the binary information sequence is grouped into message blocks of fixed length, each message block denoted by 'u', consists of 'k' information digits. Therefore there are 2^k distinct messages. The encoder, according to certain algorithm, transforms each input message 'u' into a binary n-tuple with $n > k$. This binary n-tuple is called the code word or code vector of message 'u'. Therefore corresponding to the 2^k possible message, there are 2^k code words. This set of 2^k code words is called a **BLOCK CODE**. For the block code to be useful, the 2^k code words must be distinct. Therefore there should be a one-to-one correspondence between a message 'u' and its code word 'v'.

For the Block Code, to be realizable in practice, it should have certain special structure; otherwise the encoding and decoding hardware (circuitry) would be prohibitively complex for large values of 'k' and 'n', as it has to store the 2^k code words of length 'n' each. A desirable structure for a code word to possess is **Linearity**. With this structure the encoding complexity will be greatly reduced.

Definition : A block code of length 'n' and 2^k code words is called a Linear (n, k) Code, C, if and only if its 2^k code words form a k-dimensional subspace of the vector space of all n-tuples over the field GF(2).

4.9.1. Generator Matrix

"In fact a binary block code is linear iff the modulo-2 sum of two code words is also a code word and it should contain the zero vector". Linear Block codes are often referred as Group Codes or Generalized Parity Check Codes. Since, an (n, k) linear code 'C' is a k-dimensional subspace of the vector space V_n of all the binary n-tuples, there exists 'k' linearly independent code words, $g_0, g_1, g_2, \ldots\ldots, g_{k-1}$ in C such that these code words spans the entire subspace of 2^k code words. Each code word 'v' in C is a linear combination of these k linearly independent code words

i.e., $$v = u_0 g_0 + u_1 g_1 + u_2 g_2 + \ldots\ldots\ldots\ldots\ldots + u_{k-1} g_{k-1} \text{ where } u_i \in GF(2).$$

The 'k' linearly independent code words can be arranged as the rows of a '$k \times n$' matrix as follows:

$$G = \begin{bmatrix} g_0 \\ g_1 \\ \cdot \\ \cdot \\ \cdot \\ g_{k-1} \end{bmatrix} = \begin{bmatrix} g_{0,0} & g_{0,1} & g_{0,2} & \cdot & \cdot & \cdot & g_{0,n-1} \\ g_{1,0} & g_{1,1} & g_{1,2} & \cdot & \cdot & \cdot & g_{1,n-1} \\ \cdot & \cdot & \cdot & \cdot & \cdot & \cdot & \cdot \\ \cdot & \cdot & \cdot & \cdot & \cdot & \cdot & \cdot \\ \cdot & \cdot & \cdot & \cdot & \cdot & \cdot & \cdot \\ g_{k-1,0} & g_{k-1,1} & g_{k-1,2} & \cdot & \cdot & \cdot & g_{k-1,n-1} \end{bmatrix} \tag{4.4}$$

where $g_i = (g_{i,0}\ g_{i,1}\ g_{i,2} \cdots g_{i,n-1})$ for $0 \le i < k$.

If $u = (u_0\ \ u_1\ \ u_2\ \ldots\ldots\ldots\ldots\ldots\ u_{k-1})$ is the message to be encoded, the corresponding code word can be given as $v = u.\ G$,

$$v = (u_0\ u_1\ u_2 \ldots\ldots\ldots\ldots\ldots u_{k-1}) \begin{bmatrix} g_0 \\ g_1 \\ \cdot \\ \cdot \\ \cdot \\ g_{k-1} \end{bmatrix} \tag{4.5}$$

$$v = u_0 g_0 + u_1 g_1 + \cdot\ \cdot\ \cdot\ \cdot\ + u_{k-1} g_{k-1}. \tag{4.6}$$

Clearly the rows of G generate (or span) the (n, k) linear code C. For this reason, the matrix G is called a **Generator Matrix** for the linear block code 'C'. Any 'k' linearly independent code words of an (n, k) linear code can be used to form a generator matrix for the code. Therefore an (n, k) linear code is completely specified by the 'k' rows of a generator matrix G. Therefore the encoder has only to store the 'k' rows of G and to form a linear combination of these 'k' rows based on the input message 'u'.

4.9.2. Systematic Format of a Code Word

Another desirable property for a linear block code to possess is the Systematic structure of the code words as shown below:

Parity Checking Part ($n-k$) digits	Message Part (k) digits.

Here the code word is divided into two parts, the message part and the parity checking part. The message part consists of 'k' unaltered information (or message) bits and the parity checking part consists of $(n-k)$ parity check digits, which are linear sums of the information digits. A linear block code with this structure is referred to as a **Linear Systematic Block Code**. A linear systematic (n, k) code is completely specified by ($k \times n$) matrix G of the following form.

$$G = \begin{bmatrix} g_0 \\ g_1 \\ \cdot \\ \cdot \\ g_i \\ \cdot \\ \cdot \\ g_{k-1} \end{bmatrix} =$$

$$\begin{bmatrix} p_{00} & p_{01} & \cdot & \cdot & p_{0j} & \cdot & \cdot & p_{0,n-k-1} & 1 & 0 & 0 & . & . & . & 0 \\ p_{10} & p_{11} & \cdot & \cdot & p_{1j} & \cdot & \cdot & p_{1,n-k-1} & 0 & 1 & 0 & . & . & . & 0 \\ \\ p_{i0} & p_{i1} & \cdot & \cdot & p_{ij} & \cdot & \cdot & p_{i,n-k-1} & 0 & 0 & 0 & . & 1 & . & 0 \\ \\ p_{k-1,0} & p_{k-1,1} & \cdot & \cdot & p_{k-1,j} & \cdot & \cdot & p_{k-1,n-k-1} & 0 & 0 & 0 & . & . & . & 1 \end{bmatrix}$$

i.e., G =[(k × n–k) P matrix | (k × k) identity Matrix, I_k] (4.7)

where 'p_{ij}'= 0 or 1. i.e., G = [P I_k].

Let $u = (u_0 \ u_1 \ u_2 \ \ldots\ldots\ldots\ldots \ u_{k-1})$ be the message to be encoded. The corresponding code word

$$v = (v_0 \ v_1 \ v_2 \ \ldots\ldots\ldots\ldots \ v_{n-1}) = (u_0 \ u_1 \ u_2 \ \ldots\ldots\ldots\ldots \ u_{k-1}).\ G$$

Thus the digits of the code word 'v' are

$$v_{n-k+i} = u_i \quad \text{for } 0 \le i < k \tag{4.8}$$

$$v_j = u_0 p_{0j} + u_1 p_{1j} + u_2 p_{2j} + \ldots\ldots\ldots + u_{k-1} p_{(k-1)j} \text{ for } 0 \leq j < (n-k) \qquad (4.9)$$

From the above it is clear that the right most 'k' digits of a code word 'v' are identical to the information digits ($u_0\ u_1\ u_2 \ldots\ldots\ldots\ldots\ u_{k-1}$) to be encoded, and the left most $(n - k)$ parity digits are linear sums of the information digits. The $(n - k)$ equations given by equation (4.9) are called **the Parity Check Equations** of the code.

4.9.3. Encoder Circuit

Using the above parity check equation (4.9) and equation (4.8) the encoder circuit for an (n, k) Linear Systematic Code can be implemented easily using shift registers as shown in Fig. 4.6. below. The complexity is linearly proportional to the word length 'n' of the code.

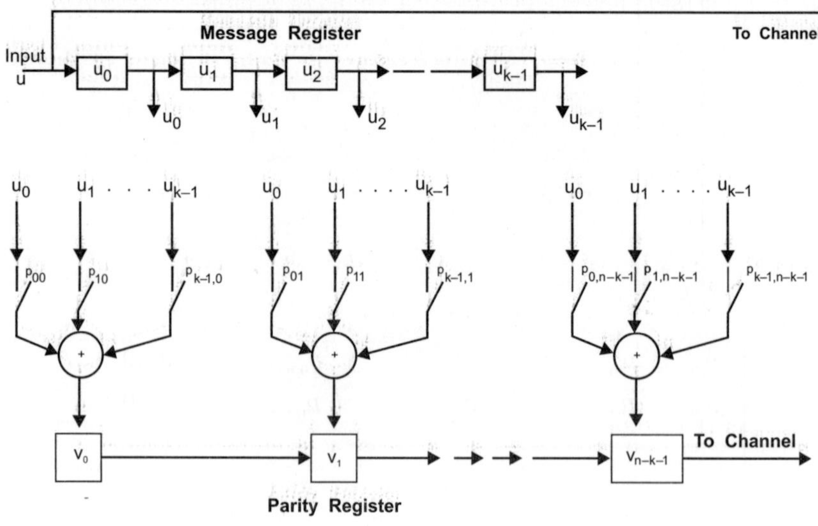

Fig. 4.6. Encoder Circuit for a Linear Systematic (n, k) Code.

The blocks in the circuit diagram represents the units as given below

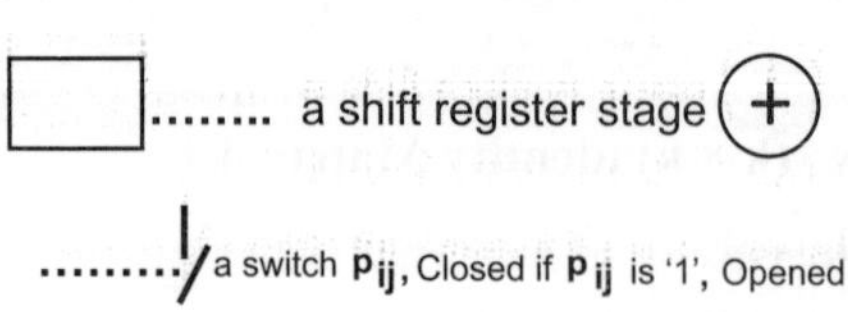

.... a shift register stage .... 'Exclusive OR' Gate.

.... a switch p_{ij}, Closed if p_{ij} is '1', Opened if p_{ij} is '0'

***Example* 4.1:** For a (7, 4) Linear block code

$$v = u \,.\, G = u \,.\begin{pmatrix} p_{00} & p_{01} & p_{02} & 1 & 0 & 0 & 0 \\ p_{10} & p_{11} & p_{12} & 0 & 1 & 0 & 0 \\ p_{20} & p_{21} & p_{22} & 0 & 0 & 1 & 0 \\ p_{30} & p_{31} & p_{32} & 0 & 0 & 0 & 1 \end{pmatrix}$$

$$v = v_0\, v_1\, v_2\, v_3\, v_4\, v_5\, v_6 = (u_0\, u_1\, u_2\, u_3).\begin{pmatrix} 1 & 1 & 0 & 1 & 0 & 0 & 0 \\ 0 & 1 & 1 & 0 & 1 & 0 & 0 \\ 1 & 1 & 1 & 0 & 0 & 1 & 0 \\ 1 & 0 & 1 & 0 & 0 & 0 & 1 \end{pmatrix}$$

$$v_0 = u_0 + u_2 + u_3$$
$$v_1 = u_0 + u_1 + u_2$$
$$v_2 = u_1 + u_2 + u_3$$
$$v_3 = u_0;\ v_4 = u_1;\ v_5 = u_2;\ v_6 = u_3.$$

The above four equations are the parity check equations.

and $\quad v = v_0\, v_1\, v_2\, v_3\, v_4\, v_5\, v_6 = v_0\, v_1\, v_2\, u_0\, u_1\, u_2\, u_3$

The code word for the message u = (1011) is v = (100 1011).

Using the parity check equations above the encoder circuit is as given in Fig. 4.7. below :

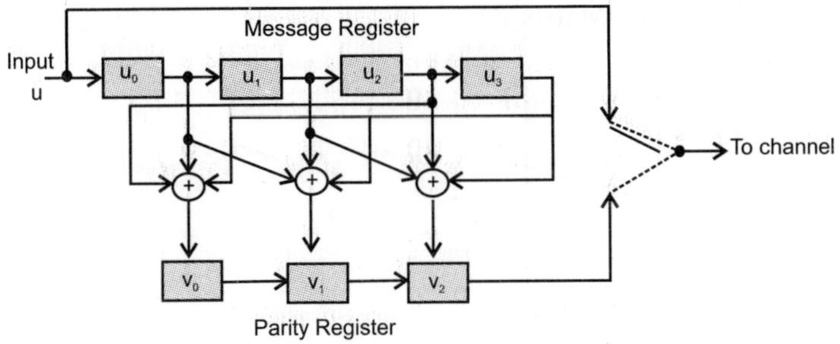

Fig. 4.7. Encoder Circuit for (7,4) Linear Systematic Code.

The 2^4 possible messages and the corresponding code words for the (7, 4) linear code is given in table (4.1) below:

Table 4.1: (7, 4) Linear Code

Message (u)	Code word (v)	Message (u)	Code word (v)
0000	000 0000	1000	110 1000
0001	101 0001	1001	011 1001
0010	111 0010	1010	001 1010
0011	010 0011	1011	100 1011
0100	011 0100	1100	101 1100
0101	110 0101	1101	000 1101
0110	100 0110	1110	010 1110
0111	001 0111	1111	111 1111

4.9.4. Parity Check Matrix

For any $(k \times n)$ matrix G with 'k' linearly independent rows there exists an $(n-k)$. n matrix H with $(n-k)$ linearly independent rows such that any vector in the row space of G is orthogonal to the rows of H and any vector that is orthogonal to the rows of H is in the row space of G .i.e., G. $H^T = 0$. Hence we can also describe the (n, k) linear code generated by G as: An n-tuple 'v' is a code word in the code generated by G iff 'v. $H^T = 0$'. This matrix **H** is called a **Parity Check Matrix** of the code.

4.9.5. Dual Code

The 2^{n-k} linear combination of the rows of matrix H form an $(n, n-k)$ linear code $\mathbf{C_d}$ which is the null (or Dual) space of the (n, k) linear code **C**, generated by the matrix **G**. i.e., for any '$v \in C$' '$w \in C_d$', $v.w = 0$. $\mathbf{C_d}$ **is called the dual of C**. Therefore a Parity Check Matrix for a linear code 'C' is a generator matrix for its Dual Code 'C_d'.

If the Generator matrix of an (n, k) linear block code is in the systematic form i.e., $G = [PI_k]$ then the Parity Check Matrix can be of the form $H = [I_{n-k}\ P^T]$, where P^T is the transpose of the Matrix P.

$$H = \begin{bmatrix} h_0 \\ h_1 \\ \cdot \\ \cdot \\ h_i \\ \cdot \\ \cdot \\ h_{n-k-1} \end{bmatrix} = \begin{bmatrix} 1 & 0 & 0 & \cdot & \cdot & \cdot & 0 & p_{00} & p_{10} & \cdot & \cdot & p_{j0} & \cdot & p_{k-1,0} \\ 0 & 1 & 0 & \cdot & \cdot & \cdot & 0 & p_{01} & p_{11} & \cdot & \cdot & p_{j1} & \cdot & p_{k-1,1} \\ \cdot & \cdot & \cdot & \cdot & \cdot & \cdot & \cdot & \cdot & \cdot & \cdot & \cdot & \cdot & \cdot & \cdot \\ \cdot & \cdot & \cdot & \cdot & \cdot & \cdot & \cdot & \cdot & \cdot & \cdot & \cdot & \cdot & \cdot & \cdot \\ 0 & 0 & 0 & \cdot & 1 & \cdot & 0 & p_{0i} & p_{1i} & \cdot & \cdot & p_{ji} & \cdot & p_{k-1,i} \\ \cdot & \cdot & \cdot & \cdot & \cdot & \cdot & \cdot & \cdot & \cdot & \cdot & \cdot & \cdot & \cdot & \cdot \\ \cdot & \cdot & \cdot & \cdot & \cdot & \cdot & \cdot & \cdot & \cdot & \cdot & \cdot & \cdot & \cdot & \cdot \\ 0 & 0 & 0 & 0 & 0 & 0 & 1 & p_{0,n-k-1} & p_{1,n-k-1} & \cdot & \cdot & p_{j,n-k-1} & \cdot & p_{k-1,n-k-1} \end{bmatrix} \quad (4.10)$$

i.e., H =[$(n-k) \times (n-k)$ identity matrix, I_{n-k} | $(n-k) \times k$ Matrix, $P_{(n-k).k}$]

The inner product of any i^{th} row in G and any j^{th} row in H is $g_i.h_j = p_{ij} + p_{ij} = 0$ for $0 \le i < k$ & $0 \le j < (n-k)$. This implies that $G.H^T = 0$. Therefore the H matrix given above is the parity check matrix of the (n, k) linear code generated by the matrix G.

The parity check equations which determine the parity bits obtained using G can also be obtained from the parity check matrix H.

Let $u = (u_0 \ u_1 \ u_2 \ .... \ u_{k-1})$ be the message to be encoded and the corresponding code word

$v = (v_0 \ v_1 \ v_2 \ .... \ v_{n-k-1} \ u_0 \ u_1 \ u_2 \ .... \ u_{k-1})$; using the fact that $v \,.\, H^T = 0$ we get

$$v_j + u_0 p_{0j} + u_1 p_{1j} + u_2 p_{2j} + \ .... + u_{k-1} p_{(k-1)j} = 0$$

$$v_j = u_0 p_{0j} + u_1 p_{1j} + u_2 p_{2j} + \ .... + u_{k-1} p_{(k-1)j} \quad \text{for } 0 \le j < (n-k) \quad (4.11)$$

Therefore an (n, k) code is completely specified by its parity check matrix. Accordingly the parity check matrix H for the (7,4) linear code of example (4.1) is $H = [\, I_3 \ P_{3.4}]$

$$H = \begin{pmatrix} 1 & 0 & 0 & 1 & 0 & 1 & 1 \\ 0 & 1 & 0 & 1 & 1 & 1 & 0 \\ 0 & 0 & 1 & 0 & 1 & 1 & 1 \end{pmatrix}$$

With the above discussions it can be concluded that for any (n, k) linear block code C, there exists a (k, n) matrix G whose row space gives C. Also there exists an $(n-k).n$ matrix H such that an n-tuple 'v' is a code word in C iff $v.H^T = 0$. If G is of the form $G = [\, PI_K \,]$ then H is of the form $H = [I_{n-k} \ P^T]$ and vice versa.

4.9.6. Decoding of Linear Block Code

Let us consider an (n, k) linear code with generator matrix G and parity matrix H. Let the transmitted code word over the noisy channel be $v = (v_0 \ v_1 \ v_2 \ .... \ v_{n-1})$. Let the corresponding received code be $r = (r_0 \ r_1 \ r_2 \ .... \ r_{n-1})$. If there is no error introduced by the channel to the code word 'v' then $r = v$. If error was introduced by the channel, then'$r = v + e$'. Where $e = (e_0 \ e_1 \ e_2 \ .... \ e_{n-1})$ is, an n-tuple, called the error vector or error pattern. But the receiver does not have any knowledge of either 'v' or 'e'.

On receiving 'r', the decoder must first determine whether 'r' contains any transmission error or not. If the presence of error is detected, the decoder will either take action to locate the error and correct them, in case of Forward Error Correction (FEC) or request for a retransmission of 'v', in case of Automatic Request for Retransmission (ARQ) scheme.

4.9.6.1. Syndrome and error detection

When 'r' is received, the decoder first computes the following $(n - k)$ tuple

S = $\mathbf{r.H^T} = (S_0 \; S_1 \; S_2 \; .... \; S_{n-k-1})$ where 'S' is called the Syndrome of the received vector 'r'.

But '$r = v + e$', therefore $S = (v + e).H^T = v.H^T + e.H^T = 0 + e.H^T$

$$\mathbf{S = e.H^T} \tag{4.12}$$

If **there is no error,** that is if the error a zero vector then **S = 0**. If, there is an error then **S ≠ 0**.Therefore, when the decoder computes the $(n - k)$ tuple syndrome vector S, it will be equal to zero iff 'r' is a code word in C, and S ≠ 0 iff 'r' is not a code word. Therefore when **S ≠ 0**, the received vector 'r' is not a code word and **the presence of error is detected.** When **S = 0,** 'r' is a code word and the **receiver accepts 'r' as the transmitted code word.** Therefore, the presence of 'S' is an indication of the presence of error. Hence the name **Syndrome,** which means symptom of decease. It is possible that the errors in **certain error vectors are not detectable** (i.e., 'r' contains errors but $S = eH^T = 0$). This happens when the error pattern **'e' is identical to a non-zero code word.** In this case, 'r' is the sum of two code words, which is another code word, and consequently $r \,.\, H^T = 0$. Error patterns of this type are called the **Undetectable Error Patterns.** Since there are $(2^k - 1)$ non-zero code vectors, there are $(2^k - 1)$ **undetectable error patterns**. When these error patterns occur the decoder makes a **Decoding Error.**

4.9.6.2. Syndrome computation

Based on $H = [\, I_{n-k} \; P^T \,]$ and $S = r \,.\, H^T$ the syndrome digits are computed as follows:

$$S = [\, r_0 \; r_1 \; r_2 \; .... \; r_{n-1} \,] \,.\, [\, I_{n-k} \; P^T \,]^T \tag{4.13}$$

$$S_0 = r_0 + r_{n-k}p_{00} + r_{n-k+1}p_{10} + r_{n-k+2}p_{20} + ... + r_{n-1}p_{(k-1),0}$$

$$S_1 = r_1 + r_{n-k}p_{01} + r_{n-k+1}p_{11} + r_{n-k+2}p_{21} + ... + r_{n-1}p_{(k-1),1}$$

.

.

$$S_j = r_j + r_{n-k}p_{0j} + r_{n-k+1}p_{1j} + r_{n-k+2}p_{2j} + ... + r_{n-1}p_{(k-1),j}$$

.

.

$$S_{n-k-1} = r_{n-k-1} + r_{n-k}\,p_{0,n-k-1} + r_{n-k+1}\,p_{1,n-k-1} + r_{n-k+2}\,p_{2,n-k-1} + ... + r_{n-1}p_{(k-1),n-k-1}$$

Close examination of the above equations reveals that the syndrome S is simply the vector sum of the received parity digits $[\, r_0 \; r_1 \; r_2 \; .... \; r_{n-k-1} \,]$ and the parity check digits recomputed from the received message digits $r_{n-k}, r_{n-k+1}, r_{n-k+2}, ..., r_{n-1}$. Therefore, the syndrome can be formed by a circuit similar to the encoding circuit.

4.9.6.3 Syndrome computation for the (7, 4) linear block code

Let the received vector '$r = r_0 \; r_1 \; r_2 \ldots r_6$'

Therefore, $S = r.\, H^T$

$$S = [r_0 \; r_1 \; r_2 \; \ldots r_6] . \begin{pmatrix} 1 & 0 & 0 \\ 0 & 1 & 0 \\ 0 & 0 & 1 \\ 1 & 1 & 0 \\ 0 & 1 & 1 \\ 1 & 1 & 1 \\ 1 & 0 & 1 \end{pmatrix}$$

i.e., $S = [\, S_0 \; S_1 \; S_2 \,]$

Therefore, the Syndrome digits are

$$\mathbf{S_0} = r_0 + r_3 + r_5 + r_6 \,;\; \mathbf{S_1} = r_1 + r_3 + r_4 + r_5 \,;\; \mathbf{S_2} = r_2 + r_4 + r_5 + r_6$$

With these equations the syndrome digits can be computed using the circuit shown in Fig. 4.8. below:

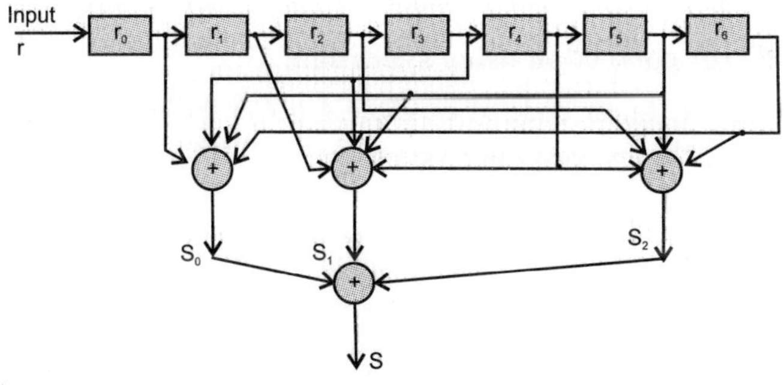

Fig. 4.8. Syndrome Circuit for (7, 4) Linear Block Code

4.9.6.4. Relation between the syndrome and error pattern

We have seen that $S = e.\, H^T$, that is the syndrome actually depends on the error pattern 'e'and not actually on the transmitted/received vector (v/r). Then if the parity check matrix H is expressed in the systematic form $H = [\, I_{n-k} \; P^T]$ then the relation between syndrome digits and the error digits are as given below. (i.e., using $S = e.H^T$).

i.e., $S = [\, e_0 \; e_1 \; e_2 \ldots e_{n-1} \,] . [\, H \,]^T$ and the digits of 'S' are

$$S_0 = e_0 + e_{n-k}\, p_{00} + e_{n-k+1} p_{10} + e_{n-k+2}\, p_{20} + \ldots\ldots\ldots + e_{n-1} p_{(k-1),0}$$

$$S_1 = e_1 + e_{n-k}\, p_{01} + e_{n-k+1} p_{11} + e_{n-k+2}\, p_{21} + \ldots\ldots\ldots + e_{n-1} p_{(k-1),1} .$$

$$S_j = e_j + e_{n-k}\,p_{0j} + e_{n-k+1}\,p_{1j} + e_{n-k+2}\,p_{2j} + \dots\dots\dots + e_{n-1}\,p_{(k-1),j}.$$

$$S_{n-k-1} = e_{n-k-1} + e_{n-k}\,p_{0,n-k-1} + e_{n-k+1}\,p_{1,n-k-1} + e_{n-k+2}\,p_{2,n-k-1} + \dots\dots\dots + e_{n-1}\,p_{(k-1),n-k-1}$$

Therefore the syndrome digits are the linear combination of the error digits, and hence syndrome digits provide information about the error digits, and therefore, can be used for error location and hence error correction.

So, one may feel that by just solving of the above $(n - k)$ linear equations of the error digits the error correction scheme is achieved. Once the error pattern 'e' is computed the vector sum $(r + e)$ gives the transmitted code vector, 'v'. But computing the true error vector is not that easy. Because the above $(n - k)$ linear equations do not have unique or single solution, they have 2^k solutions. That is there are 2^k error patterns that will result for the same syndrome, and the true error pattern is one among them. Therefore, the decoder has to identify the true error from the set of 2^k patterns.

For this in order to minimize the probability of decoding error, the most probable error pattern that satisfies the equation between the syndrome and error pattern is chosen as the error vector. If the channel is Binary Symmetric Channel (BSC), the most probable error pattern is the one that has the least (smallest) non-zero digits.

4.9.6.5. Error correction using syndrome

Consider the (7, 4) linear code of example 4.1. Let the transmitted code vector $v = 100\ 1011$ and the received code vector '$r = 100\ 1001$'. Then

$$S = r.H^T = [100\,1001].\begin{pmatrix} 1 & 0 & 0 \\ 0 & 1 & 0 \\ 0 & 0 & 1 \\ 1 & 1 & 0 \\ 0 & 1 & 1 \\ 1 & 1 & 1 \\ 1 & 0 & 1 \end{pmatrix} = [1\ 1\ 1]$$

Let the error pattern for the above syndrome be '$e = (e_0\ e_1\ e_2\ e_3\ e_4\ e_5\ e_6)$'. Therefore, from the relation, $S = e.H^T$ we get

$\mathbf{S_0} = 1 = e_0 + e_3 + e_5 + e_6$., $\mathbf{S_1} = 1 = e_1 + e_3 + e_4 + e_5$., $\mathbf{S_2} = 1 = e_2 + e_4 + e_5 + e_6$.

The following $2^4 = 16$ error patterns will satisfy the above equations.

(0000010), (1101010), (0110110), (1011110),
(1110000), (0011000), (1000100), (0101100),
(1010011), (0111011), (1100111), (0001111),
(0100001), (1001001), (0010101), (1111101),

The error vector e = (0000010) has the smallest number of non-zero components. If the channel is a BSC then e = (0000010) will be the most probable error vector that satisfies the above. Therefore, the decoder takes this error pattern and computes the transmitted code word as:

$$v = r + e = 1001001 + 0000010 = 1001011.$$

The (7, 4) linear block code is a single error correcting (SEC) code.

4.9.6.6. Maximum likelihood decoding of linear block codes

Assuming that all the code words are equally likely to be transmitted, the best decision at the receiver is to decode into a code word v_j such that the Hamming Distance, HD, between the received vector r_j and v_j is minimum, such a decoding principle is called the ***Maximum Likelihood Decoding.***

Hamming Weight: A terminology often used is the Hamming Weight (H.W) or simply the weight of the code word 'v', denoted by $w(v)$ and is defined as the number of non-zero digits of 'v'. For a Binary code word $w(v)$ will be the number of binary ones in 'v'.

e.g., $w(v)$ of 1100101 is 4 and HD between 1100101 and 1000110 is 3.

For a Linear block code the HD between two n-tuples v_1 and v_2 is equal to the HW of the sum of v_1 & v_2, that is $d(v_1, v_2) = w(v_1 + v_2)$

Minimum Distance and Minimum Weight

The minimum distance of Block Code 'C' denoted by d_{min} is defined as

$$d_{min} = \min \{ d(v_1, v_2) : v_1, v_2 \in C, v_1 \neq v_2 \} \tag{4.14}$$

When C is a linear block code, the sum of two code vectors is also a code word (vector). Therefore, the HD between two vectors in a linear code C is equal to the HW of a third code in C.

Therefore $d_{min} = \min \{ w(v_1 + v_2) : v_1, v_2 \in C, v_1 \neq v_2 \}$

$= \min \{ w(v_3) : v_3 \in C, v_3 \neq 0 \}$ where $(v_3 = v_1 + v_2)$

$$d_{min} = w_{min} \ldots\ldots\ldots\ldots \tag{4.15}$$

The parameter w_{min} is called the **Minimum Weight** of the linear code C.

Therefore, the minimum distance of a linear block code is equal to the minimum weight of its non-zero code words and hence for determining d_{min} it is enough to compute the w_{min}.

The minimum distance of a linear block code can be determined by observing the Generator Matrix *G*, and Parity Check matrix *H*. Since the rows of the *G* matrix are valid code vectors, the weight of the row in the *G* matrix, with minimum weight gives the minimum weight of the code and hence the minimum distance of the code.

Also, it can be shown that **the minimum weight (or the minimum distance) of the linear code is equal to the smallest number of columns of 'H' that sum to zero.**

4.9.7. Error Detecting Capabilities of a Block Code

Let *C* be a block code with a minimum distance d_{min}. i.e., any two distinct code vectors of *C* differ in atleast d_{min} places. When a code word '$v \in C$' is transmitted over a noisy channel, an error pattern of 'j' errors will result in a received vector 'r', which differs from the transmitted code vector 'v' in 'j' places i.e., $d(v, r) = j$.

Therefore, for a block code *C* of minimum distance d_{min}, no error pattern of $(d_{min} - 1)$ or fewer errors can change one code vector into another. Hence any error pattern of $(d_{min} - 1)$ or fewer errors will result in a received vector 'r' that is not a code word in *C*. When the receiver detects that the received vector 'r' is not a code word of *C* i.e., $r \notin C$, it is said that errors are detected.

Therefore, a block code C with minimum distance d_{min} is capable of detecting all the error patterns of $(d_{min} - 1)$ or few errors.

However, it cannot detect all the error patterns of d_{min} errors, because there exists at least one pair of code words that differ in d_{min} places, such that an error pattern of d_{min} errors will carry one code word into another code word belonging to *C*. The same thing will happen to error patterns of more than d_{min} errors.

Therefore, the Random Error Detecting Capability of a Block Code with minimum distance d_{min} is "$(d_{min} - 1)$".

Even though a block code with minimum distance d_{min} guarantees detecting all the error patterns of $(d_{min} - 1)$ or fewer errors, it is also capable of detecting a large fraction of error patterns with d_{min} or more errors.

The number of error patterns with $(d_{min}-1)$ errors or fewer errors is

$$= {}^nC_0 + {}^nC_1 + {}^nC_2 + \dots\dots\dots\dots\dots\dots + {}^nC_{(d_{min}-1)}$$

$$= \sum_{i=0}^{i=d_{min}-1} {}^nC_i \tag{4.16}$$

For (7, 4) linear block code $d_{min} = 3$ and the number of error patterns that can be detected is,

$${}^7C_0 + {}^7C_1 + {}^7C_2 = 1 + 7 + 21 = 29.$$

An (n, k) linear block code is capable of detecting $(2^n - 2^k)$ error patterns of length ' n '.

Proof: Possible non-zero error patters are $(2^n–1)$. Possible non-zero code vectors are $(2^k–1)$. Therefore number of error patterns identical to code words $= (2^k–1)$. When these identical error pattern occurs they alter the transmitted code word 'v_1' into another code word 'v_2' which also belong to C (by linearity property), and decoder accepts v_2 as the transmitted code word and thus commits an erroneous decoding.

Therefore there are $(2^k–1)$ undetectable error patterns (4.17)

When an error pattern is not identical to any of the code word then the received vector 'r' will not be belonging to C and hence, error will be detected.

Therefore, the number of non-identical error patterns that an (n, k) linear block code can detect

$$= (2^n - 1) - (2^k - 1) = (2^n - 2^k) \quad (4.18)$$

For large value of 'n', $(2^k - 1)$ will be much smaller than 2^n and hence only a small fraction of error patterns will pass through the decoder without being detected.

e.g.: for (7, 4) code , the number of detectable error patterns $= 2^7 - 2^4 = 128 - 16 = 112$.

Number of undetectable error patterns $= 2^k - 1 = 2^4 - 1 = 15$.

Probability of undetectable error = probability that the error vector resembles non-zero codes.

$$= P_u[E] = \sum_{i=1}^{n} A_i p^i (1-p)^{n-i} \quad (4.19)$$

Where A_i = number of codes with weight 'i' ; p = the channel transition probability.

***Example* 4.2:** For C with minimum distance d_{min}, A_1 to $(A_{d_{min}-1})$ are zero.

For (7, 4) linear code, for which $d_{min} = 3$, $A_0 = 1$, $A_1 = A_2 = 0$, $A_3 = A_4 = 7$, $A_5 = A_6 = 0$, $A_7 = 1$(can be verified from the Table 4.1).

Therefore, $P_u(E) = 7p^3(1-p)^4 + 7p^4(1-p)^3 + p^7 = 7\times10^{-6}$ for $p = 1 \times 10^{-2}$.

That is for a transmission of one million bits over the channel (BSC) with $p = 1 \times 10^{-2}$ there are on the average seven erroneous code words will pass through the decoder without being detected. This $P_u(E)$, can be proved to be $\leq 2^{-(n-k)}$ and decreases exponentially with $(n - k)$.

4.9.8. Error Correcting Capabilities of a Block Codes

The error correcting capabilities of Linear Block Code C with minimum distance d_{min}:

The minimum distance d_{min} can either be odd or even. Let 't' be a positive integer such that $(2t+1) \leq d_{min} \leq (2t+2)$. Then it can be proved that C is capable of correcting all the error patterns of 't' or fewer errors.

Proof: Let 'v' be the transmitted code vector and 'r' be the corresponding received code word. Let 'w' be another code word in C. The H.D among v, r and w satisfy the triangular inequality

$$d(v,r) + d(w,r) \geq d(v,w) \tag{4.20}$$

Let an error pattern of 't'' errors occur during the transmission of the code vector v. Then, 'r' differs from 'v' in 't'' places, i.e., $d(v,r) = t'$. Since 'v' and 'w' are code vectors of C we have,

$$d(w,r) \geq d_{min} \geq 2t+1 \tag{4.21}$$

From equations (4.20) and (4.21) we have $d(w,r) \geq d(v,w) - d(v,r)$

i.e., $d(w,r) \geq (2t+1) - t'$ and if $t' \geq t$, then $d(w,r) > t$.

Therefore when an error pattern of "t" or fewer errors occur, the received vector r is closer (in HD) to the transmitted code word v, than to any other code word w in C. Based on the maximum likelihood decoding scheme, r is decoded into v, which is the actual transmitted code word. This results in a correct decoding and thus errors are corrected.

Therefore for a Linear Block Code, to correct "t" errors, its minimum distance d_{min} should be such that $d_{min} \geq 2t+1$

or $$t = [(d_{min} - 1)/2] \tag{4.22}$$

where $[\times]$ means the largest integer which is $< x$.

On the other hand, the code is not capable of correcting all the error patterns of $(t+1)$ and above, because there is atleast one case where an error pattern of $(t+1)$ errors or more results in a received code vector which is closer to an incorrect code vector than to the actual transmitted code vector.

Therefore, a block code with minimum distance "d_{min}" guarantees correcting all the error patterns of "$t = [(d_{min} - 1)/2]$" or fewer errors.

The parameter "$t = [(d_{min} - 1)/2]$" is called the random error correcting capability of the block code.

4.9.9. Probability of Erroneous Decoding

If a "t" error correcting block code is used strictly for error correction on a BSC with transition probability "p", the probability that the decoder commits an erroneous decoding is upper bounded by,

$$P_d(E) \leq \sum_{i=t+1}^{\lambda} {}^nC_i p^i (1-p)^{n-i} \tag{4.23}$$

i.e., the Probability that more than " t " errors occur.

4.9.10. Simultaneous Error Correction and Error Detection

In practice, a code is often used for correcting λ or fewer errors and simultaneously detecting l ($l > \lambda$) or fewer errors. That is, when λ or fewer errors occurs the code is capable of correcting them, when more than λ but fewer than $(l + 1)$ errors occur, the code is capable of detecting their presence without making a decoding error. For this, the d_{min} of a code should be atleast "$(\lambda + l + 1)$". Thus for a block code, with $d_{min} = 10$, is capable of correcting 3 or fewer errors and simultaneously detecting 6 or fewer errors.

Thus the random error detecting and random error correcting capability of a block code are determined by the minimum distance of the code. Clearly for given n and k, one would like to construct a block code with minimum distance as large as possible, in addition to the implementation considerations.

4.9.11. Singleton Bound

The minimum distance of any linear (n, k) code satisfies

$$d_{min} \leq n-k+1. \tag{4.24}$$

Proof : A systematic code will have only one non-zero information symbol. Hence such a code cannot have minimum weight larger than $(n - k) + 1$. Hence, the minimum weight and hence, the minimum distance cannot be larger than $(n - k + 1)$. Any code whose minimum distance satisfies $d_{min} = n - k + 1$ is called a **Maximum Distance Separable (MDS)** code. e.g Reed Solomon Codes.

The probability of undetectable errors $P_u(E) \leq 2^{-(n-k)}$ and $P_u(E)$ decreases exponentially with $(n-k)$.

4.9.12. Standard Array and Syndrome Decoding

A method of partitioning the 2^n possible received vectors into 2^k disjoint subsets such that each subset contains one and only one code vector for maximum likelihood decoding is described here. The partitioning is based on the linear structure of the code.

First the 2^k code vectors of C are placed in a row with the "all zero" code vector as the first (left most) element. From the remaining $(2^n - 2^k)$ n-tuples, an n-tuple e_2 is chosen and is placed under the zero vector v_1. Now we form a second row by adding e_2 to each code word v_j in the first row and placing the sum $e_2 + v_j$ below v_j. Having completed the second row, an unused n-tuple e_3 is chosen from the remaining n-tuples and is placed under v_1. Then the third row is formed by adding e_3 to each code word v_j in the first row and placing $e_3 + v_j$ under v_j. This process is repeated till all the n-tuples are used. This results in an array of rows and columns as in Fig. 4.9. This array is called the **Standard Array** of the given Linear block code C.

$$\begin{matrix} v_1 = 0 & v_2 & v_3 & \cdot\ \cdot & v_j & \cdot\ \cdot & v_{2^k} \\ e_2 & (e_2 + v_2) & (e_2 + v_3) & \cdot\ \cdot & (e_2 + v_j) & \cdot\ \cdot & (e_2 + v_{2^k}) \\ e_3 & (e_3 + v_2) & (e_3 + v_3) & \cdot\ \cdot & (e_3 + v_j) & \cdot\ \cdot & (e_3 + v_{2^k}) \\ \cdot & \cdot & \cdot & \cdot\ \cdot & \cdot & \cdot\ \cdot & \cdot \\ \cdot & \cdot & \cdot & \cdot\ \cdot & \cdot & \cdot\ \cdot & \cdot \\ e_i & (e_i + v_2) & (e_i + v_3) & \cdot\ \cdot & (e_i + v_j) & \cdot\ \cdot & (e_i + v_{2^k}) \\ \cdot & \cdot & \cdot & \cdot\ \cdot & \cdot & \cdot\ \cdot & \cdot \\ e_{2^{n-k}} & (e_{2^{n-k}} + v_2) & (e_{2^{n-k}} + v_3) & \cdot\ \cdot & (e_{2^{n-k}} + v_j) & \cdot\ \cdot & (e_{2^{n-k}} + v_{2^k}) \end{matrix}$$

Fig. 4.9. Standard Array for an (n, k) Linear Block Code.

4.9.12.1. Properties of standard array

- From the construction rule of a standard array the sum of any two vectors in the same row is a code vector in C.
- No two n-tuple in a standard array are identical.
- Every n-tuple appears in one and only one row.
- There are $2^n / 2^k = 2^{n-k}$ disjoint rows in the Standard array. Each row consists of 2^k distinct vectors (elements).

The 2^{n-k} rows are called the **Coset** of the code C and the first n-tuple e_j of each coset is called a **Coset Leader.** Any element of the coset can be used as its coset leader, this does not change the elements of the coset; it simply permutes them (i.e., simply changes the position in the same row).

4.9.12.2. Decoding with the help of standard array

A Standard array of an (n, k) linear code C consists of 2^k disjoint columns. Each column consists of 2^{n-k} 'n-tuples' with the top most one as a code vector in C. Let D_j denote the jth column of the standard array. Then $D_j = \{v_j, e_2 + v_j, e_3 + v_j, \ldots, e_i + v_j, \ldots, e_{2^{n-k}} + v_j\}$ where v_j is a code vector in C and $e_2, e_3, \ldots e_i, \ldots, e_{2^{n-k}}$ are the coset leaders of the Standard array. The 2^k disjoint columns D_1, D_2, ..., D_j,......., D_{2^k} can be used for decoding the code C. As D_j consists of only one code vector of C , there is a **one-to-one correspondence to a code vector v_j and the disjoint column D_j.** Therefore, if the received vector 'r' is found in the column D_j, then r is decoded into v_j. But correct decoding will be done iff the received vector r is in the column D_j that corresponds to the actual transmitted code vector v_j.

This will happen only when the error pattern caused by the channel is a coset leader. In this case the received vector r will be decoded correctly into the transmitted code vector v_j. On the other hand if the error pattern caused by the channel is not a coset leader, an erroneous decoding will result.

e.g., : Let 'v_i' is the transmitted code vector and 'x' be the error pattern which is not equal to any coset leader. Then 'x' may be in same coset and under some non-zero code vector, say in the pth coset and under a non-zero code code vector 'v_j'.

Then $x = e_p + v_j$ and the received vector is

$$r = v_i + x = v_i + e_p + v_j = e_p + (v_i + v_j)$$

$$r = e_p + v_s \text{ (i.e., } v_i + v_j = v_s)$$

Thus the received vector r is in D_s and is decoded as v_s, which is not the transmitted code vector and this results in erroneous decoding. Therefore, the decoding is correct iff the error pattern caused by the channel is a coset leader. Therefore, the 2^{n-k} coset leaders (including the zero vector) are called the **Correctable Error Patterns.**

Therefore every (n, k) linear block code is capable of correcting 2^{n-k} error patterns.

4.9.12.3. Selection of coset leaders

In order to minimize the probability of decoding error, the error pattern those are most likely to occur for a given channel should be chosen as the coset leaders. For a BSC, an error pattern of smaller weight is more probable than error pattern of larger weight. Therefore, when a Standard array is formed, **each coset leader should be chosen to be a vector of least weight** from the remaining available vectors. As a result, the decoding based on the Standard array is the **minimum distance decoding** (that is Maximum Likelihood Decoding—MLD).

4.9.13. Probability of Decoding Error

Let'α_i' denote the number of coset leaders, of weight i. The numbers $\alpha_0, \alpha_1, \alpha_2, \ldots\ldots \alpha_n$ are called the weight distribution of the coset leaders. By knowing all α_i's we can compute the probability of decoding error. Since decoding error occurs when the error patterns are not the coset leaders, the error probability for a BSC with transition probability 'p' is

P(E) =1– (probability of correct decoding) = 1 – (probability of occurrences of the coset leaders).

$$P(E) = 1 - \sum_{i=0}^{n} \alpha_i p^i (1-p)^{n-i} \tag{4.25}$$

e.g., For a (6 , 3) code $\alpha_0 = 1, \alpha_1 = 6, \alpha_2 = 1, \alpha_3 = \alpha_4 = \alpha_5 = \alpha_6 = 0$.

Therefore, $P(E) = 1 - \{ p^0(1-p)^{6-0} + 6p(1-p)^{6-1} + 1.p^2(1-p)^{6-2} \}$

$$= 1 - \{(1-p)^6 + 6p(1-p)^5 + p^2(1-p)^4\}$$

For $p = 10^{-2}$, we have $P(E) \approx 1.37 \times 10^{-3}$

An (n, k) linear code is capable of detecting $(2^n - 2^k)$ error patterns; however can correct only 2^{n-k} error patterns.

For large n, 2^{n-k} is small fraction of $2^n - 2^k$. Therefore probability of decoding error is much more than probability of undetected error. $P_D(E) << P_u(E)$.

For an (n, k) linear code C with minimum distance d_{min}, all the n-tuples of weight of '$t = [\, (d_{min} - 1) / 2 \,]$' or less can be used as coset leaders of a Standard array of C.

A Standard Array has an important property to simplify the decoding process.

All the 2^k n-tuples of coset have the same syndrome. The syndromes for different cosets are different.

Proof : Let e_j be the coset leader of a coset in a Standard Array. Any vector in this coset is the sum of e_j and some code vector v_i in C. The syndrome of this vector is

$$(e_j + v_i).H^T = e_j \,.\, H^T + v_i \,.\, H^T$$

$$= e_j \,.\, H^T + 0 \;= e_j \,.\, H^T$$

Therefore, the syndrome of all the vectors in a coset is the same, (as each coset is having one coset leader) i.e., **all the vectors of a coset have the same syndrome.**

Let e_j and e_p be the coset leaders of the two j^{th} and p^{th} cosets respectively and $j < p$. Suppose the syndrome of each of these two are same then $e_j \,.\, H^T = e_p \,.\, H^T$ i.e., $(e_j + e_p)H^T = 0$.

Therefore $e_j + e_p$ should be a code vector in C say v_i thus $e_j + e_p = v_i$ and $e_p = e_j + v_i$.

Therefore, e_p is in the jth coset, but it is not so, because a coset leader should be previously unused. **Therefore, no two cosets have the same syndrome.**Therefore, there is a one-to-one correspondence between a coset leader (a correctable error pattern) and a syndrome.

Using this correspondence relationship, we can form a decoding table which is much simpler to use than a standard array. The table will consist of 2^{n-k} coset leaders (the correctable error patterns) and their corresponding syndromes. This table is either stored or wired in the receiver.

4.9.14. Decoding Table Using Coset Leaders and Syndromes

Decoding of the received vector using the standard array is known as Syndrome Decoding or Look-up Table Decoding. There are three steps involved in this method. They are,

Step **1:** Computation of the syndrome of 'r' i.e., $S = r \,.\, H^T$.

Step **2:** Identifying the coset leader 'e_j' whose syndrome is equal to $r.H^T$. Then e_j is assumed as the error pattern.

***Step* 3:** Decoding the received vector ' r ' into the code vector $v = r + e_j$.

The decoding method explained above is called the **Syndrome Decoding or Table Look-up Decoding**. This method can be used for any (n, k) linear code. It results in **Minimum Decoding Delay and Minimum error probability**.

However, for large $(n - k)$ the implementation of this method becomes difficult and either a large storage or a complicated logic circuitry is needed. There are many other decoding techniques developed based on this principle (with variations). But they require additional properties in a code other than the linear structure.

4.9.15. Implementation of Table Look-up Decoding of an (*n*, *k*) Linear Block Code

The decoding table can be thought of as the truth table of 'n' switching functions :

$$e_0 = f_0 (S_0, S_1, \ldots\ldots S_{n-k-1})$$

$$e_1 = f_1 (S_0, S_1, \ldots\ldots S_{n-k-1})$$

$$e_2 = f_2 (S_0, S_1, \ldots\ldots S_{n-k-1})$$

$$\cdot$$

$$e_{n-1} = f_{n-1} (S_0, S_1, \ldots\ldots S_{n-k-1})$$

where $S_0, S_1, \ldots\ldots, S_{n-k-1}$ are the syndrome digits, which are regarded as the switching variables, and $e_0, e_1, \ldots\ldots e_{n-1}$ are the estimated error digits. Once these ' n ' switching functions are derived and simplified, a combinational logic circuit with $(n - k)$ syndrome digits as input and the estimated error digits as outputs can be realized. The general decoder, using Look-up table decoding, for (n, k) linear code is shown in Fig. 4.10 below. The cost of this scheme depends on the complexity of the combinational logic circuits.

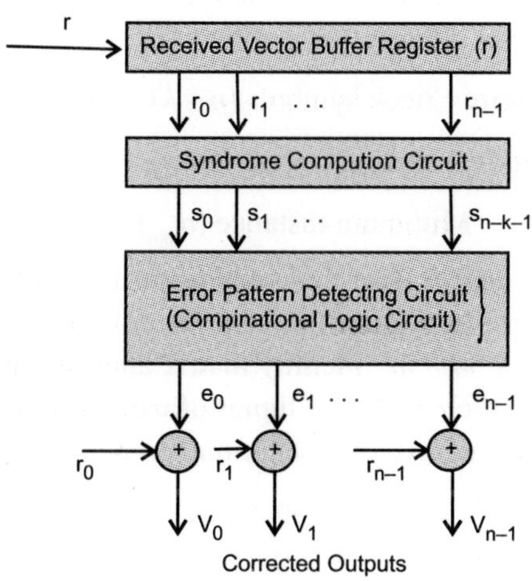

Fig. 4.10. General Decoder for a Linear Block Code.

***Example* 4.3:** Consider the (6, 3) Linear Code C, whose generator matrix is

$$G = \begin{pmatrix} 0 & 1 & 1 & 1 & 0 & 0 \\ 1 & 0 & 1 & 0 & 1 & 0 \\ 1 & 1 & 0 & 0 & 0 & 1 \end{pmatrix}$$

Standard Array for this code generated using the procedure of section 4.9.12 is given below:

Table 4.2: Standard Array for (6, 3) Linear Block Code.

Coset Leader v_1= 000000	v_2= 110001	v_3= 101010	v_4= 011100	v_5= 011011	v_6= 101101	v_7= 110110	v_8= 000111	Syndrome 000
000001	110000	101011	011101	011010	101100	110111	000110	110
000010	110011	101000	011110	011001	101111	110100	000101	101
000100	110101	101110	011000	011111	101001	110010	000011	011
001000	111001	100010	010100	010011	100101	111110	001111	001
010000	100001	111010	001100	001011	111101	100110	010111	010
100000	010001	001010	111100	111011	001101	010110	100111	100
001001	111000	100011	010101	010010	100100	111111	001110	111

4.9.16. Hamming Codes

The most widely used codes for error control in digital communication and data storage systems are the Hamming codes and its variations. For every positive integer $m \geq 3$, there exists a Hamming Code with the following parameters:

$$\text{Code length } (n) = 2^m - 1.$$

$$\text{Number of information symbols } (k) = 2^m - m - 1$$

$$\text{Number of Parity Check symbols } (n - k) = m$$

$$\text{Error Correcting Capability } (t) = 1$$

$$\text{Minimum distance } (d_{min}) = 3$$

The parity-check matrix H of this code consists of all the $(2^m - 1)$ non-zero m-tuples as its columns. In systematic form the columns of H are arranged as H = [I^m Q], where I_m is a '$m \times m$' identity matrix and the sub matrix Q consists of $(2^m - m - 1)$ columns which are the m-tuples of weight 2 or more. For example let $m = 3$. The parity-check matrix of a Hamming code of length $n = 2^m - 1 = 7$, can be put in the form

$$H = \begin{pmatrix} 1 & 0 & 0 & 1 & 0 & 1 & 1 \\ 0 & 1 & 0 & 1 & 1 & 1 & 0 \\ 0 & 0 & 1 & 0 & 1 & 1 & 1 \end{pmatrix}$$

which is the parity check matrix of the (7, 4) linear block code of Example:4.1 Hence, the (7, 4) linear block code is a Hamming code. The columns of Q may be arranged in any order without affecting the distance property and weight distribution of the code. Therefore, in systematic form the generator matrix of the code is of the form $G = [\,Q^T\, I2^m - m{-}1] = [\,P_{4\times3}\;\; I_4\,]$, where Q^T is the transpose of the sub matrix Q in H and Q^T $I2^m - m{-}1]$ is a $(2^m - m - 1) \times (2^m - m - 1)$ identity matrix and Q is a $(2^m - m - 1) \times m$ matrix , i.e., $k \times (n{-}k)$ matrix. Since the columns of H are non-zero and distinct no two columns add to zero. Therefore the minimum distance of a Hamming code is atleast 3.

Since H consists of all non-zero m-tuples as its columns, the vector sum of any two columns say h_i and h_j must be a column in H, say h_p. Thus $h_i + h_j + h_p = 0$. The minimum distance of a code C is the minimum number of columns of H which add to zero. Therefore the minimum distance of the Hamming code is exactly equal to three. Hence the code is capable of correcting all the Single error patterns or capable of detecting all the error patterns of two or fewer errors.

4.9.16.1. Standard array of hamming code

If we form the standard array for the Hamming code of length 'n', all the n-tuples of weight one can be used as coset leaders. The number of n-tuples of weight one is n. Since $n - k = m$, the code has 2^m cosets (i.e 2^{n-k} rows in the standard array), i.e the zero vector 0 and n tuples of weight one form all the coset leaders of the standard array. That is the Hamming code corrects only the error patterns with single error and no others. This is a very interesting structure.

Referring to the code mapping, in the receiver, the n-dimensional space is partitioned into 2^{n-k} distinct spheres with radius $= t$; then all the received code vectors with 't' errors will lie within the spheres and the correct decoding is done. The vectors with more errors will fall outside the concerned sphere or in wrong spheres. This, results in wrong decoding. Based on this a class of **Perfect** and **Quasi-Perfect Codes** are defined.

4.9.17. Perfect Code

A "t" error correcting code is called a **"Perfect Code"** if its Standard Array has all the error patterns of "t" or fewer errors and no others as coset leaders. That is the number of coset leaders $= \sum_{i=0}^{t} ({}^nC_i)$. It can also be defined as - **A 't' error correcting code is said to be Perfect Code iff it is capable of correcting error patterns with 't' or less than 't' errors.**

4.9.18. Quasi-Perfect Code

A code is said to be Quasi Perfect, if its coset leaders contain all error patterns with 't' or less errors and some patterns with $(t + 1)$ errors, but none with $(t + 2)$ or more errors.

$$\text{i.e., } \sum_{i=0}^{t} ({}^nC_i) < \text{Number of coset leaders} \leq \sum_{i=0}^{t+1} ({}^nC_i) \tag{4.26}$$

Thus ***Hamming Codes form a class of Single Error Correcting (SEC) Perfect Codes.***

Perfect codes are rare. Besides Hamming Codes, the only other non-trivial Binary **Perfect Code** is the three error correcting (23,12) **Golay Code;** (11,6) ternary Golay code is another perfect code.

But (6, 3) **Group Code,** (15, 5) **BCH (Bose-Chaudhuri-Hocquenghem) Codes** and (8, 4) **Hamming Codes** are **Quasi Perfect Codes.**

4.9.19. Decoding of Hamming Codes

Syndrome decoding (Table Look-up) method can be used for decoding the Hamming codes. The decoder for Hamming code of length $(2^m - 1)$ can be implemented in the same way as that for (7, 4) Hamming Code (using Logic gates).

4.9.20. Hamming Bound

For any binary (n, k) linear code with minimum distance (or minimum weight) '$2t+1$' or greater, the number of parity-check digits satisfies the following inequality:

$$(n-k) \geq \log_2 [1 + ({}^nC_1) + ({}^nC_2) + \ldots\ldots\ldots\ldots + ({}^nC_t)] \tag{4.27}$$

The inequality above gives an upper bound on the random error-correcting capability 't' of an (n, k) linear code. This bound is known as the **Hamming Bound.**

***Proof*:** For any binary (n, k) linear code with minimum distance (or minimum weight) '$2t + 1$' or greater, all the n-tuples of weight 't' or less can be used as coset leaders in a standard array. The number of coset leaders in a standard array are $2^{(n-k)}$.

i.e., $$2^{n-k} \geq [1 + ({}^nC_1) + ({}^nC_2) + \ldots\ldots\ldots\ldots + ({}^nC_t)].$$

$$\therefore (n-k) \geq \log_2[1 + ({}^nC_1) + ({}^nC_2) + \ldots\ldots\ldots\ldots + ({}^nC_t)].$$

Before concluding this chapter, let us take up an example and show that the Shannon's second theorem can be achieved and also show that by going for error control coding we can reduce the overall or residual probability of error of the system.

***Example* 4.4:** Let us consider the transmission of data over a channel (like telephone line) having bandwidth of B = 3 KHz, signal to noise ratio S/N =13 dB, at a rate of 1200 bits/sec with probability of error less than 10^{-4}. Let us assume that we have to transmit the data using a modem operating at a speed of 3600 bits/sec with error probability $q_c = 8\times10^{-4}$. We are supposed to design an error control scheme which will provide an overall probability of error $P_e < 1\times10^{-4}$.

Given: r_b = 1200 bits/sec, B = 3000 Hz, S/N =13 dB, r_c = 3600 bits/sec and $q_c = 8\times10^{-4}$

$$(S/N) = 10 \log_{10}(S/N) = 13 \text{ dB}.$$

$$\therefore (S/N) = 10^{1.3} = 19.953.$$

The channel capacity as per Shannon-Hartley theorem is,

$$C = B \log_2(1 + \frac{S}{N})$$

$$= 3000 \log_2(1 + 19.953)$$

i.e., C = 13167 bits/sec

Therefore, r_b and r_c are less than the channel capacity C, and hence, according to Shannon's second theorem, it is possible to have data transmission with an arbitrarily small probability of error. That is it is possible to reduce the probability of error P_e to as small a value as possible using appropriate coding technique.

When we use an (n, k) error control code, the operational speed of the modem is given by,

$$r_c = r_b\,(n/k) \text{ bits/sec}$$

$$\therefore (n/k) = (r_c/r_b) = (3600/1200) = 3$$

For a value of $k = 1$, $n = 3$,

That is the number of parity check bits = $n - k = 2$ bits. The minimum distance of the code as per Singleton bound of equation (4.24) is $d_{min} \le n - k + 1 = 3$. Hence, the code can detect and correct single error.

The overall or residual probability of error of the system using the above error control code is given by,

P_e = Probability that the decoded message block is not equal to the transmitted message block.

= Probability that the number of errors $t \ge 2$.

i.e., $P_e = P(t \ge 2)$

i.e., $P_e = \sum_{t=2}^{n} {}^nC_t\, p^t (1-p)^{n-t}$

$$= \sum_{t=2}^{3} {}^3C_t (q_c)^t (1-q_c)^{3-t}$$

$$= 3(q_c)^2 (1-q_c) + (q_c)^3$$

$$= (q_c)^2 (3-2\,q_c)$$

$$= (8\times10^{-4})^2[3-(2)8\times10^{-4})]$$

$\therefore$ $P_e \cong 0.02 \times 10^{-4}$, which is much less than 10^{-4}.

That is the residual or the overall probability error P_e of the system using error control code is about 2% of the desired value 10^{-4}. From the above it is clear that without error control coding the overall probability is equal to that of the modem $= 8\times10^{-4}$ and the use of error control coding reduced it to 0.02×10^{-4}. Hence, the use of error control coding reduces the residual or the overall probability of the system.

PROBLEMS

4.1. Show that if a linear binary code has a minimum distance that is odd, extending the code by appending an overall parity-check symbol increases the minimum distance by 1.

4.2. Consider a systematic (8, 4) code whose parity-check equations are

$$v_0 = u_1 + u_2 + u_3$$

$$v_1 = u_0 + u_1 + u_2$$

$$v_2 = u_0 + u_1 + u_3$$

$$v_3 = u_0 + u_2 + u_3$$

where u_0, u_1, u_2 and u_3 are message digits and v_0, v_1, v_2, v_3 are parity-check digits.

(*a*) Find the Generator and Parity-check Matrices for this code.

(*b*) Show that the minimum distance of this code is 4.

(*c*) Find the error detecting capability of this code.

4.3. Construct an encoder circuit and a syndrome circuit for the code specified in problem 4.2.

4.4. For a linear code with both even-weight and odd-weight code vectors, show that the number of even-weight code vectors is equal to the number of odd-weight code vectors.

4.5. Show that the 'Dual Code' of the (8, 4) linear code C given in problem 4.2 is identical to the code C. [In this case C is said to be Self-Dual].

4.6. Form a parity-check matrix for a (15, 11) Hamming Code. Also design a decoder circuit.

4.7. Show that the Hamming Codes achieve the Hamming bound.

4.8. Compute the probability of undetected error for a (15, 11) Hamming Code on a Binary Symmetric Channel (BSC) with a transition probability $p = 10^{-2}$.

4.9. The Generator matrix of a linear code is given by

$$G = \begin{pmatrix} 1 & 0 & 1 & 0 & 1 & 1 \\ 0 & 1 & 1 & 1 & 0 & 1 \end{pmatrix}$$

Find the generator matrix and the parity-check matrix for an equivalent systematic code.

4.10. Form the standard array for the systematic code in problem 4.9.

CHAPTER 5

BINARY CYCLIC CODES

5.0 Introduction

Cyclic Codes are a class of linear codes first introduced by Prange in 1957 as an alternative to Linear Block codes. They are used for both random error correction and burst error correction. They became popular because of the facts that;

- Encoding and decoding (syndrome computation) can be implemented easily using Linear Sequential Circuits (i.e., shift registers with feedback connections),
- They possess inherent algebraic structure which makes it possible to find different practical decoding methods.

Cyclic Codes:

An (n, k) linear code C is called a ***Cyclic Code*** if every cyclic shift of a code in C is also a code vector in C. When the elements of the code word are binary digits then C is called a ***Binary Cyclic Code***.

A Cyclic Shift:

If the elements i.e., digits of an n-tuple $v = (v_0, v_1, v_2, \ldots\ldots\ldots, v_{n-1})$ are cyclically shifted one place to the right we get another n-tuple

$$v^{(1)} = (v_{n-1}, v_0, v_1, v_2, \ldots\ldots\ldots, v_{n-2}).$$

This shifted n-tuple is called a cyclic shift of 'v'. If the digits of v are cyclically shifted 'i' places to the right, the resultant n-tuple would be

$$v^{(i)} = (v_{n-i}, v_{n-i+1}, \ldots\ldots\ldots, v_{n-1}, v_0, v_1, v_2, \ldots\ldots\ldots, v_{n-i-1}).$$

The above shows that, cyclically shifting 'v' by 'i' places to the right is equivalent to cyclically shifting by $(n - i)$ places to the left. In order to study the algebraic properties of a cyclic code the digits of a code vector 'v' are considered as the coefficients of a polynomial as follows:

$$v(X) = (v_0 + v_1X + v_2X^2 + \ldots\ldots\ldots + v_{n-1}X^{n-1}) \quad (5.1)$$

Therefore each code vector corresponds to a polynomial of degree $(n - 1)$ or less. If $v_{n-1} \neq 0$, the degree of $v(X)$ is $(n - 1)$ otherwise it is less than $(n - 1)$. As the code vectors are unique there is a one-to-one correspondence between the vector 'v' and the polynomial '$v(X)$' and the polynomial '$v(X)$' is called the Code Polynomial of 'v'. The code polynomial corresponding to the code vector $v^{(i)}$ is given by

$$v^{(i)}(X) = (v_{n-i} + v_{n-i+1} X + \ldots + v_{n-1} X^{i-1} + v_0 X^i + v_1 X^{i+1} + v_2 X^{i+2} + \ldots + v_{n-i-1} X^{n-1}) \tag{5.2}$$

$v(X)$ and $v^i(X)$ exhibits an interesting feature between them. By multiplying equation (5.1) by X^i we have

$$X^i v(X) = (v_0 X^i + v_1 X^{i+1} + v_2 X^{i+2} + \ldots\ldots + v_{n-i-1} X^{n-1} + v_{n-i} X^n + + v_{n-i+1} X^{n+1} + \ldots\ldots + v_{n-1} X^{n+i-1}).$$

$$= (v_{n-i}, + v_{n-i+1} X \ldots\ldots, v_{n-1} X^{i-1} + v_0 X^i + v_1 X^{i+1} + \ldots\ldots, v_{n-i-1} X^{n-1}) + [v_{n-i} (X^{n+1}) + v_{n-i+1} X(X^{n+1}) + \ldots\ldots + v_{n-1} X^{i-1}(X^n + 1)]$$

$$= v^{(i)}(X) + q(X)\,(X^{n+1})$$

$$\mathbf{X^i v(X) = q(X)\,(X^n+1) + v^{(i)}(X)} \tag{5.3}$$

where, $q(X) = (v_{n-i} + v_{n-i+1} X \ldots\ldots\ldots\ldots + v_{n-1} X^{i-1})$

Equation (5.3) is called the **Euclid's Division algorithm and $q(X)$ is called the Quotient Polynomial.**

That is **the code polynomial $v^{(i)}X$ is the remainder obtained from dividing the polynomial $X^i v(X)$ by (X^n+1).**

5.1 Algebraic Structures of Cyclic Codes

Next we consider a number of important algebraic properties of the Cyclic Codes which simplifies the implementation of encoding and syndrome computation.

Property 1: The non-zero code polynomial of minimum degree in a cyclic code C is unique.

Property 2: Let $g(X) = g_0 + g_1 X + g_2 X^2 + \ldots\ldots + g_{r-1} X^{r-1} + X^r$ be the non-zero code polynomial of minimum degree in an (n, k) Cyclic Code C. Then the constant term 'g_0' must be equal to 1.

From this property the non-zero code polynomial of minimum degree in an (n, k) cyclic code C is of the form:

$$\mathbf{g(X) = 1 + g_1 X + g_2 X^2 + \ldots\ldots\ldots\ldots + g_{r-1} X^{r-1} + X^r} \tag{5.4}$$

For example the non-zero code polynomial of minimum degree for a (7, 4) Cyclic code is $\mathbf{g(X) = 1 + X + X^3}$.

Property 3: Consider the polynomials $X.g(X)$, $X^2.g(X)$ and $X^{n-r-1}.g(X)$, which have degrees $(r + 1)$, $(r + 2)$ and $(n-1)$ respectively. Then from equation (5.3) we have

$$X.g(X) = q(X)(X^n + 1) + g^{(1)}(X).$$

X.g(X) is of degree $(r + 1)$ which is less than 'n'. Therefore $q(X) = 0$.

i.e., $X.g(X) = g^{(1)}(X)$.

Similarly $\{X^2.\, g(X) = g^{(2)}(X)\}, \{X^3.g(X) = g^{(3)}(X)\}, ..., \{X^{n-r-2}.g(X)$

$$= g^{(n-r-2)}(X)\} \text{ and } \{X^{n-r-1}.g(X) = g^{(n-r-1)}(X)\}.$$

i.e., they are the cyclic shifts of code polynomial '$g(X)$' and therefore they are code polynomials in C. Since C is linear, a linear combination of $g(X)$, $X.g(X)$, $X^2.g(X)$,..., and $X^{n-r-1}.g(X)$, along with the digits of the message 'u' gives the code polynomial.

i.e., $v(X) = u_0 g(X) + u_1 X.g(X) + u_2 X^2.g(X) + ... + u_{n-r-1} X^{n-r-1}.g(X)$,

i.e., $$\mathbf{v(X) = (u_0 + u_1 X + u_2 X^2 + ... + u_{n-r-1} X^{n-r-1}) . g(X)} \qquad \textbf{(5.5)}$$

where $u_i = 0$ or 1

Therefore it can be stated that ***"a binary polynomial of degree* (n–1) *or less is a code polynomial if and only if it is a multiple of g (X)"***

Property 4: In an (n, k) cyclic code, their exists one and only one code polynomial of degree $(n - k)$, which is the minimum degree code polynomial

$$\mathbf{g(X) = 1 + g_1 X + g_2 X^2 + ... + g_{n-k-1} X^{n-k-1} + X^{n-k}} \qquad \textbf{(5.6)}$$

i.e., Every binary polynomial of degree $(n - 1)$ or less and multiple of $g(X)$ is a code polynomial.

The code polynomial $v(X)$ in an (n, k) Cyclic Code can be expressed in the form $v(X) = u(X) . g(X)$

$$= (u_0 + u_1 X + u_2 X^2 + ...+ u_{k-1} X^{k-1}) . g(X)$$

If the coefficients of $u(X)$ namely $u_0, u_1, ..., u_{k-1}$ are the 'k' information digits to be encoded, $v(X)$ is the corresponding code polynomial. Hence, the encoding can be achieved by multiplying the message polynomial $u(X)$ by $g(X)$. Therefore, a (n, k) cyclic code is completely specified by its non-zero code polynomial of minimum degree, given by equation (5.6).

Hence, the polynomial $g(X)$ is called the Generator Polynomial of the Code.

The degree of $g(X)$ is equal to the number of Parity Check digits $(n - k)$, of the code.

***Example* 5.1:** The g(X) of the (7, 4) linear cyclic code is $g(X) = 1 + X + X^3$. It can be verified that the (7, 4) cyclic code generated is the same as the (7, 4) linear block code of example (4.1) and it can also be verified from table (4.1) that each code polynomial is a multiple of $g(X)$.

Property 5: The generator polynomial $g(X)$ of an (n, k) cyclic code is a factor of $(X^n + 1)$. $g(X)$ is of degree $(n - k)$, therefore $X^k g(X)$ is of degree 'n'. Dividing $X^k g(X)$ by $(X^n + 1)$, we get

$$X^k g(X) = (X^n+1) + g^{(k)}(X) \tag{5.7}$$

that is $q(X)$ the quotient is 1 and $g^{(k)}(X)$ is the remainder. The polynomial $g^{(k)}(X)$ is the code polynomial obtained by cyclically shifting $g(X)$ by 'k' places to the right. Hence $g^{(k)}(X)$ is a multiple of $g(X)$, say $g^{(k)}(X) = a(X)\, g(X)$.

$\therefore$ from equation (5.7) $X^n + 1 = \{ X^k + a(X) \}\, g(X)$. Hence $g(X)$ is a factor of (X^n+1).

Property 6: If $g(X)$ is a polynomial of degree $(n - k)$ and is a factor of $(X^n + 1)$ then $g(X)$ generates an (n, k) Cyclic Code.

This property assures that any factor of (X^n+1) with degree $(n - k)$ generates an (n, k) cyclic code. For large n, (X^n+1) may have many factors with degree $(n-k)$. Some of these polynomials generate good codes and some generate bad codes. The selection of polynomials which generate good codes is a difficult task.

***Example* 5.2:** $(X^7 + 1)$ can be factorised into $(1 + X)(1 + X + X^3)(1 + X^2 + X^3)$ and there are two factors with degree three each generating a (7, 4) cyclic code. (7, 4) cyclic code taken as the example 5.1 is generated by $g(X) = 1 + X + X^3$. This code has $d_{min} = 3$ and is a SEC Code.

5.2 Systematic Format of Cyclic Codes

The code generated by multiplying the message polynomial and the generator polynomial $g(X)$ is not in the systematic form. Given the generator polynomial $g(X)$ of an (n, k) cyclic code, the code can be put in the systematic form as follows:

Suppose the message polynomial of u is $(u_0 + u_1 X + u_2 X^2 + \dots\dots\dots + u_{k-1} X^{k-1})$,

- Multiply $u(X)$ by X^{n-k} to get a polynomial of degree $(n - 1)$ or less.

i.e., $X^{n-k}.u(X) = (u_0 X^{n-k} + u_1 X^{n-k+1} + u_2 X^{n-k+2} + \dots\dots\dots + u_{k-1} X^{n-1})$

- Divide $X^{n-k}.u(X)$ by the generator polynomial $g(X)$ to get $X^{n-k}.u(X) = a(X)\, g(X) + b(X)$ where $a(X)$ is the quotient polynomial and $b(X)$ is the remainder polynomial.

Since, $g(X)$ is of degree $(n - k)$, $b(X)$ must be of degree $(n - k - 1)$ or less.

i.e., $b(X) = b_0 + b_1 X + b_2 X^2 + \dots\dots\dots + b_{n-k-1} X^{n-k-1}$

or $b(X) + X^{n-k}.u(X) = a(X)\, g(X)$

Since the R.H.S is a multiple of $g(X)$, it is a valid code polynomial in the cyclic code. Therefore, $b(X) + X^{n-k}.u(X)$ is also a code polynomial. This in the expanded form is

$$= b_0 + b_1X + b_2X^2 + ... + b_{n-k-1}X^{n-k-1} + u_0X^{n-k} + u_1X^{n-k+1} + u_2X^{n-k+2} + ... + u_{k-1}X^{n-1}$$

which corresponds to the code polynomial in the systematic form. The coefficients of b(X) are the Parity check digits.

To summarise, the 3 steps involved in obtaining Systematic Cyclic Code are:

Step 1: Premultiply u(X) by X^{n-k}

Step 2: Obtain the remainder b(X), the parity check digits, by dividing $X^{n-k}.u(X)$ by g(X)

Step 3: Combine b(X) and $X^{n-k}.u(X)$ to get the code polynomial.

i.e., $b(X) + X^{n-k}.u(X)$ is the cyclic code in systematic format.

5.3 Generator Matrix of Cyclic Codes

Let $g(X) = g_0 + g_1X + g_2X^2 + ... + g_{n-k}X^{n-k}$ be the generator polynomial of an (n, k) Cyclic Code C. The cyclic code is spanned by the k, code polynomials g(X), X g(X), $X^2g(X)$, ... , $X^{k-1}g(X)$. If the k, n-tuples corresponding to these k code polynomials are used as the rows of a (k, n) matrix we obtain the Generator Matrix for the Cyclic Code C as given below:

G =

$$\begin{bmatrix} g_0 & g_1 & g_2 & . & . & . & . & . & g_{n-k} & 0 & 0 & 0 & . & . & 0 \\ 0 & g_0 & g_1 & g_2 & . & . & . & . & g_{n-k-1} & g_{n-k} & 0 & 0 & . & . & 0 \\ 0 & 0 & g_0 & g_1 & g_2 & . & . & . & . & g_{n-k-1} & g_{n-k} & 0 & . & . & 0 \\ . & . & . & . & . & . & . & . & . & . & . & . & . & . & 0 \\ 0 & 0 & 0 & 0 & 0 & g_0 & g_1 & g_2 & . & . & . & . & . & g_{n-k-1} & g_{n-k} \end{bmatrix}$$

In this $g_0 = g_{n-k} = 1$. The above Generator Matrix G is not in the Systematic form. But it can be put in the systematic form by row operations.

Example **5.3:** The (7, 4) cyclic code with generator polynomial $g(X) = 1 + X + X^3$ has the following Generator Matrix G, which is not in the systematic form. If the first row is added to the third row and the sum of first two rows are added to the fourth row we obtain the generator matrix G' which is in the systematic form.

$$G = \begin{pmatrix} 1 & 1 & 0 & 1 & 0 & 0 & 0 \\ 0 & 1 & 1 & 0 & 1 & 0 & 0 \\ 0 & 0 & 1 & 1 & 0 & 1 & 0 \\ 0 & 0 & 0 & 1 & 1 & 0 & 1 \end{pmatrix}; \quad G' = \begin{pmatrix} 1 & 1 & 0 & 1 & 0 & 0 & 0 \\ 0 & 1 & 1 & 0 & 1 & 0 & 0 \\ 1 & 1 & 1 & 0 & 0 & 1 & 0 \\ 1 & 0 & 1 & 0 & 0 & 0 & 1 \end{pmatrix}$$

5.4 Parity Check Matrix of Cyclic Codes

It was shown that $g(X)$ is a factor of (X^n+1), say $(X^n+1) = g(X).h(X)$, where $h(X)$ is a polynomial of degree 'k' and is of the form

$$h(X) = h_0 + h_1X + h_2X^2 + \dots\dots + h_kX^k,$$
with $h_0 = h_k = 1$.

We can get the Parity Check Matrix using $h(X)$. Let $v = v_0, v_1, v_2, \dots\dots, v_{n-1}$ be the code vector in the Cyclic Code C. Then $v(X) = a(X)\,g(X)$. Now multiplying $v(X)$ by $h(X)$ we have

$$v(X).h(X) = a(X)g(X).h(X)$$
$$= a(X)(X^n + 1) \qquad \text{because } g(X)h(X) = (X^n + 1)$$
$$= a(X) + X^n.a(X)$$

Since, $a(X)$ is of degree $(k - 1)$ or less, the terms with $X^k, X^{k+1}, \dots\dots, X^{n-1}$ do not appear in $a(X) + X^n\,a(X)$.

Therefore, the coefficients of the terms with $X^k, X^{k+1}, \dots\dots, X^{n-1}$ in the expansion of $v(X).h(X)$ should be $= 0$. Therefore we obtain the following $(n - k)$ equalities:

$$\sum_{i=0}^{k} h_i v_{n-i-j} = 0, \text{ for } 1 \le j \le (n-k) \qquad (5.8)$$

The reciprocal of the polynomial $h(X)$ is defined as

$$X^k\,h(X^{-1}) = h_k + h_{k-1}X + h_{k-2}X^2 + \dots\dots + h_0X^k$$

It can be easily seen that $X^kh(X^{-1})$ is also a factor of $(X^n + 1)$.

The polynomial $X^kh(X^{-1})$ generates an $(n, n-k)$ cyclic code with the following $(n-k) \times n$ matrix as a generator matrix.

$$H = \begin{pmatrix} h_k & h_{k-1} & . & . & . & . & . & . & h_0 & 0 & 0 & . & 0 \\ 0 & h_k & h_{k-1} & . & . & . & . & . & . & h_0 & 0 & . & 0 \\ 0 & 0 & h_k & h_{k-1} & . & . & . & . & . & . & h_0 & . & 0 \\ 0 & . & . & . & . & . & . & . & . & . & . & . & 0 \\ 0 & 0 & 0 & 0 & 0 & . & 0 & h_k & h_{k-1} & . & . & h_1 & h_0 \end{pmatrix}$$

From the $(n - k)$ equalities of equation (5.8) it follows that any code vector 'v' in C is orthogonal to every row of H. Therefore H is the parity check matrix of the Cyclic Code C, and the row space of H is the dual code of C. Since the parity check matrix is obtained from $h(X)$, it is called the Parity Check Polynomial of C. Hence a cyclic code is also uniquely specified by its parity check polynomial.

***Example* 5.4:** Since $g(X).h(X) = (X^n+1)$; $h(X) = (X^n+1) / g(X)$

The Parity Check Polynomial of the (7, 4) cyclic code is $= (X^7+1) / 1+X+X^3$

$$= 1 + X + X^2 + X^4$$

Property 7: The Dual code of an (n, k) cyclic code C is also cyclic and is generated by the polynomial $X^k\, h(X^{-1})$, where $h(X) = (X^n+1) / g(X)$

5.5 Generator and Parity Check Matrices of Cyclic Codes in Systematic Format

The Generator Matrix G in Systematic format can also be formed easily. Dividing (X^{n-k+i}) by the generator polynomial $g(X)$, for $i = 0, 1\ 2, \ldots\ldots\ldots\ldots, (k-1)$, we obtain

$$X^{n-k+i} = a(X)\, g(X) + b_i(X)$$

Where the remainder $b_i(X) = b_{i0} + b_{i1} X + b_{i2} X^2 + \ldots\ldots\ldots + b_{i,n-k-1} X^{n-k-1}$

Since, $b_i(X) + X^{n-k-1}$ for $i = 0, 1, 2, \ldots\ldots\ldots\ldots\ldots\ldots, (k-1)$ are multiples of $g(X)$, they are code polynomials in C. Arranging these 'k' code polynomials as rows of a $(k \times n)$ matrix we obtain

$$G = \begin{pmatrix} b_{00} & b_{01} & b_{02} & \cdots\cdots & b_{0,n-k-1} & 1 & 0 & 0 & 0 & \cdots\cdots & 0 \\ b_{10} & b_{11} & b_{12} & \cdots\cdots & b_{1,n-k-1} & 0 & 1 & 0 & 0 & \cdots\cdots & 0 \\ b_{20} & b_{21} & b_{22} & \cdots\cdots & b_{2,n-k-1} & 0 & 0 & 1 & 0 & \cdots\cdots & 0 \\ . & . & . & & . & 0 & . & & & & 0 \\ . & . & . & & . & 0 & . & & & & 0 \\ b_{k-1,0} & b_{k-1,1} & b_{k-1,2} & \cdots\cdots & b_{k-1,n-k-1} & 0 & 0 & 0 & 0 & \cdots\cdots & 1 \end{pmatrix}$$

which is the generator matrix in the Systematic form. The corresponding parity check matrix of the cyclic code in the systematic form is as given below:

$$H = \begin{pmatrix} 1 & 0 & 0 & 0 & \cdots\cdots & 0 & b_{00}\ b_{10} & b_{20} & \cdots\cdots & b_{k-1,0} \\ 0 & 1 & 0 & 0 & \cdots\cdots & 0 & b_{01}\ b_{11} & b_{21} & \cdots\cdots & b_{k-1,1} \\ 0 & 0 & 1 & 0 & \cdots\cdots & 0 & b_{02}\ b_{12} & b_{22} & \cdots\cdots & b_{k-1,2} \\ . & . & . & . & .. & . & & & & \\ . & . & . & . & .. & . & . & & & \\ 0 & 0 & 0 & 0 & \cdots\cdots & 1 & b_{0,n-k-1} & b_{1,n-k-1} & \cdots\cdots & b_{k-1,n-k-1} \end{pmatrix}$$

***Example* 5.5:** For a (7, 4) Cyclic code the generator matrix G in the systematic form can be obtained by dividing X^{7-4+i}, by the generator polynomial $g(X) = 1 + X + X^3$ for $i = 0, 1, 2, 3$.

Accordingly $b_i(X) = b_{i0} + b_{i1} X + b_{i2} X^2 + b_{i,3} X^3$ and the Generator matrix is

$$G = \begin{pmatrix} b_{00} & b_{01} & b_{02} & 1 & 0 & 0 & 0 \\ b_{10} & b_{11} & b_{12} & 0 & 1 & 0 & 0 \\ b_{20} & b_{21} & b_{22} & 0 & 0 & 1 & 0 \\ b_{30} & b_{31} & b_{32} & 0 & 0 & 0 & 1 \end{pmatrix} = \begin{pmatrix} 1 & 1 & 0 & 1 & 0 & 0 & 0 \\ 0 & 1 & 1 & 0 & 1 & 0 & 0 \\ 1 & 1 & 1 & 0 & 0 & 1 & 0 \\ 1 & 0 & 1 & 0 & 0 & 0 & 1 \end{pmatrix}$$

5.6 Linear Switching Circuits for Polynomial Multiplication and Division

The digital logic circuits namely the shift registers can be used for the cyclic shifts and polynomial arithmetic used in the description of cyclic codes. The Encoder and Decoder circuits of error correction schemes are usually constructed using adders (modulo-2), memory elements and multipliers. For digital circuits modulo-2 adders can be constructed with Logical 'Exclusive OR' gates and memory devices may be either delay lines or binary shift registers (SR's). Multiplication of two polynomials and division of the one by the other are also realized using modulo-2 adders and Shift Registers.

As a convention the high-order coefficients of polynomials are transmitted first. For example;

$$B(X) = b_0 + b_1X + b_2 X^2 + \dots\dots\dots\dots + b_n X^n$$

The coefficients are either binary 1's or 0's; the right most bit is transmitted first in any operation (multiplication or division).

5.6.1. Multiplication of Two Polynomials A(X) and B(X)

$$C(X) = A(X).B(X)$$

$$= (a_0 + a_1 X + a_2 X^2 + \dots\dots + a_k X^k).(b_0 + b_1 X + b_2 X^2 + \dots\dots + b_n X^n)$$

$$= a_0b_0 + (a_0b_1 + a_1b_0)X + (a_0b_2 + a_1b_1 + a_2b_0)X^2 + \dots + (a_{k-1}b_n + a_kb_{n-1}) X_{k+n-1} + (a_kb_n) X^{k+n}$$

This may be realized with the circuit given in Fig. 5.1.

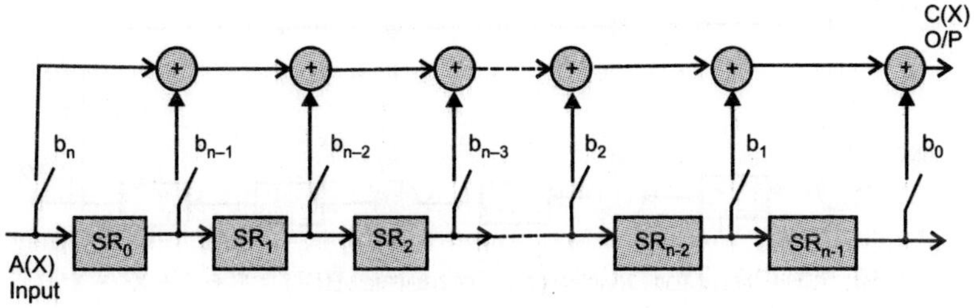

Fig. 5.1. A Circuit for Multiplying A(X) by B(X)

Here A(X) is the input and the coefficients of B(X) are given as the weighting factors connected to the mod-2 adders. Since higher order coefficients are transmitted first, the highest order coefficient '$a_k b_n$' of the product polynomial C(X) is obtained first at the output. Then the coefficients of X^{k+n-1} is obtained as the sum of $(a_{k-1}b_n + a_k b_{n-1})$, the first term directly and the second term through the storage device SR_1. Lower order coefficients are then generated through successive SR's and Mod-2 adders. The switch in the circuit shown will be in closed or open position based on $b_i = 1$ or 0.

After $(k + n - 1)$ shifts, the shift registers contain $(0, 0, \ldots, a_0, a_1)$ and the output is $a_0 b_1 + a_1 b_0$ which is the coefficient of X. After $(k + n)$ shifts the shift registers contain $(0, 0, \ldots, a_0)$ and the output is $a_0 b_0$. The product is now complete and the shift register content is zero.

5.6.2. Division of Polynomial A(X) by B(X)

The division of A(X) by B(X) may be implemented by using SR's and mod-2 adders as shown in Fig. 5.2. below:

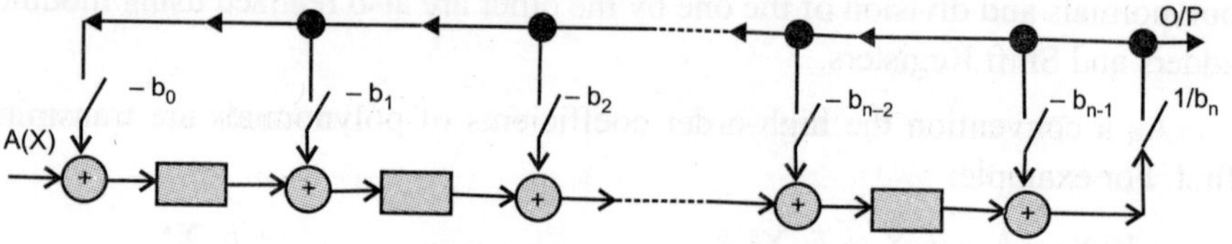

Fig. 5.2. Divider Circuit

In a division circuit, say for dividing A(X) by B(X), the first coefficient of the quotient is $a_k / b_n = q_j$, $k \geq n$, and q_j .B(X) is to be subtracted from A(X). This subtraction is done by the feed-back connections shown. After '*k*' shifts, that is by the degree of A(X), the entire quotient will appear at the output and the remainder is stored in the shift registers.

***Example* 5.6:** Consider dividing A (X) by B(X) = 1 + X + X² + X⁴ by the circuit shown in figure below: Consider now that polynomial A(X) = 1 + X⁷ ; then the output Q(X) is zero for the 4 shifts till X⁷ reaches the right most SR. In the next shift Q(X) will be 1; and this will be added to the first, second and third SR.

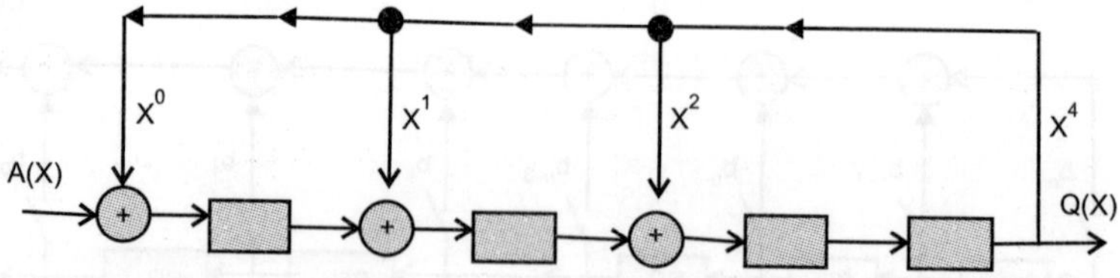

Fig. 5.3. Circuit for Dividing by Polynomial B(X) = 1 + X + X² + X⁴

In the next shift, X^6 of A(X) = 0, and the output is also zero; but at the same time the stored 1's are shifted into 2nd, 3rd and 4th SR's. The successive input to the divider, the SR content and the output Q(X) are shown in the table (5.1). Thus the quotient is $(1 + X + X^3)$ and the remainder is zero, as may be checked by long division. If, however, $A(X) = 1 + X + X^2 + X^7$ then the SR content after the 8th shift will be 0110; thus the remainder $R(X) = X + X^2$ with $Q(X) = 1 + X + X^3$

Table 5.1: Division Operation

Number of Shifts	Input a_j's of A(X)	SR Contents	Output symbols q_j's of Q(X)
0	0	0000	0
1	1 (X^7)	1000	0
2	0 (X^6)	0100	0
3	0 (X^5)	0010	0
4	0 (X^4)	0001	0
5	0 (X^3)	1110	1
6	0 (X^2)	0111	0 Q(X)
7	0 (X)	1101	1
8	1 (X^0)	0000	1
9	0	0000 {R(X)=0}	$Q(X)=1+X+X^3$

5.6.3. Simultaneous Multiplier and Divider

It is possible to combine a divider circuit with a multiplier circuit, giving thereby a composite multiplier-divider circuit which is useful in various coder circuits. For example, suppose A(X) is to be multiplied by $(1 + X^2 + X^4)$ and divided by $(1 + X + X^2 + X^4)$; then the division circuit of figure (5.3), can be further extended to include a multiplication as shown in Fig. 5.4, shown below:

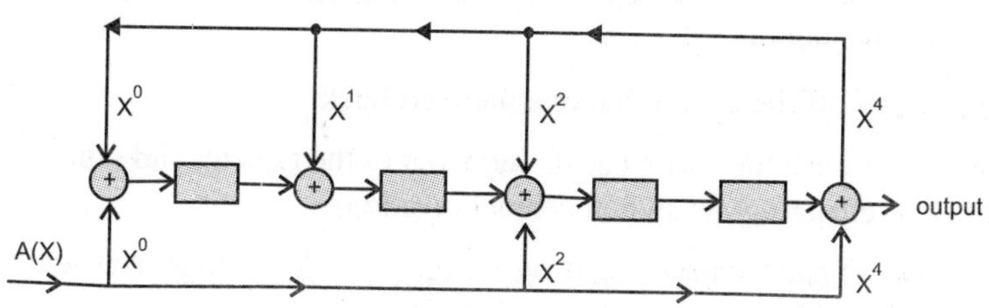

Fig. 5.4. Circuit for Simultaneous Multiplication and Division.

For $A(X) = (1 + X^7)$ the output is $(1 + X + X^2 + X^4 + X^7)$, as may be checked by long division.

5.7 Cyclic Code Encoder

The cyclic code encoder can be established by implementing the three steps mentioned in section (5.2) previously, namely

Step 1: Multiply the message polynomial u (X) by X^{n-k}.

Step 2: Divide the above product, $X^{n-k}.u(X)$ by the generator polynomial $g(X)$, to obtain the remainder polynomial, $b(X)$ and

Step 3: Form the code polynomial by combining $b(X) + X^{n-k}.u(X)$.

The above three steps can be carried out with a division circuit consisting of a linear $(n - k)$ stage shift register with feed-back connections based on the generator polynomial g(X). The circuit for the above implementation is shown in Fig. 5.5 below:

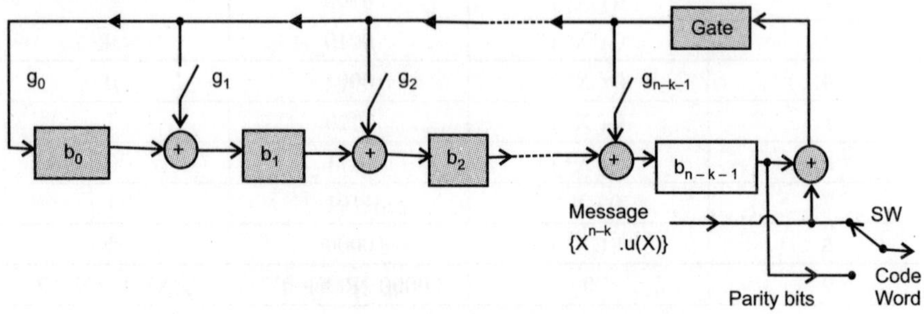

Fig. 5.5. An (n, k) Cyclic Code Encoder.

The steps involved in the encoding procedure are explained below:

Step 1: Turn on the gate; Shift the ' k ' information digits namely u_0, u_1, u_2, ..., u_{k-1} into the encoder circuit as well as into the channel simultaneously. Note: Shifting the message $u(X)$ into the circuit from the front end (as shown) is equivalent to pre-multiplying $u(X)$ by X^{n-k}. As the complete information digits has entered the circuit, the $(n - k)$ digits in the shift register form the remainder which are the parity check digits.

Step 2: Turn off the gate to remove the feed-back.

Step 3: Now shift the parity check digits out of the register and send them into the channel by changing the switch contact position.

These $(n - k)$ parity check digits b_0, b_1, b_2, ..., b_{n-k-1}, together with the ' k ' information digits form the complete code vector.

Example **5.7:** Let us consider the (7, 4) cyclic code whose $g(X) = 1 + X + X^3$. The encoder circuit for this code, based on g(X) is as shown in Fig. 5.6.

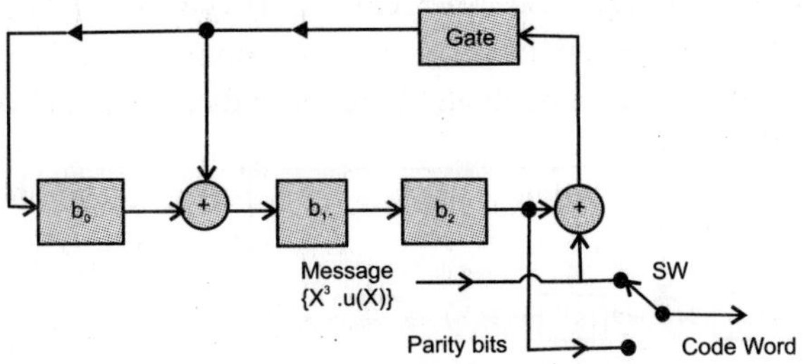

Fig. 5.6. (7, 4) Cyclic Code Encoder.

Let the message to be encoded is $u = (1\ 0\ 1\ 1)$. The contents of the register as the message digits are shifted into the register are as shown in table (5.2) below:

Table (5.2): (7,4) Cyclic Code Encoding Steps

Input	Contents of Register
-	0 0 0 (Initial position)
1	1 1 0 (1st Shift)
1	1 0 1 (2nd Shift)
0	1 0 0 (3rd Shift)
1	1 0 0 (4th Shift)

After k, (= 4), shifts the contents of the register are (1 0 0). The complete code vector is (1 0 0 1 0 1 1) and the code polynomial is $1 + X^3 + X^5 + X^6$.

5.8 Decoding of Cyclic Codes-Syndrome Computation and Error Detection

The first step in decoding a cyclic code is computation of the syndrome. For a linear systematic code, the syndrome is the vector sum of the received parity digits and the parity check digits recomputed from the received information digits. For a cyclic code in systematic form the syndrome can be computed easily. The received vector 'r' is considered as a polynomial of degree $(n-1)$ or less. That is

$$r(X) = r_0 + r_1 X + r_2 X^2 + \dots\dots + r_{n-1} X^{n-1}$$

Dividing $r(X)$ by the generator polynomial $g(X)$, we have

$$\mathbf{r(X) = a(X).g(X) + S(X)} \qquad \textbf{(5.9)}$$

where S(X), the remainder polynomial, is of degree 'n–k–1' or less. Then the $(n$–$k)$ coefficients of S(X) form the syndrome S. S(X) is identical to zero iff the received polynomial $r(X)$ is a code polynomial. The syndrome computation can be accomplished with a division circuit as shown in Fig. (5.7), which is identical to the $(n-k)$ stage encoding circuit except for the received polynomial is shifted into the register from the left end. The received polynomial $r(X)$ is shifted into the register with all stages initially set to zero. As soon as the entire $r(X)$ has been shifted into the register; the contents in the register form the syndrome $S(X)$.

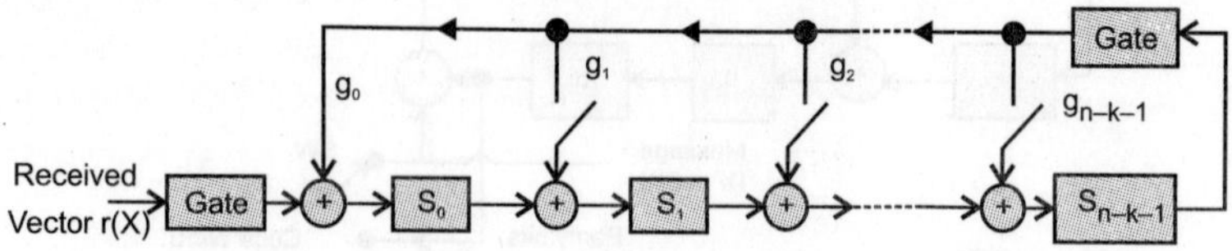

Fig. 5.7. An (n–k) Stage Syndrome Circuit with input from the left end.

***Example* 5.8:** The syndrome circuit for the (7, 4) cyclic code having $g(X) = 1 + X + X^3$ is shown in figure 5.8 below. The contents of the syndrome register with received vector ' $r = 001\ 0110$ ' as input is given in the table (5.3) below.

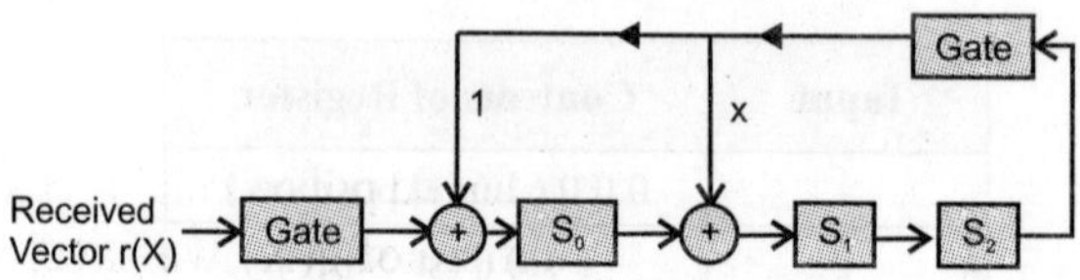

Fig. 5.8. Syndrome Circuit for the (7, 4) Cyclic Code Generated by $g(X) = 1 + X + X^3$

Table 5.3: Contents of Syndrome Register of (7, 4) Cyclic Code

Shift	Input	Register Contents
0	0	000 (Initial)
1	0	000
2	1	100
3	1	110
4	0	011
5	1	011
6	0	111
7	0	**101 (Syndrome S)**
8	-	100 (Syndrome $S^{(1)}$)
9	-	010 (Syndrome $S^{(2)}$)

Because of the cyclic structure of the code, the syndrome $S(X)$ has the following property:

***Property* 8:** For a Linear Cyclic code C having $S(X)$ as the syndrome for a received polynomial, $r(X) = r_0 + r_1 X + r_2 X^2 + ... + r_{n-1} X^{n-1}$, the remainder $S^{(1)}(X)$

resulting from dividing X.S(X) by the generator polynomial g(X) is the syndrome of $r^{(1)}$(X), which is a cyclic shift of r(X).

From the above property it can be stated that the remainder $S^{(i)}$(X) resulting from dividing $X^{(i)}$S(X) by the generator polynomial g(X) is the syndrome of $r^{(i)}$(X) which is the ith cyclic shift of the received polynomial r(X). This property is useful in decoding the cyclic codes.

The syndrome $S^{(1)}$(X) of $r^{(1)}$(X) can be obtained by shifting the syndrome register once with S(X) as the initial content and with the input gate disabled. This is because of the fact that shifting the syndrome register once with S(X) as the initial content is equivalent to dividing XS(X) by g(X). Thus after the shift, the register contains $S^{(1)}$(X). To obtain $S^{(i)}$(X) of $r^{(i)}$(X), it is sufficient to shift the syndrome register 'i' times with S(X) as the initial content.

5.9 Error Detection and Correction

Let v(X) be the transmitted code word and let "$e(X) = e_0 + e_1 X + e_2 X^2 + ... + e_{n-1} X^{n-1}$ "be the error pattern. Then the received polynomial is $r(X) = v(X) + e(X)$. Since the code polynomial v(X) is a multiple of g(X) we have, the error pattern given by

$$\begin{aligned} e(X) &= \{r(X)\} + \{v(X)\}. \\ &= \{a(X)g(X) + S(X)\} + \{b(X)g(X)\} \\ &= [a(X) + b(X)].g(X) + S(X) \end{aligned}$$

This means that the syndrome is actually equal to the remainder obtained by dividing the error pattern by the generator polynomial g(X).

The syndrome can be computed from the received vector; however, the error pattern e(X) is unknown to the decoder. Therefore the decoder has to estimate e(X) based on the syndrome S(X).

5.10 Standard Array Decoding of Cyclic Codes

If e(X) is a coset leader in the Standard Array and if table-lookup decoding is used, e(X) can be correctly determined from the syndrome. From the above expression for the error pattern it is clear that ***the syndrome S(X) is identical to zero iff either the error pattern e(X) = 0, or it is identical to a code polynomial. If e(X) is identical to a code polynomial it is an undetectable error pattern.***

Cyclic codes are very effective in detecting errors both random and burst. The error detection circuit is just a syndrome circuit with an OR gate, with the syndrome digits as inputs. If the syndrome is not zero, the output of OR gate will be '1' and the presence of errors has been detected.

5.11 Error Detecting Capabilities of an (n, k) Cyclic Code

An (n, k) Cyclic code is capable of detecting error bursts of length (n – k) or less.

Let the error pattern is a burst of length $(n-k)$ or less, that is the errors are confined to $(n-k)$ or fewer consecutive positions. Then the burst error pattern can be expressed as $\boldsymbol{e(X) = X^j.B(X)}$ where $0 \leq j \leq (n-1)$ and B(X) is a polynomial of degree $(n-k-1)$ or less, that is less than the degree of the generator polynomial g(X), and with b_0 non-zero. ***B(X) describes the burst pattern and X^j describes the starting location of the burst.***

Since the degree of B(X) is less than that of g(X), B(X) is not divisible by g(X). Since g(X) is a factor of $(X^n + 1)$ and X is not a factor of g(X) and also $(X^n + 1)$, we have g(X) and X^j must be relatively prime. Therefore $\boldsymbol{e(X) = X^j.B(X)}$ is not divisible by g(X), that is e(X) is not a multiple of g(X). Therefore the syndrome caused by this e(X) is not equal to zero.

This implies that, an (n, k) Cyclic Code is capable of detecting any error burst of length (n – k) or less.

***Example* 5.9:** In a (7, 4) code, if the error vector $e = 0011100$, i.e. having an error burst of length 3 $(= n - k)$, then the error polynomial $e(X) = X^2 + X^3 + X^4 = X^2(1 + X + X^2)$,

i.e. $j = 2$ and $B(X) = 1 + X + X^2$

5.11.1. End-around Burst

For a cyclic code, an error pattern with errors confined to 'i' higher order positions, and $(l - i)$ lower-order positions is also regarded as a burst of length "l" or less. Such a burst is called **"End-around burst".** For example 'e = 101000001101' is an end-around error burst of length 7. An (n, k) Cyclic code is also capable of detecting all the end-around error bursts of length $(n-k)$ or less. Summarizing we have the following property:

***Property* 9:** An (n, k) Cyclic code is capable of detecting any error burst of length $(n-k)$ or less, including the end around bursts.

***Property* 10:** "The minimum weight of a cyclic code is equal to the number of terms in the generator polynomial".

Since the generator polynomial g(X) is the minimum degree code polynomial of the cyclic code, the number of terms in the corresponding code vector, i.e., the number of terms in g(X) gives the minimum weight and hence the minimum distance of the cyclic code.

***Example* 5.10:** The minimum weight of the (7, 4) cyclic code with $g(X) = 1 + X + X^3$, is equal to the number of terms in g(X)= 3.

5.12 Decoding of Cyclic Codes: Meggitt Decoder

With the above discussions it is clear that any cyclic code decoder will mainly have first a **Buffer Register to store the received vector**, second a **Syndrome Register** and third an **Error Pattern Detector.** The table look up method of implementing the error pattern detector mentioned in section 5.10 was practical for syndromes of length $(n-k)$ up to 12 or 14 bits. However the lookup table decoder can be simplified by using the strong algebraic structure of the cyclic code to find the relation between the syndromes. Using such relations it is sufficient to store the error patterns related to few typical syndromes. Simple methods can then be used to estimate the remaining required error patterns. One such decoder is the Meggitt decoder. This decoder tests the syndrome only for those error patterns that have an error in the high-order component, that is error patterns in which $e_{n-1} = 1$. Errors in other locations can be decoded latter by using the cyclic properties of the code. As a result the syndrome evaluator table need to contain only those syndromes corresponding to an error pattern in which $e_{n-1} = 1$(or non-zero). If the computed syndrome is one such syndrome then e_{n-1} is corrected. Next the received vector is cyclically shifted and the process repeated to test for a possible error in the next to higher order place ($e_{n-2} = 1$). This procedure is repeated digit by digit; each digit of the received vector is tested in turn for error and is corrected whenever an error is detected.

The basic Meggitt decoder circuit of an (n, k) cyclic code is shown in figure (5.9). This Meggitt decoder can be used, in principle, to any cyclic code.The received code word (polynomial) is shifted into the syndrome register from the left end. The effect of an error digit on the syndrome is removed by feeding the error digit into the syndrome (shift) register from the left end through an Exclusive –OR gate as shown. The decoder operation is explained below:

1. The received vector is shifted into the buffer register and syndrome register simultaneously. The syndrome is formed in the syndrome register when all the digits of the received vector are moved in.
2. The syndrome formed at the output of the syndrome register is read into the error pattern detection circuit and is tested for a corresponding error pattern. The error detection circuit is so designed to give an output '1' only when the syndrome formed corresponds to a correctable error pattern with an error at the highest order position X^{n-1}. That is iff $e_{n-1} = 1$ in the error polynomial. That is, if a '1' appears at the output of the detection circuit, the received symbol (digit) in the right most stage of the buffer register is assumed to be erroneous and must be corrected; if a '0' appears at the output or the detector, the received symbol at the right most stage of the buffer register is assumed to be correct and no correction is necessary. Thus the output of detector is the estimated error value for the symbol to come out of the buffer. The error pattern detection circuit can be a combinational logic circuit.

3. The first received symbol is read out of the buffer. At the same time, the syndrome register is shifted once. If the first received symbol is detected to be erroneous, it is then corrected by the output of the detector. The output of the detector is also fed back to the syndrome register to modify the syndrome (i.e., to remove the error effect from the syndrome). This result in a new syndrome which corresponds to the altered received vector shifted one place to the right.
4. The new syndrome formed in the above step is used to detect whether or not the second received symbol (now at the right most stage of the buffer register) is an erroneous symbol. The decoder repeats step 2 and step 3. The second received symbol is corrected in exactly the same manner as the first received symbol was corrected.
5. The decoder decodes the received vector symbol by symbol in the same manner as explained above until all the symbols of the received vector is read out of the buffer register.

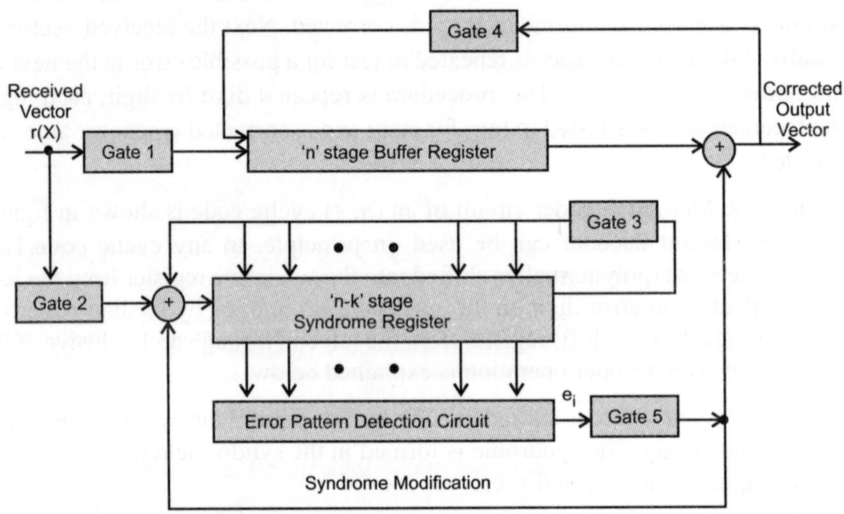

Fig. 5.9. Generic Cyclic Code Decoder Circuit {"Meggitt Decoder"}

The decoding stops after a total of '*n*' shifts. If *e*(X) is a correctable error pattern the contents of the syndrome register should be zero at the end of the decoding operation, and the received vector *r*(X) has been correctly decoded. On the other hand, if the syndrome register does not contain all 0's at the end of the decoding process, an uncorrectable error pattern has been detected.Whether the Meggitt decoder is practical or not, depends entirely on its error pattern detection circuit. Also the decoding of a received vector takes '*n*' extra clock period after the receipt of that code vector. Of course this extra '*n*' clock cycles can be at a faster rate than the rate at which the code vector is received.

***Example* 5.11:** The Meggitt decoder circuit for the (7, 4) cyclic code whose $g(x) = 1 + X + X^3$ is shown in Fig. 5.10 below:

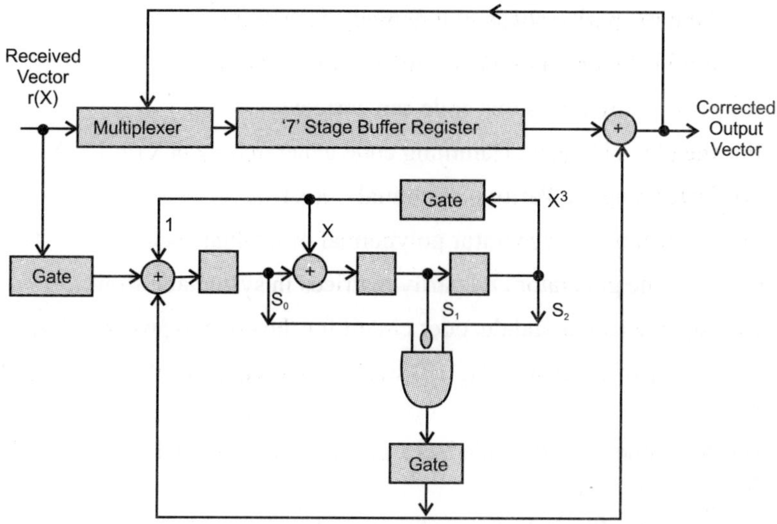

Fig. 5.10. Meggitt Decoder for (7, 4) Cycle Code Generated by $g(x) = 1 + X + X^3$.

The properties of the Generator Polynomial g(X) are summarized below:

1. $g(X)$ is an irreducible and primitive polynomial (i.e., a factoı of X^n+1)
2. Every code polynomial is a multiple of $g(X)$
3. The degree of $g(X)$ is equal to
   - The number of parity check digits (symbols)
   - The number of Shift Register stages in the encoder.
   - The length of burst errors the code can detect.
4. The number of terms in $g(X)$ is equal to the 'minimum weight' of the code.

PROBLEMS

5.1. (*a*) Find the product of the polynomials (1+X) and $(1 + X + X^3)$ using modulo-2 addition.

(*b*) Divide $(X^6 + X^5 + X^3)$ by $(X^3 + X + 1)$

(*c*) If $f(X) = (X^4 + X + 1)$, then show that $[f(X)]^2 = f(X^2)$.

(*d*) Find $(1+X)(1 + X + X^2)(1 + X + X^4)(1 + X^3 + X^4)$

(*e*) Find the factors of (X^7+1)

5.2. Draw a shift register circuit to divide an incoming polynomial by the polynomial $(X^3 + X^2 + 1)$

5.3. The polynomial $g(X) = 1 + X^4 + X^6 + X^7 + X^8$ is the generator polynomial of a cyclic code with block length $n = 15$.

(*a*) Find the parity-check polynomial of the code.

(*b*) Find the Generator Matrix in systematic format.

(*c*) How many errors this code can correct?

5.4. For the (15, 11) cyclic Hamming code generated by $g(X) = 1 + X + X^4$

(*a*) Determine the Parity polynomial '$h(X)$'

(*b*) Determine the generator polynomial of its dual code.

(*c*) Find the generator and parity matrices in systematic form.

5.5. Devise an encoder and decoder circuit for the code in problem 5.4.

5.6. Find $g(X)$ of a binary (21, 12) cyclic code capable of correcting double error.

5.7. A (15, 5) linear cyclic code has a generator polynomial $g(X) = 1 + X + X^2 + X^4 + X^5 + X^8 + X^{10}$.

(*a*) Draw the block diagram of an encoder and syndrome evaluator for this code.

(*b*) Find the code polynomial in systematic form for the message $u(X) = 1 + X^2 + X^4$.

(*c*) Is the received polynomial $r(X) = 1 + X^4 + X^6 + X^8 + X^{14}$ a code polynomial? If not find the syndrome for this polynomial.

5.8. Prove that $g(X) = 1 + X + X^3$ is the generator polynomial of a (7, 4) cyclic code.

5.9. For a (7, 4) cyclic code having $g(X) = 1 + X^2 + X^3$

(*a*) Draw the syndrome computation circuit.

(*b*) Find the syndrome for the following received vectors with single error: (*i*) 0011000, (*ii*) 1110110, (*iii*) 0011111 and (*iv*) 0001110. List the states of the shift register after each received bit input.

(*c*) Correct the errors and get the transmitted code words.

(*d*) Verify the answer by direct hand computation.

5.10. Let $g(X)$ be the generator polynomial of an (n, k) cyclic code.

(*a*) Show that $g(X^\mu)$ generates a (μ_n, μ_k) cyclic code.

(*b*) Show that the code generated by $g(X^\mu)$ has the same minimum weight as that of the code generated by $g(X)$.

CHAPTER

CONVOLUTIONAL CODES

6

6.0 Introduction

In case of block codes, the data is grouped into "k" bit words, and coded into "n" bit code words after adding $(n - k)$ parity check bits. Decoding is done word by word and there is no coupling between successive words.

Contrast with this, in an alternative technique, an encoder transforms "k" bits of data into n-bit codes, but the encoder output also depends on the earlier blocks of data that have been processed by the encoder. Thus the encoder possess memory and this memory couples the currently processed data with few (say 'm') earlier data blocks. The size of the encoder memory is indicated by the ***constraint length*** "m" and the code is designated by (n, k, m) along with the code rate $R = k / n$. Such codes are known as ***Tree Codes*** or ***Recurrent Codes***. An important class of Tree Codes are known as ***Convolutional Codes***. Convolutional Codes were first introduced by Elias in 1955 as a substitute to block codes.

Convolutional codes also posses Linear Time Invariance property. An (n, k, m) convolutional code can be implemented with a "k" input and a "n" output linear sequential circuit with input memory "m". Typically, "n" and "k" are small integers with $k < n$, but the memory order "m" can be made large to achieve low error probabilities. Several practical procedures have been developed for decoding.

In Convolutional Code, as the encoding of a message depends on previous message blocks also, a message block may be encoded into different code blocks at different instances of time. As decoding also depends on previous code vectors error propagates from code vector to code vector as the decoding proceeds. Therefore, this is called sometimes as an error propagating code. In the important special case when $k = 1$, the information sequence is not divided into blocks and can be processed continuously. In that case code length depends on message length. The encoding of convolutional codes can be carried out using both time domain approach and transform domain approach.

6.1 Encoding of Convolutional Codes using Time Domain Approach

The encoding of convolutional code using time domain approach is explained using the following example:

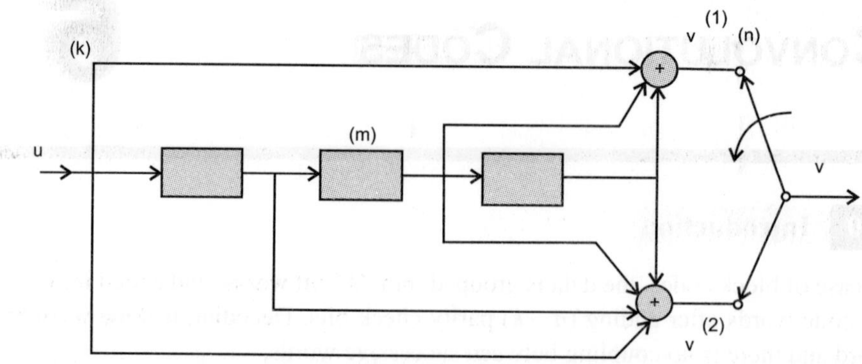

Fig. 6.1. (2,1,3) Binary Convolutional Encoder

A typical (n, k, m) for $n = 2$, $k = 1$, $m = 3$ is shown in Fig. 6.1 above. The encoder consists of an "m" (= 3) stage shift register together with "n" (= 2) modulo-2 adders and a multiplexer for serialising the encoder outputs. Since modulo-2 addition is a linear operation, the encoder is a linear feed forward shift register (LFF SR). All convolutional encoders can be implemented using a linear feed forward shift register of this type.

The information sequence $u = (u_0, u_1, u_2, \text{..........})$ enter the encoder one bit at a time. Since, the encoder is a linear system the two encoder output sequences

$$v^{(1)} = [v_0^{(1)}, v_1^{(1)}, v_2^{(1)} \text{........}] \text{ and } v^{(2)} = [v_0^{(2)}, v_1^{(2)}, v_2^{(2)} \text{.....}] \tag{6.1}$$

can be obtained as the convolution of the input sequence 'u' with the n (= 2) encoder "impulse responses". These impulse responses define the time domain behaviour of the (binary) convolutional encoder.

The impulse responses are obtained by letting $u = (1\ 0\ 0 \text{ ....})$ and observing the two output sequences $v^{(1)}$ and $v^{(2)}$. Since the encoder has a 'm' - time unit memory, the impulse responses can last at most $(m + 1)$ time units, and are written as:

$$g^{(1)} = [g_0^{(1)}, g_1^{(1)}, g_2^{(1)} \text{........} g_m^{(1)}]$$

and $$g^{(2)} = [g_0^{(2)}, g_1^{(2)}, g_2^{(2)} \text{........} g_m^{(2)}] \tag{6.2}$$

For the encoder of Fig. (6.1) $g^{(1)} = (1011)$ and $g^{(2)} = (1111)$

The impulse responses $g^{(1)}$ and $g^{(2)}$ are called the **Generator Sequences** of the code. The **Encoding Equations** can be written as:

$$v^{(1)} = u * g^{(1)}$$

$$v^{(2)} = u * g^{(2)} \tag{6.3}$$

where "$*$" denotes 'discrete convolution' and all operations are modulo-2. The convolution operation implies that for all "$l \geq 0$",

$$v_l^{(j)} = \sum_{i=0}^{m} u_{l-i} g_i^{(j)} = u_l g_0^{(j)} + u_{l-1} g_1^{(j)} + \dots\dots\dots\dots + u_{l-m} g_m^{(j)} \tag{6.4}$$

for $j = 1,2$ and $u_{l-i} = 0$ for all $l < i$

That is the output sequence at each output terminal is the sum of the product of $(m + 1)$ input bits and the response in the corresponding time impulse response.

Hence, for the encoder of the Fig. 6.1 above

$$v_1^{(1)} = u_l + u_{l-2} + u_{l-3}$$

$$v_1^{(2)} = u_l + u_{l-1} + u_{l-2} + u_{l-3}$$

This can easily be verified by inspection of the encoding circuit. After encoding, the two output sequences are multiplexed into a single sequence, called the code word, for transmission over the channel. The code word is given by

$$v = [v_0^{(1)} v_0^{(2)}, v_1^{(1)} v_1^{(2)}, v_2^{(1)} v_2^{(2)} \dots\dots\dots\dots] \tag{6.5}$$

Encoding terminates when the last non-zero input element (bit) comes out of the encoder.

***Example* 6.1:** For the encoder of figure (6.1), let the information sequence $u = (10111)$ then the output sequences are $v^{(j)} = u * g^{(j)}$

$$v^{(1)} = (10111) * (1011) = (1000\ 0001)$$

$$v^{(2)} = (10111) * (1111) = (11011101)$$

and the code word is $v = (11, 01, 00, 01, 01, 01, 00, 11)$.

We can see that the length of the output code $= n\ (m + \text{L})$ where 'L' is the length of the input sequence. The steps involved in getting the output sequence at terminal 1 is as given below:

$$v_0^{(1)} = u_0 g_0^{(1)} + u_1 g_1^{(1)} + u_2 g_2^{(1)} + u_3\, g_3^{(1)}$$

$$v_0^{(1)} = u_0 g_0^{(1)} + 0 + 0 + 0 = 1 \tag{6.6}$$

Similarly $v_1^{(1)} = u_1 g_0^{(1)} + u_0 g_1^{(1)} + 0 + 0 = 0$

$$v_2^{(1)} = u_2 g_0^{(1)} + u_1 g_1^{(1)} + u_0 g_2^{(1)} + 0 = 0$$

$$v_3^{(1)} = u_3 g_0^{(1)} + u_2 g_1^{(1)} + u_1 g_2^{(1)} + u_0 g_3^{(1)} = 0$$

$$v_4^{(1)} = u_4 g_0^{(1)} + u_3 g_1^{(1)} + u_2 g_2^{(1)} + u_1 g_3^{(1)} = 0$$

$$v_5^{(1)} = 0g_0^{(1)} + u_4g_1^{(1)} + u_3g_2^{(1)} + u_2\,g_3^{(1)} = 0$$

$$v_6^{(1)} = 0g_0^{(1)} + 0g_1^{(1)} + u_4g_2^{(1)} + u_3g_3^{(1)} = 0$$

$$v_7^{(1)} = 0g_0^{(1)} + 0g_1^{(1)} + 0g_2^{(1)} + u_4g_3^{(1)} = 1$$

Similarly the outputs at terminal (2) can also be computed.

6.2 Generator Matrix of a Convolutional Code

The encoding of a convolutional code can be done by using a Generator Matrix. Method of forming the generator matrix and encoding is as explained below:

The generator sequences $g^{(1)}$ and $g^{(2)}$ can be interlaced and then arranged in the matrix form as below:

$$G = \begin{pmatrix} g_0^1 & g_0^{(2)} & g_1^{(1)} & g_1^{(2)} & g_2^{(1)} & g_2^{(2)} & \cdots\cdots & g_m^{(1)} & g_m^{(2)} & & & & \\ & & g_0^{(1)} & g_0^{(2)} & g_1^{(1)} & g_1^{(2)} & \cdots\cdots & g_{m-1}^{(1)} & g_{m-1}^{(2)} & g_m^{(1)} & g_m^{(2)} & & \\ & & & & g_0^{(1)} & g_0^{(2)} & \cdots\cdots & g_{m-2}^{(1)} & g_{m-2}^{(2)} & g_{m-1}^{(1)} & g_{m-1}^{2} & g_m^{(1)} & g_m^{(2)} \end{pmatrix} \tag{6.7}$$

Where the blank areas are zeros, the encoding equations can be rewritten in the matrix form as

$$v = u\,G \tag{6.8}$$

Where all operations are Modulo-2. "G" is called the (time domain) **Generator Matrix** of the encoder. Note that each row of G is identical to the previous row but shifted by $n = 2$ places to the right, and G is a semi-infinite matrix, corresponding to the fact that the information sequence 'u' is of arbitrary length. If 'u' has finite length "L", then **G** has **L rows** and $\boldsymbol{n(m+L)} = 2(m + L)$ columns, and '$\boldsymbol{v}$' has length $\boldsymbol{n(m + L)} = 2(m + L)$.

***Example* 6.2:** For the encoder of Fig. (6.1) If $u = (\ 10111\ )$ then,

$$\boldsymbol{v = u\,G}$$

$$= (10111).\begin{bmatrix} 11 & 01 & 11 & 11 & & & & \\ & 11 & 01 & 11 & 11 & & & \\ & & 11 & 01 & 11 & 11 & & \\ & & & & 11 & 01 & 11 & 11 \end{bmatrix}$$

$$= 11, 01, 00, 01, 01, 01, 00, 11$$

The result is the same as we have obtained using discrete convolution in example (6.1).

***Example* 6.3:** Consider a (3, 2, 1) Convolutional encoder shown in figure (6.2) below :

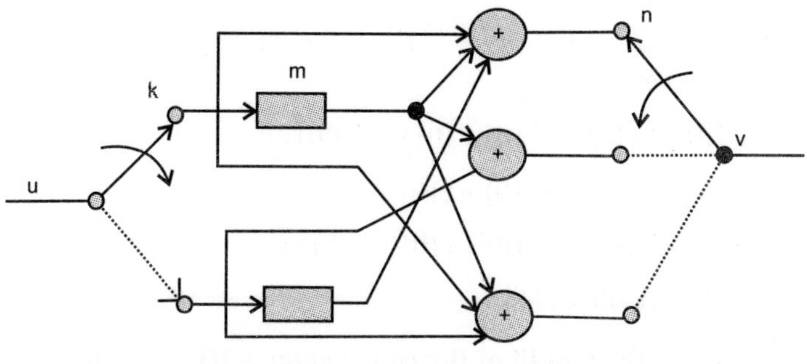

Fig. 6.2. Rate R = (2/3), (3,2,1) Covolutional Encoder

Since $k = 2$, the encoder consists of two m (= 1) stage shift registers together with n (= 3) Mod-2 adders and two multiplexers, one at the input and the other at the output. The information sequence enters the encoder k (= 2) bits at a time, and can be written as

$$u = (u_0^{(1)}\, u_0^{(2)},\, u_1^{(1)}\, u_1^{(2)},\, u_2^{(1)}\, u_2^{(2)}) \text{ or as two input sequences}$$

$$u^{(1)} = (u_0^{(1)}, u_1^{(1)}, u_2^{(1)} \ldots\ldots\ldots)$$

$$u^{(2)} = (u_0^{(2)}, u_1^{(2)}, u_2^{(2)} \ldots\ldots\ldots) \qquad (6.9)$$

There are three generator sequences corresponding to each input sequence.

Let $g_i^{(j)} = g_{(i,0)}^{(j)}\, g_{(i,1)}^{(j)}\, g_{(i,2)}^{(j)} \ldots\ldots\ldots\ldots g_{(i,m)}^{(j)}$ represent, the generator sequence corresponding to input 'i' and output 'j'. Therefore, there will be $(k \times n)$ generator sequences and corresponding $(k \times n)$ generator polynomials. The generator sequence of Fig. 6.2 are:

$$g_1^{(i)} = (11);\ g_1^{(2)} = (01)\ ;\ g_1^{(3)} = (11);$$

$$g_2^{(1)} = (01);\ g_2^{(2)} = (10)\ ;\ g_2^{(3)} = (10);$$

and the encoding equations can be written as

$$v^{(1)} = u^{(1)} * g_1^{(1)} + u^{(2)} * g_2^{(1)}$$

$$v^{(2)} = u^{(1)} * g_1^{(2)} + u^{(2)} * g^{2}_{(2)}$$

$$v^{(3)} = u^{(1)} * g_1^{(3)} + u^{(2)} * g_2^{(3)} \qquad (6.10)$$

The convolution equation implies that

$$v_l^{(1)} = u_l^{(1)} \times 1 + u_{(l-1)}^{(1)} \times 1 + u_l^{(2)} \times 0 + u_{(l-1)}^{(2)} \times 1$$

$$v_l^{(2)} = u_l^{(1)} \times 0 + u_{(l-1)}^{(1)} \times 1 + u_l^{(2)} \times 1 + u_{(l-1)}^{(2)} \times 0$$

$$v_l^{(3)} = u_l^{(1)} \times 1 + u_{(l-1)}^{(1)} \times 1 + u_l^{(2)} \times 1 + u_{(l-1)}^{(2)} \times 0$$

This can be seen from the encoding circuit. After multiplexing, the code word is given by

$$v = (v_0^{(1)}\, v_0^{(2)}\, v_0^{(3)},\ v_1^{(1)}\, v_1^{(2)}\, v_1^{(3)},\ v_2^{(1)}\, v_2^{(2)}\, v_2^{(3)}) \qquad (6.11)$$

***Example* 6.4:** If $u^{(1)} = 101$, $u^{(2)} = (110)$ is the input to the encoder of example (6.3) the output sequences are

$$v^{(1)} = (101) * (11) + (110) * (01) = (1001)$$

$$v^{(2)} = (101) * (01) + (110) * (10) = (1001)$$

$$v^{(3)} = (101) * (11) + (110) * (10) = (0011)$$

$$v = (110, 000, 001, 111).$$

When k > 1 the length of the code vector = [(L / k) + m] n (6.12)

In this case L, length of the input sequence 'u' = 6, $k = 2$, $m = 1$, $n = 3$; Hence Length of code = **[(L / k) + m] n** = 12. We can also form the Generator Matrix and carry out encoding using this matrix.

In a convolutional encoder with $k > 1$, any set of 'kn' generator sequences can be used to realize a rate 'R = k/n' feed forward convolutional encoder as shown in figures (6.1) and (6.2) with 'k' shift registers to store the input sequences and 'n' modulo-2 adders to form the output sequences. This type of feed forward convolutional encoder realization in which the 'k' input sequences enter the left end of each shift register, and the 'n' output sequences are produced by modulo-2 adders external to the shift registers, is called **Controller Canonical Form.**

6.3 Encoder Memory Order

It is evident that when $k > 1$, the encoding and the representing notation for a convolutional code are more complicated. i.e., the encoder consists of k shift registers, each may be of different length (stages). If K_i is the length of the ith shift register then the encoder memory "m" is defined as:

$$m \triangleq \max (K_i) \text{ for } 1 \le i \le k \qquad (6.13)$$

(i.e., the maximum length, out of all 'k' shift registers)

For the rate R =1/2 encoder of figure (6.1), $m = k_1 = 3$; and for the rate R = 2/3, encoder of figure (6.2), $m = \max(k_1, k_2) = \max(1, 1) = 1$.

6.4 Overall Constraint Length of Convolutional Encoder

The ***overall constraint length*** of a convolutional encoder or the convolutional code is defined as the sum of the lengths of all 'k' shift registers. This is also some times called as the ***Total Encoder memory*** and is given by

$$K = \sum_{1 \le i \le k} K_i \qquad (6.14)$$

For the rate R = 1/2 encoder of figure 6.1, $K = K_1 = 3$; and for the rate R = 2/3, encoder of Fig. 6.2, $K = K_1 + K_2 = 1 + 1 = 2$.

***Example* 6.5:** The encoder diagram of a (4, 3, 2) Convolutional code is given below:

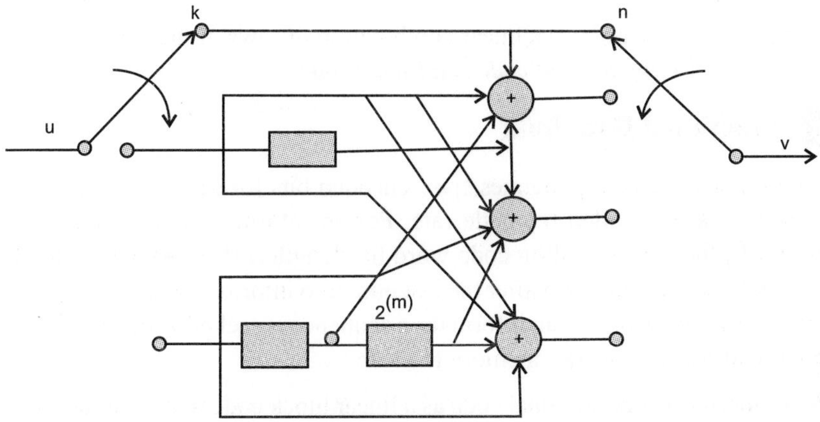

Fig. 6.3: (4, 3, 2) Binary Convolutional Encoder

6.5 General Form of Generator Matrix of a Convolutional Code

A (n, k, m) Convolutional Code is a linear code. In general case of an (n, k, m) code, the generator matrix is

$$G = \begin{pmatrix} G_0 & G_1 & G_2 & \cdots\cdots & G_m & & \\ & G_0 & G_1 & \cdots\cdots & G_{m-1} & G_m & \\ & & G_0 & \cdots\cdots & G_{m-2} & G_{m-1} & G_m \\ & & & . & & . & . \quad . \\ & & & . & . & . & . \end{pmatrix} \tag{6.15}$$

where each G_l is a " $k \times n$ " sub matrix whose entries are:

$$G_l = \begin{pmatrix} g_{1.l}^{(1)} & g_{1.l}^{2} & \cdots\cdots & g_{1.l}^{(n)} \\ g_{2.l}^{(1)} & g_{2.l}^{(2)} & \cdots\cdots & g_{2.l}^{(n)} \\ . & . & & . \\ . & . & & . \\ g_{k.l}^{(1)} & g_{k.l}^{(2)} & \cdots\cdots & g_{k.l}^{(n)} \end{pmatrix} \tag{6.16}$$

It can be noted that each set of " k " rows of G is identical to the previous set of " k " rows but shifted by n places to the right.

For an information sequence

$$u = (u_0, u_1, ...) = u_0^{(1)}\, v_0^{(2)} \,........\, u_0^{(k)}, u_1^{(1)}\, u_1^{(2)} \,........\, u_1^{(k)}......$$

the code word $v = (v_0 , v_1,....) = v_0^{(1)}\, v_0^{(2)} \,.....\, v_0^{(n)}, v_1^{(1)}\, v_1^{(2)} \,.....\, v_1^{(n)}...$ is again given by $\boldsymbol{v} = \boldsymbol{u}\,\mathbf{G}$

Since the code word v is a linear combination of rows of the generator matrix G, an (n, k, m) ***Convolutional code is a linear code.***

6.6 Fractional Rate Loss

A convolutional encoder generates " n "encoded bits for each " k " information bits and $\mathbf{R} = \boldsymbol{k/n}$ is called the code rate. For an information sequence of finite length $\boldsymbol{k}$ **. L,** the corresponding code word has length $\boldsymbol{n}$ **(L +** $\boldsymbol{m}$**),** where the final "$n * m$" outputs are generated after the last non-zero information block has entered the encoder. In other words, an information sequence is sacked with all zero blocks in order to allow the encoder memory to clear.

Considering a convolutional code as a linear block code with generator matrix G, the block code rate is given by $\boldsymbol{k}$**L/**$\boldsymbol{n}$**(L +** $\boldsymbol{m}$**)** i.e., the ratio of the number of information bits to the length of the code word. If $L >> m$, then $(L\,/\,L + m) \cong 1$, and the block code rate and convolutional code rate are approximately equal, that is R $= k/n$. Therefore, the rate of the convolution code and the rate of convolution code when it is viewed as a block code are the same.

If L were small, the rate $\boldsymbol{k}$**L/**$\boldsymbol{n}$ **(L+**$\boldsymbol{m}$ **)** which is the effective rate of information transmission, would be reduced below the code rate by a fractional amount.

$$\frac{\left(\frac{k}{n}\right)-\left(\frac{k\mathrm{L}}{n(\mathrm{L}+m)}\right)}{\left(\frac{k}{n}\right)} = \frac{m}{\mathrm{L}+m} \tag{6.17}$$

This fraction is called the ***Fractional Rate Loss***. Hence, to keep the fractional rate loss small (i.e., near to zero), L is always kept much larger than m.

***Example* 6.6:** For the information sequence in example (6.1); L = 5, and the fractional rate loss is 3/8 = 37.5 %. However, for an information sequence of length L = 1000 the fractional rate loss is only 3/1003 = 0.3 %.

6.7 Encoding of Convolutional Codes using Transform Domain Approach

As per linear filter theory the convolutional integral, which describes the linear filtering operation in the time domain, can be replaced by multiplication of Fourier Transform in frequency domain. Since, the convolutional encoder is a linear system, the adder outputs of the convolutional encoder can be computed by applying an appropriate transform. Each sequence in the encoding equations can be replaced by a corresponding polynomial and the convolution operation can be replaced by polynomial multiplication. In the polynomial representation of a binary sequence, the sequence itself can be represented by the coefficients of the polynomial. The concept is explained below:

For a (2, 1, m) code, the encoding equations become

$$v^{(1)}(D) = u(D)\, g^{(1)}(D)$$

$$v^{(2)}(D) = u(D)\, g^{(2)}(D),$$

where $u(D) = u_0 + u_1 D + u_2 D^2 + \ldots..$ (6.18)

is the information polynomial.

$v^{(1)} D = v_0^{(1)} + v_1^{(1)} D + v_2^{(1)} D^2 + \ldots..$ and

$v^{(2)} D = v_0^{(2)} + v_1^{(2)} D + v_2^{(2)} D^2 + \ldots.$ (6.19)

are the encoder or output polynomials and

$$g^{(1)}(D) = g_0^{(1)} + g_1^{(1)}(D) + \ldots\ldots\ldots\ldots\ g_m^{(1)}(D)^m$$

$$g^{(2)}(D) = g_0^{(2)} + g_1^{(2)}(D) + \ldots\ldots\ldots\ldots\ g_m^{(2)}(D)^m \quad (6.20)$$

are the **Generator polynomials** of the code, and all operations are Modulo-2. After multiplexing the code word becomes, v **(D)** = $v^{(1)}$ **(D²) + D** $v^{(2)}$ **(D²)**. The indeterminate 'D' can be interpreted as a delay operator, the power of D denoting the number of time units a bit is delayed w.r.t. the initial bit in the sequence.

The generic form of the generator polynomial is of the form:

$$g^{(j)}(D) = g_1^{(j)} + g_2^{(j)} D + g_3^{(j)} D^2 + \ldots\ldots\ldots + g_{m+1}^{(j)} D^m \quad (6.21)$$

The adder output will be of the form $\mathbf{v^{(j)}(D) = u(D)\, g^{(j)}(D)}$

The final encoder output is obtained by multiplexing the adder outputs and is of the form

$$v(D) = v^{(1)}(D^n) + Dv^{(2)}(D^n) + D^2 v^{(3)}(D^n) + \ldots\ldots + D^{n-1} v^{(n)}(D^n) \quad (6.22)$$

***Example* 6.7:** For the (2, 1, 3) code of Fig.(6.1), the generator polynomials are

$$g^{(1)}(D) = 1 + D^2 + D^3 = [1\ 0\ 1\ 1]$$

$$g^{(2)}(D) = 1 + D + D^2 + D^3 = [1\ 1\ 1\ 1]$$

The information polynomial for the information sequence 10111 is $u(D) = 1 + D^2 + D^3 + D^4$

The encoding equations are:

$$v^{(1)}(D) = (1 + D^2 + D^3 + D^4)(1 + D^2 + D^3) = 1 + D^7$$

$$v^{(2)}(D) = (1 + D^2 + D^3 + D^4)(1 + D + D^2 + D^3) = 1 + D + D^3 + D^4 + D^5 + D^7$$

and the code word is

$$v(D) = 1 + D + D^3 + D^7 + D^9 + D^{11} + D^{14} + D^{15}$$

The result is the same as we have obtained using convolution as well as matrix multiplication. The generator polynomials of an encoder can be determined directly from its circuit diagram.

6.7.1. Transfer Function Matrix

Since the convolutional code encoder is a Linear Time Invariant system, the generator polynomial $\mathbf{g}_i^{(j)}(\mathbf{D})$ can be interpreted as the encoder transfer fuction relating the ith input to the jth output. For a 'k' input and 'n' output linear convolutional code there will be '$k.n$' Transfer fuctions. These transfer functions can be put in the form of a '$k \times n$' matrix called the **Transfer Function Matrix** as given below:

$$G(D) = \begin{pmatrix} g_1^{(1)}(D) & g_1^{(2)}(D) & \ldots\ldots & g_1^{(n)}(D) \\ g_2^{(1)}(D) & g_2^{(2)}(D) & \ldots\ldots & g_2^{(n)}(D) \\ \cdot & \cdot & \cdot & \cdot \\ \cdot & \cdot & \cdot & \cdot \\ g_k^{(1)}(D) & g_k^{(2)}(D) & \ldots\ldots & g_k^{(n)}(D) \end{pmatrix} \quad (6.23)$$

Using G(D) the encoding equation can be written as

$$\mathbf{V(D) = U(D).G(D)} \quad (6.24)$$

where $U(D) = [\, u^{(1)}(D), u^{(2)}(D), \ldots\ldots u^{(k)}(D)\,]$ is the k-tuple of input sequences and $V(D) = [v^{(1)}(D), v^{(2)}(D), \ldots\ldots v^{(n)}(D)]$ is the n-tuple of output sequences.

After multiplexing the code word becomes

$$v(D) = v^{(1)}(D^n) + Dv^{(2)}(D^n) + D^2v^{(3)}(D^n) + \ldots\ldots + D^{n-1}v^{(n)}(D^n). \quad (6.25)$$

***Example* 6.8:** The transfer function matrix of the (3, 2, 1) code of figure (6.2) is given by

$$G(D) = \begin{pmatrix} 1+D & D & 1+D \\ D & 1 & 1 \end{pmatrix}$$

For the input sequence $u^{(1)}(D) = 1 + D^2$ and $u^{(2)}(D) = 1 + D$, the encoding equations are:

$$V(D) = [v^{(1)}(D), v^{(2)}(D), v^{(3)}(D)] = [\,1 + D^2, 1 + D\,] \begin{pmatrix} 1+D & D & 1+D \\ D & 1 & 1 \end{pmatrix}$$

$$= [\,1 + D^3, 1 + D^3, D^2 + D^3\,]$$

And the code word $v(D) = 1 + D + D^8 + D^9 + D^{10} + D^{11} = 110\ 000\ 001\ 111$.

Again the result is the same as obtained with convolution and matrix multiplication methods.

6.8 Systematic Convolutional Code

An important sub-class of convolutional codes is the class of systematic codes. In a systematic code, the first "k" output sequences are exact replicas of the "k" input sequences, i.e.,

$$v^{(i)} = u^{(i)} \text{ for } i = 1, 2, \ldots\ldots\ldots\ldots k$$

and the generator sequences satisfy $g_i^{(j)} = 1$ if $j = i$ and $g_i^{(j)} = 0$ if $j \neq i$; for $i = 1, 2, \ldots.k$.

The generator matrix is given by

$$G = \begin{pmatrix} I & P_0 & 0 & P_1 & 0 & P_2 & \cdots\cdots & 0 & P_m & & & & \\ & & I & P_0 & 0 & P_1 & \cdots\cdots & 0 & P_{m-1} & 0 & P_m & & \\ & & & & I & P_0 & \cdots\cdots & 0 & P_{m-2} & 0 & P_{m-1} & 0 & P_m \\ & & & & & \cdot & & \cdot & & & \cdot & & \cdot \\ & & & & & & \cdot & & \cdot & & & \cdot & \cdot \\ & & & & & & & \cdot & & \cdot & & \cdot & \cdot \end{pmatrix} \quad \ldots.(6.26)$$

where, 'I' is the $k * k$ identify matrix, 0 is the $k * k$ all zero matrix, and P_l is the $k * (n - k)$ matrix and is given by

$$P_l = \begin{pmatrix} g_{1,l}^{(k+1)} & g_{1,l}^{(k+2)} & \cdots\cdots & g_{1,l}^{(n)} \\ g_{2,l}^{(k+1)} & g_{2,l}^{(k+2)} & \cdots\cdots & g_{2,l}^{(n)} \\ \cdot & \cdot & & \cdot \\ \cdot & \cdot & & \cdot \\ g_{k,l}^{(k+1)} & g_{k,l}^{(k+2)} & \cdots\cdots & g_{k,l}^{(n)} \end{pmatrix} \quad (6.27)$$

and the transfer function matrix becomes

$$G(D) = \begin{pmatrix} 1 & 0 & \cdots\cdots & g_1^{(k+1)}D & \cdots\cdots & g_1^{(n)}(D) \\ 0 & 1 & \cdots\cdots & g_2^{(k+1)}D & \cdots\cdots & g_2^{(n)}D \\ \cdot & \cdot & \cdot & \cdot & & \cdot \\ 0 & 0 & & 1\ g_k^{(k+1)}D & \cdots\cdots & g_k^{(n)}D \end{pmatrix} \quad (6.28)$$

Since the first k output sequences equal the input sequences, they are called ***information sequences*** and the last $(n–k)$ output sequences are called **Parity**

sequences. In a non-systematic Convolutional code in general $k * n$ generator sequences must be specified to define an (n, k, m) Convolutional code. Whereas in the case of a systematic code only $k\,(n - k)$ sequences must be specified to define a Systematic code. (Hence, Systematic codes represent a sub class of the set of all possible codes).

***Example* 6.9:** Consider the (2,1,3) Systematic Convolutional Code whose encoder is shown in Figure (6.4), below:

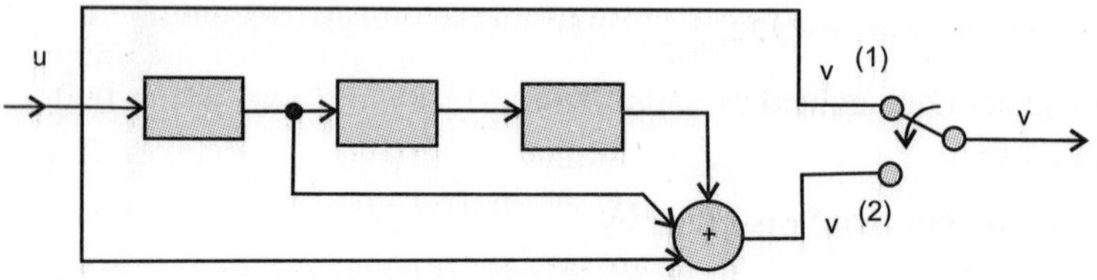

Fig. 6.4. (2, 1, 3) Systematic Encoder .

The generator sequences of the encoder are $g^{(1)} = (1000)$ and $g^{(2)} = (1101)$, and the generator matrix is

$$G = \begin{pmatrix} 11 & 01 & 00 & 01 & & & \\ & 11 & 01 & 00 & 01 & & \\ & & 11 & 01 & 00 & 01 & \\ & & . & . & . & . & \\ & & & . & . & . & . \\ & & & . & . & . & . \end{pmatrix}$$

and the transfer function matrix is

$$G(D) = [1 \;\; g_1^{(k+1)}(D)\,] = [1 \;\; (1 + D + D^3)] \text{ (because } k = 1)$$

For an input sequence $u(D) = 1 + D^2 + D^3$ (i.e., u =1011) the information sequence at the output terminal $v^{(1)}$ is,

$$v^{(1)}(D) = u(D)\, g^{(1)}(D) = (1 + D^2 + D^3) \bullet 1$$

$$= (1 + D^2 + D^3) = 1011\;000$$

[the last three 0's are appended to the input sequence to see that the last information digit comes out of the encoder]

and the parity sequence is $v^{(2)}(D) = u(D)g^{(2)}(D)$

$$= (1 + D^2 + D^3)\,(1 + D + D^3)$$

$$= 1 + D + D^3 + D^2 + D^3 + D^5 + D^3 + D^4 + D^6$$

$= 1 + D + D^2 + D^3 + D^4 + D^5 + D^6 = 1111111$

$\therefore \quad v = 11, 01, 11, 11, 01, 01, 01$.(obtained by interleaving $v^{(1)}(D)$ and $v^{(2)}(D)$)

6.8.1. Advantages of Systematic Convolutional Codes

Encoding is simpler (because of less hardware).

- For example, the (2, 1, 3) non-systematic codes of Fig 6.1 requires two modulo-2 adders with a total of seven inputs, whereas the (2, 1, 3) Systematic Code of Fig. 6.4 requires only one Modulo-2 adder with only 3 inputs.
- The number of generator sequences and shift registers required will be less.
- Also, for an (n, k, m) systematic code with $k > (n - k)$ there exists a modified encoding circuit which normally requires fewer than K shift register stages (as shown in the following example 6.10).

***Example* 6.10:** Consider the (3, 2, 2) systematic code with transfer function Matrix

$$G(D) = \begin{pmatrix} 1 & 0 & 1+D+D^2 \\ 0 & 1 & 1+D^2 \end{pmatrix}$$

The straight forward canonical form realisation of the encoder requires a total of $k = k_1 + k_2 = 4$ shift register stages, as shown in Fig. 6.5a. On the other hand since the information sequences are given by

$v^{(1)}(D) = u^{(1)}(D)$ and

$v^{(2)}(D) = u^{(2)}(D)$ and the parity sequence is given by

$v^{(3)}(D) = u^{(1)}(D)g_1^2(D) + u^{(2)}(D)g_2^2(D)$

The encoder can also be modified as shown in Fig 6.5b. This requires only 2 stages of encoder memory rather than 4. In most cases the straight forward realization is more efficient. However for an (n, k, m) systematic code with $k > n - k$, a simpler realization usually exists, as noted in Fig. 6.5(b) of example 6.10.

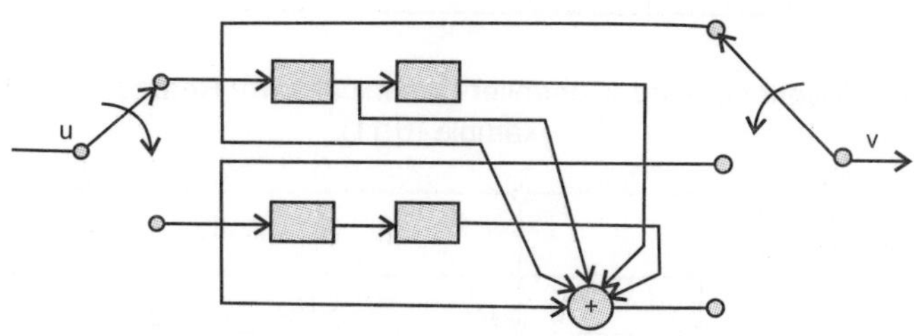

Fig. 6.5(a). Controller Canonical Form of Convolutional Encoder.

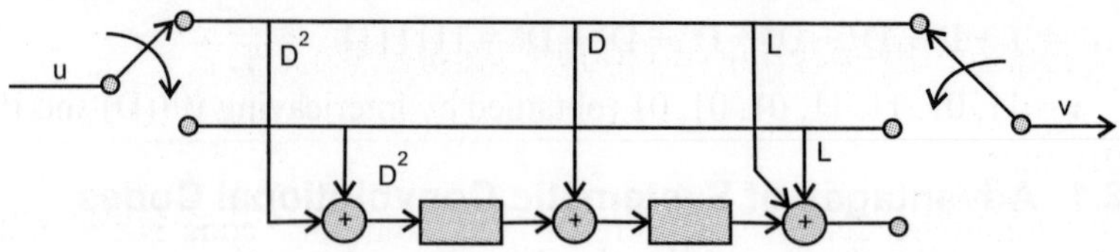

Fig. 6.5(b). Observer Canonical Form of Convolutional Encoder.

The modified encoder realization is called **'Observer Canonical form'.** In this form of realization, there is one shift register corresponding to each of the '*n*' output sequences, the '*k*' input sequences enter modulo-2 adders internal to the shift register, and the outputs at the right end of each shift register form the '*n*' output sequences. Also the lowest-degree (D^0, i.e., the constant) terms in the generator (parity-check) polynomials represent the connections at the right ends of the shift registers, whereas the highest-degree terms represent the connection at the left ends of the shift registers. It can be noted that this is exactly the opposite in the case of a controller canonical form realization. Hence, when an encoder is realized in observer canonical form it is common to write the parity check polynomials with highest degree term first (for example '$D^2 + D + 1$' as in Fig. 6.5b).

***Example* 6.11:** Transfer Function Matrix, Encoder circuit and Modified encoder circuit of (3, 2, 3) Systematic Code are given below in Figs. 6.6(*a*) and (*b*).

$$G(D) = \begin{pmatrix} 1 & 0 & 1+D^2+D^3 \\ 0 & 1 & 1+D+D^3 \end{pmatrix}$$

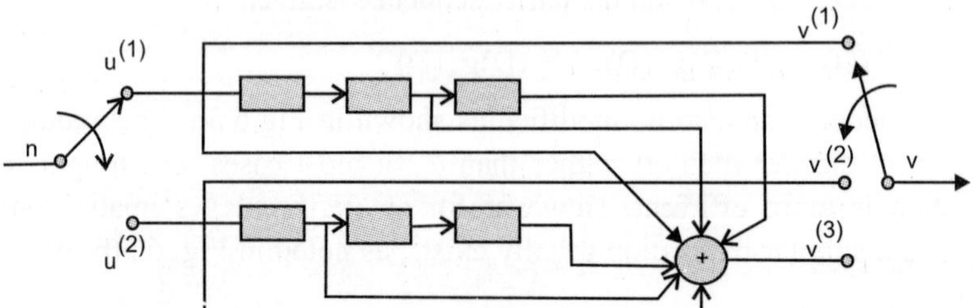

Fig. 6.6 (a). Straight Forward (Controller Canonical form) Realization of Encoder for example (6.11)

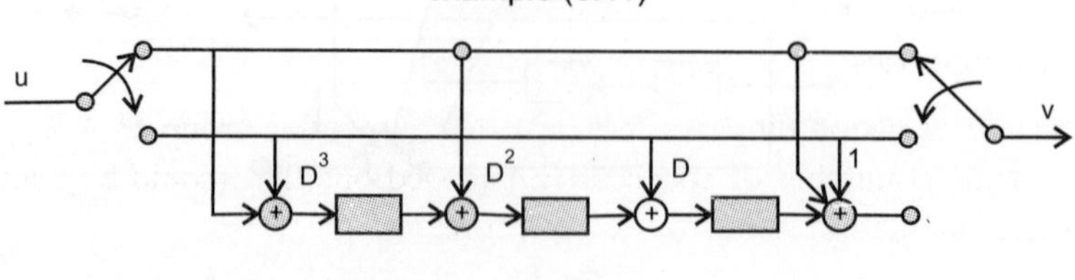

Fig. 6.6 (b). Modified (Observer canonical form) Encoder of example (6.11)

6.9 Distance Properties of Convolutional Codes

Free Distance

The most important distance measure of a convolutional code is the minimum "free distance" d_{free} defined as:

$$d_{free} \triangleq \min \{d(v', v'') : u' \neq u''\} \tag{6.29}$$

where v' and v'' are the code words corresponding to the information sequence u' and u'' respectively.

Hence d_{free} is the minimum distance between any two code words in the code. Since a convolutional code is a linear code,

$$\begin{aligned} d_{free} &= \min \{w(v' + v'') : u' \neq u''\} \\ &= \min \{w(v) : u \neq 0\} \\ &= \min \{w(u\,G) : u \neq 0\} \end{aligned} \tag{6.30}$$

where "v" is the code word corresponding to the information sequence 'u'. Hence d_{free} is the weight of minimum weight code word of any length produced by a non-zero information sequence.

***Example* 6.12:** d_{free} = 6 for the (2, 1, 3) code of Fig. 6.1; d_{free} = 3 for the (3, 2, 1) code of Fig. 6.2.

The d_{free} depends on K [the overall constraint length or the total encoder memory defined by equation (6.14)], i.e., the total number of stages in the shift register. It is difficult to find d_{free} for larger value of K. Algorithms are available for values of K upto about 20. No general solution to the problem of finding d_{free} for large values of K has yet been discovered.

Column Distance Function (CDF)

The column distance function of order 'i' is defined as

$$\begin{aligned} d_l &\triangleq \min \{d([v']_l, [v'']_l) : [u']_0 \neq [u'']_0\} \\ &= \min \{w[v]_l : [u]_0 \neq 0\} \end{aligned} \tag{6.31}$$

And 'd_l' is the minimum weight code word over the $(l + 1)$ time unit whose initial information block is non-zero.

Distance properties

Since each information bit can affect $(m + 1)$ consecutive group of "n" output bits, the smallest number of consecutive encoded bits that should be considered for distance calculation is $n\,(m + 1)$. Therefore, let,

1. v_1 and v_2 be two code word paths that originate at the same node of the tree or trellis (discussed in section 6.11.1) and each containing $n\,(m + 1)$ encoded bits.

2. The first information bit associated with v_1 be the complement of the first information bit associated with v_2.

Now the ***minimum distance*** d_{min} of the convolutional code can be defined as d_{min} = minimum number of places in which v_1 and v_2 differ [for any pair of code sequences satisfying the conditions 1 and 2].

Clearly we need not restrict to '$n(m+1)$' consecutive encoded bits only while defining d_{min}. We can take a length L of each code sequence, L being greater than '$n(m+1)$'. Then the resulting minimum distance is called the ***decoding minimum distance*** d_L. Now if we let L to go to infinity then d_L is called the ***minimum free distance***, d_f. It is found that when 'l' tends to infinity $d_l = d_{free}$; also

$$d_f \geq d_L \geq d_{min} \tag{6.32}$$

The minimum distance of an (n, k) convolutional code is defined as the smallest H.D = $d(C_1, C_2)$ between two initial code vectors $C_1(x)$ and $C_2(x)$ which disagree in the 0th block. The "d_{min}" can be determined by the inspection of the tree or the trellis diagram of the coder and the number of correctable errors in "n.K" bits of the received vector r_i is given by the condition $(2t+1) \leq d_{min}$.

6.10 Structural Properties of Convolutional Code

6.10.1. State Diagram

Total Encoder memory of a (n, k, m) encoder = K = $\sum_{i=1}^{k} K_i$ (6.33)

The encoder state at time unit 'l' is the binary K-tuple of inputs namely,

$$u_{l-1}^{(1)}\, u_{l-2}^{(1)}\, u_{l-3}^{(1)} \ldots\ldots u_{l-K_1}^{(1)}\, u_{l-1}^{(2)}\, u_{l-K_2}^{(2)} \ldots\ldots u_{l-1}^{(k)}\, u_{l-2}^{(k)} \ldots\ldots u_{l-K_K}^{(k)} \tag{6.34}$$

and there is 2^k different possible states in total. For $(n, 1, m)$ code K = $K_1 = m$ and the encoder state at that time unit " l " is $(u_{l-1}, u_{l-2}, \ldots\ldots u_{l-m})$. Each new block of k input causes a transition to a new state. Hence, there are 2^k branches leaving each state. ***State diagram is the graph of the possible states of the encoder and the possible transition from one state to another***. Each branch is labelled with the k-tuple formed by the 'k' inputs, $(u_l^{(1)}\, u_l^{(2)} \ldots\ldots u_l^{(k)})$, causing the transition and the n tuple formed by the corresponding outputs $(v_l^{(1)}\, v_l^{(2)} \ldots\ldots v_l^{(n)})$.

***Example* 6.13:** The state diagram for the (2, 1, 3) code of Fig. 6.1 is given in Fig. 6.7 below. The states are labelled as $S_0, S_1, \ldots\ldots S(_{2^k-1})$ where S_i represents the state whose binary K-tuple representation is equivalent to the integer 'i' (decimal value). For the encoder of Fig. 6.1, there are three stages in the shift register and hence, there are $2^3 = 8$ states namely S_0 to S_7. This can be represented in the form of a '**State Table**' as given in table 6.1.

Table 6.1: State Table for the (2,1,3) Encoder of Figure (6.1)

Encoder State	S_0	S_1	S_2	S_3	S_4	S_5	S_6	S_7
Binary Representation	000	100	010	110	001	101	011	111

The state transitions of the encoder can be put in the form of a **'State Transition Table'** as given in table (6.2) below for the (2,1,3) encoder of Fig. (6.1):

The Encoder outputs are $v^{(1)} = u_l + u_{l-2} + u_{l-3}$; $v^{(2)} = u_l + u_{l-1} + u_{l-2} + u_{l-3}$

Table 6.2: State Transition table for the (2,1,3) Encoder of Fig. (6.1)

Present State	Binary Form	Input	Next State	Binary Form	$u_l\ u_{l-1}\ u_{l-2}\ u_{l-3}$	Output $v^{(1)}\ v^{(2)}$
S_0	000	0	S_0	000	0 0 0 0	0 0
		1	S_1	100	1 0 0 0	1 1
S_1	100	0	S_2	010	0 1 0 0	0 1
		1	S_3	110	1 1 0 0	1 0
S_2	010	0	S_4	001	0 0 1 0	1 1
		1	S_5	101	1 0 1 0	0 0
S_3	110	0	S_6	011	0 1 1 0	0 1
		1	S_7	111	1 1 1 0	0 1
S_4	001	0	S_0	000	0 0 0 1	1 1
		1	S_1	100	1 0 0 1	0 0
S_5	101	0	S_2	010	0 1 0 1	1 0
		1	S_3	110	1 1 0 1	0 1
S_6	011	0	S_4	001	0 0 1 1	0 0
		1	S_5	101	1 0 1 1	1 1
S_7	111	0	S_6	011	0 1 1 1	0 1
		1	S_7	111	1 1 1 1	1 0

The state diagram for this (2,1,3) encoder of Fig. 6.1 is shown in Fig. 6.7. Assuming that the encoder is initially in state S_0 the code word corresponding to any given information sequence can be obtained by following the path in the state diagram determined by the information sequence and noting the corresponding output's on the branch labels. Following the last non-zero information block, the encoder is returned to state S_0 by a sequence of m - all zero blocks appended to the information sequence.

For $u = 11101$, from Fig. (6.7) the code word $v = (11, 10, 01, 01, 11, 10, 11, 11)$

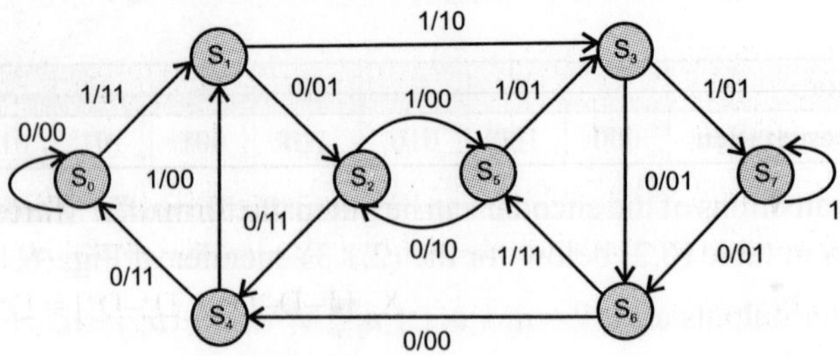

Fig. 6.7. Encoder State Diagram for (2, 1, 3) Code

***Example* 6.14:** The state diagram for a (3,1,2) Systematic Convolutional code of Fig. 6.8 whose sub generator sequence are given by $g^{(2)} = 101$; $g^{(3)} = 110$, and transfer function matrix **G (D) = [1, 1 + D², 1 + D]**, is shown in Fig. 6.9.

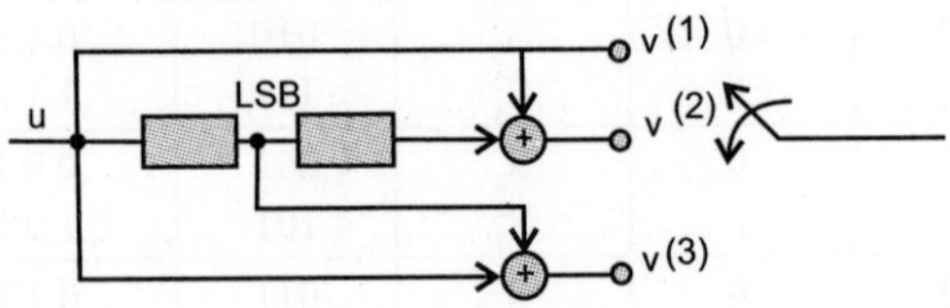

Fig. 6.8. (3, 1, 2) Systematic Convolutional Encoder

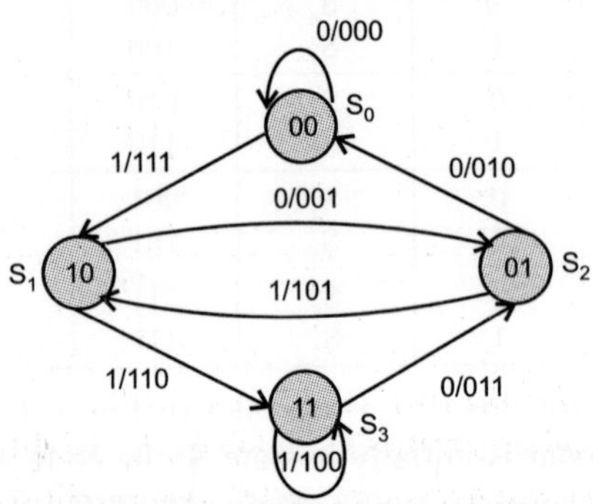

Fig. 6.9. State Diagram of (3, 1, 2) Systematic Convolutional Code

Let the message u = 1011, then the output at the three output terminals are given by,

$$v^{(1)} = u_i * g^{(1)} = (1 + D^2 + D^3)\,(1) = 1011\ 00$$

$$v^{(2)} = u_i * g^{(2)} = (1 + D^2 + D^3)\,(1 + D^2) = 1 + D^2 + D^3 + D^2 + D^4 + D^5$$

$$= 1 + D^3 + D^4 + D^5$$

$$= 1\ 0\ 0\ 1\ 1\ 1$$

$$v^{(3)} = u_i * g^{(3)} = (1 + D^2 + D^3)(1 + D) = 1 + D^2 + D^3 + D + D^3 + D^4$$
$$= 1 + D + D^2 + D^4$$
$$= 1\ 1\ 1\ 0\ 1\ 0$$

By multiplexing $v^{(1)}$, $v^{(2)}$ and $v^{(3)}$ we have v = 111, 001, 101, 110, 011, 010

The same code vector can be traced through the state diagram in Fig. 6.9.

6.10.2. Hamming Distance Computation using Modified State Diagram

Apart from obtaining the output sequence for a given input sequence, the state diagram can be modified to get the complete weight distribution of all non-zero code words. That is the state diagram is useful in determining the Hamming weight of the code words and hence the minimum distance of the code. To illustrate this let us consider the (3, 1, 2) systematic convolutional code of example 6.14, whose encoder diagram and state diagram are given in Fig. (6.8) and (6.9). The modified state diagram for this code is given in Fig. 6.9 (a) below.

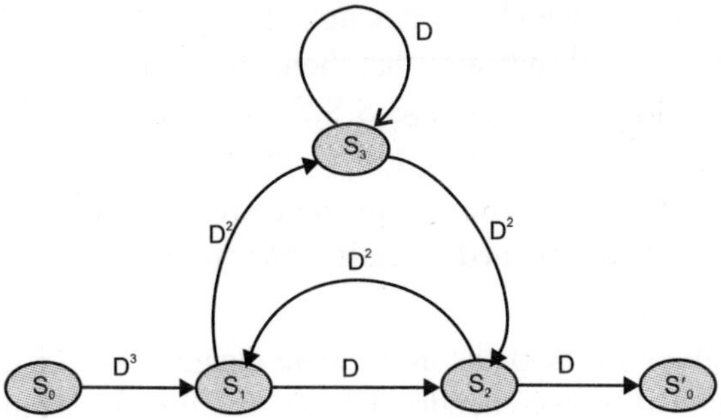

Fig. 6.9(a). Modified State Diagram of (3, 1, 2) Encoder in Figure (6.8)

In the modified state diagram the state S_0 is split into an initial state, S_0, representing the input and a final state, S'_0 representing the output. The self loop around S_0 is not included as it does not contribute to the Hamming distance (H.D). The branches are labelled as D^i, where '*i*', the exponent of D denote the H.D between the actual output sequence in that branch and the all zero out sequence. Now this diagram consists of **5 nodes** to write the following **4 state equations.**

$$X_{S_1} = D^3 X_{S_0} + D^2 X_{S_2};$$
$$X_{S_2} = D X_{S_1} + D^2 X_{S_3};$$
$$X_{S_3} = D^2 X_{S_1} + D X_{S_3};$$
$$X_{S'0} = D\, X_{S2}.$$

Now the transfer function of the encoder is given as $T(D) = [\, X_{S'_0} / X_{S_0}]$. By solving the 4 state equations above we can get the expression for T(D).

$$X_{S2} = DX_{S_1} + D^2X_{S_3} = DX_{S_1} + D^2[(D^2X_{S_1})/(1-D)]$$
$$= X_{S_1}[D + (D^4)/(1-D)] = X_{S_1}[(D-D^2 + D^4)/(1-D)]$$

Now substituting the value of X_{S_1} we have,

$$X_{S_2} = [D^3X_{S_0} + D^2 X_{S_2}] [(D) (1-D+D^3)/(1-D)]$$
$$X_{S_2} \{1- [(D^3) (1-D+D^3)/(1-D)]\} = (D^3) X_{S_0} [D(1-D+D^3)/(1-D)]$$
$$X_{S_2}[1-D-D^3 + D^4-D^6] = D^4[1-D+D^3] X_{S_0};$$

Now substituting the value of X_{S_2} in terms of $X_{S'_0}$ we have,

$$\{(X_{S'_0})/D\} [1-D-D^3 + D^4-D^6] = D^4[1-D+D^3]X_{S_0}$$
$$\therefore \quad \{(X_{S'_0})/X_{S_0}\} = \{D^5[1-D+D^3] / [1-D-D^3+D^4-D^6]\}$$
$$= [D^5 - D^6 + D^8] / [1-D-D^3 + D^4- D^6]$$
$$\therefore \quad T(D) = D^5 + 2 D^8 + D^9 + D^{10} + 4D^{11} + 3D^{12} + \ldots$$

The T(D) for this code indicates that there is a single path with H.D, $d = 5$ from the all zero path that merges with the all zero path at a given node. From the state diagram it is observed that the $d = 5$ path is "$S_0S_1S_2S'_0$". There is no other path from node S_0 to node S'_0 having a distance of $d = 5$. The second term in the expression for T(D) indicates that there are 2 paths from S_0 to S'_0 having a distance of $d = 8$; these paths are $S_0S_1S_3S_2S'_0$ and $S_0S_1S_2S_1S_2S'_0$. The third term indicates that there is a path ($S_0S_1S_3S_3S_2S'_0$) with $d = 9$ and so on. Thus the transfer function gives the distance properties of the convolutional code. The minimum distance of the code so obtained is called the "minimum free distance, d_{free}." In this case $d_{free} = 5$.

Since T(D) gives all the paths in the state diagram, it also represents all the code words generated by the encoder. Hence, this function is also called as the '***Generating Function***' of the encoder.

Also, since T(D) provides a complete description of the Hamming weights of all non-zero codewords, it is also called as the ***Weight Enumerating Function (WEF)*** of the code.

6.11 Maximum Likelihood Decoding of Convolutional Codes

Viterbi Algorithm

A popular algorithm for decoding convolutional codes is based on the maximum likelihood principle. As per this principle a ***Maximum Likelihood Decoder (MLD)*** declares a valid code vector which is very nearly equal to the received code vector '*r*', as the transmitted code vector '*v*'. Viterbi proposed an algorithm, for decoding convolutional code, based on the maximum likelihood principle in 1967. This algorithm is popularly known as the *Viterbi algorithm*. It was seen that the Viterbi algorithm is equivalent to a dynamic programming solution to the problem of finding the shortest path through a weighted graph. Viterbi algorithm can be used to produce the maximum likelihood estimate of the transmitted sequence over a band limited channel with inter symbol interference. This is obvious because the channel memory which produces the inter symbol interference is analogous to the encoder memory of the convolutional code.

6.11.1. Trellis Diagram

A better understanding of the Viterbi algorithm can be had by expanding the state diagram of the encoder in time (i.e., to represent each time unit with a separate state diagram). This expanded state diagram is called a ***Trellis Diagram***.

The trellis diagram, for the (3, 1, 2) code having Transfer Function Matrix G (D) = (1 + D, 1 + D^2, 1 + D + D^2), is shown in Fig. (6.10c); the encoder and the state diagram are shown in Figs. 6.10(a) and (b) for this code.

For an (n, k, m) convolutional code and for an information sequence length of L (= 5 in this case), the trellis diagram contains (L + m + 1) time units or levels, and these are labelled from 0 to L + m, as shown in the figure (6.10c). Assuming that the encoder always starts at state S_0 and returns to S_0; the first 'm' time units corresponds to the encoder's departure from state S_0 and the last 'm' time units corresponds to the encoder's return to state S_0. The branches of the trellis diagram corresponding to these 'm' time units during which the encoder returns to state 'S_0' is called the **'Tail of the Trellis'.**

Each time unit contains a replica of the state diagram. There are two branches leaving and entering each state. The upper branch leaving each state at time unit 'i' represents the input $u_i = 1$, while the lower branch represents $u_i = 0$. Each branch is labelled with the n corresponding outputs v_i and each of the 2^L code word of length N = (L + m) $*$ n is represented by a unique path through the trellis. That is each path in a trellis corresponds to a valid code vector. For example the code word corresponding to the information sequence u_i = (11101) is shown highlighted in Fig. (6.10c), that is v_i = 111, 010, 001, 110, 100, 101, 011.

In general case of an (n, k, m) code and an information sequence of length kL, there are 2^k branches leaving and entering each state and 2^{kL} distinct paths through the trellis corresponding to the 2^{kL} code words.

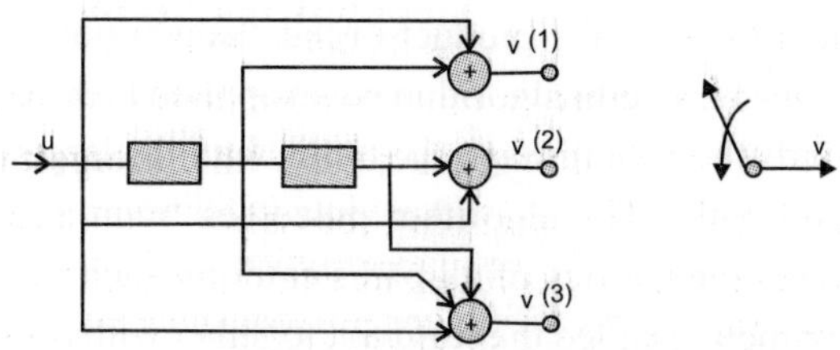

Fig. 6.10(a). (3, 1, 2) Convolutional Encoder

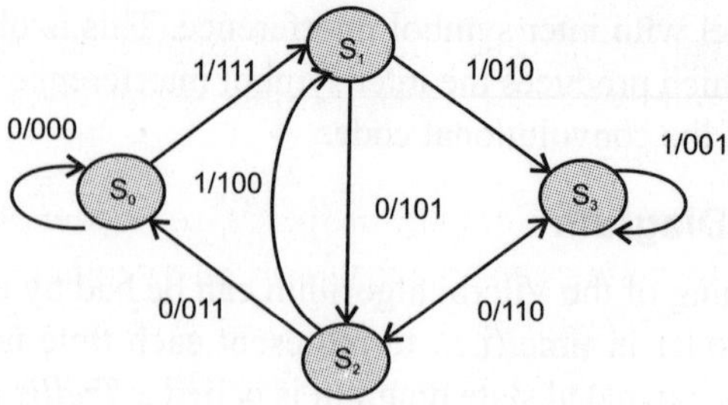

Fig: 6.10 (b). State Diagram for the Encoder of Fig. 6.10 (a)

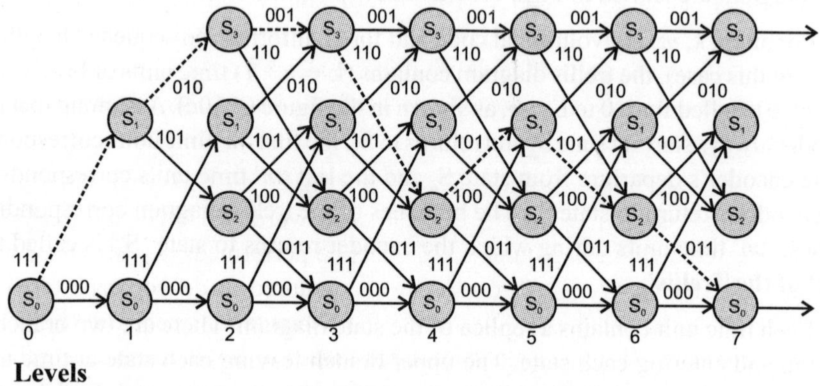

Fig. 6.10 (c). Trellis Diagram for (3, 1, 2) Code with L = 5

For u_i = 11101 the code word traced is v_i = 111, 010, 001, 110, 100, 101, 011.

6.11.2. Viterbi Decoding of Convolutional Codes using Trellis Diagram path Metrics

A maximum likelihood decoder for a Discrete Memoryless Channel (DMC) chooses a code vector '*v*', corresponding to a received vector '*r*', which maximises the ***log-likelihood function 'logP(r/v)'*** known as the ***metric*** associated with the path '*v*' in the trellis. P(*r*/*v*) is the product of the channel transition probabilities $P(r_i / v_i)$ ($i = 0$ to $L + m - 1$). Viterbi algorithm when applied to the received sequence '*r*' from a DMC, finds the path through the trellis with the largest metric (i.e., the maximum likelihood path). The algorithm processes '*r*' in an iterative way. At each step, it compares the metrics of all paths entering each state, and stores the path with the largest metric, called the survivor, together with the metric. The steps involved in the algorithm are listed below:

The Viterbi Algorithm:

Step 1. Beginning at the time unit $j = m$, compute the partial metric for the single path entering each state. Store the survivor path and its metric for each state.

Step 2. Increment j by 1. Compute the partial metric for all the paths entering a state by adding the branch metric entering that state to the metric of the connecting survivor at preceding time unit. For each state, store the path with the largest metric (the survivor), together with its metric, and eliminate all other paths.

Step 3. If $j < L + m$, repeat step 2. Otherwise, stop.

There will be 2^{km} or 2^K survivors from time unit 'm' through time unit L, one for each of the 2^K states. After time unit L there are fewer survivors, since there are fewer states while the encoder is returning to all zero state. Finally, at time unit $L + m$, there is only one state, the all zero state, and hence only one survivor, and the algorithm terminates.

Since a MLD must compute the metric ('logP(r/v)') for each code word, or path in the trellis, and then choose the code word that gives the maximum metric, the number of computations that must be performed by a MLD becomes excessively large. 2^{km} or 2^K computations are required to decode each block of 'k' information bits. Therefore, the number of computations becomes larger and larger as the value of 'm' increases. As large values of 'm' is required to achieve low error probabilities, the number of computations are really very high. *Hence, it is very difficult to achieve very low error probabilities with maximum likelihood decoding through computation of the path metrics.*

6.11.3. Decoding of Convolutional Codes using Minimum Distance Criteria of Paths in Trellis Diagram

Another method of decoding a convolutional code is by using the minimum distance principle. For a BSC the metric value for a path corresponding to a code vector 'v' is the minimum distance between 'v' and the received vector 'r' i.e., $d(r,v)$. Therefore, for decoding a convolutional received code vector, it is sufficient to find a path in the trellis which has the minimum distance, "d_{min}" for the received code vector $\{r_i\}$. Since, the trellis has all the paths corresponding to all possible transmitted sequences, the minimum distance path gives the most likely path for $\{r_i\}$, and hence the minimum distance path is taken as the transmitted sequence $\{v_i\}$. Such a decoder using minimum distance "d_{min}" as the criterion for decoding is also known as a ***Minimum Distance Decoder***.

To calculate the most-likely path, the HD between 'r_i' bits and the branches of the trellis are calculated and added to the HD's of the previous nodes in order to compute the HD of the nodes at the next stage. Those paths are said to survive which results in the minimum HD at the nodes. These paths (and their history) and the HD of the corresponding nodes are stored in a table called Survivor path and HD, as shown in table (6.3), for the encoder of Fig. 6.10(a). Similar computations are repeated for other nodes. It can be shown that the paths which have survived at any stage of the computations emerge from one old branch at the distance of 3 K or 4 K bits in the past, and thus the oldest bit of all the paths is expected to be the same. The oldest bit is then declared to be decoded data-bit. This process is continued until all the received r_i bits are decoded.

Table 6.3: Survivor Paths and their HD for the Encoder of Fig. 6.10(a)

States \ Nodes	2	3	4	5	6	7	8
S_{00}	0 (2)	00 (2)	000 (3) 100 (4)	0000 (4) 1000 (5) 1100 (6)	00000 (5) 11000 (8)	000000 (7)	0000000 (8) 1110100 (3)
S_{10}	1 (1)	01 (5)	001 (4) 101 (5)	0001 (4) 1101 (7)	00001 (6) 11101 (3)	000001 (6) 111101 (6)	0000001 (8) 1110101 (6)
S_{01}		10 (3)	110 (5) 010 (6)	1110 (3) 0010 (7)	11110 (5) 00010 (6)	111110 (7) 111010 (3)	1110110 (7) 1111010 (8)
S_{11}		11 (2)	111 (2) 011 (7)	1111 (4) 0011 (4)	11111 (5) 00011 (7)	111111 (6) 111011 (5)	1110111(6) 1111011 (7)

In this particular example, it is seen that at the nodes 6 and 7, no definite decision can be made regarding the first two bits of u_i. But at node 8, it is observed that u_i = 11101 has HD = 3 which is the least among all the HD's at the node, and further 11101 occurs in 4 probable paths, while the other data bits 11110 and 00000 occur in two paths only. Then by majority vote v_i = 11101 is decided as the correct transmitted information bits.

***Example* 6.15:** Consider the (3, 1, 2) Systematic Convolutional code and its encoder block diagram of Fig. 6.8. It has been shown that for the a given $\{u_i\}$ = {010110......}, the coder output $\{v_i\}$ = {000 111 001 101 110 011 010 000.....}. Let us assume that the received sequence is $\{r_i\}$ = {000 110 000 101 110 011 010 000 ....}.

A Trellis diagram can also drawn as shown in Fig. 6.11 below in which the upper branch leaving each state at time unit 'i' represents the input $u_i = 0$, while the lower branch represents $u_i = 1$. Referring to this trellis of Fig. (6.11), the probable paths of $\{r_i\}$ may now be traced, computing at the same time the HD between $\{r_i\}$ and the trellis path. The corresponding data bits and the HD at each node are shown in Table 6.4. In this particular example, it is seen that at the nodes

5 and 6, no definite decision can be made regarding the first two bits of u_i. But at node 7, it is observed that u_i = 101100 has HD = 2 which is the least among all the HD's at the node, and further 1011 occurs in 4 probable paths, while the other data bits 0011, occur only in two paths. Then a majority vote is taken and v_i = 1011 is decided as the correct transmitted information bits. In general, however, this ambiguity does not occur with a decoding window of say 4K or more bits. The code of this example corrects $t \leq 2$ in 9 received bits, and this may be checked by introducing two more errors, say in the 5th and 6th blocks of digits in $\{r_i\}$, thus producing 4 errors in 18 received bits. The tentative decisions regarding the initial bits of $\{r_i\}$ may be taken at the 4th / 5th node as well, and a final check is made regarding the minimum value of the HD of the accepted sequence after sufficient data bits are decoded.

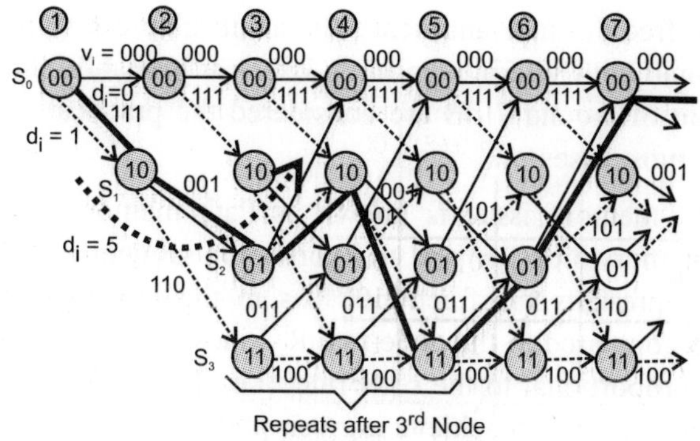

Fig. 6.11. Trellis Diagram of (3, 1, 2) Encoder of Fig. 6.8.

Table. 6.4 Survivor Paths and their HD of (3, 1, 2) Encoder of Fig. 6.8.

SR state \ nodes	2	3	4	5	6	7
00	0 (2)	00 (2)	000 (4) ~~100 (5)~~	0000 (6)		101100 (2) 001100 (3)
10	1 (1)	~~01 (5)~~	001 (3) 101 (2)			101101 (5) 001101 (6)
01		10 (2)	~~110 (5)~~	~~0010 (6)~~ ~~1010 (5)~~ ~~1110 (6)~~	10110 (2) 00110 (3)	101110 (6)
11		11 (3)	111 (4)	1011 (2) 0011 (3) ~~1111 (5)~~	10111 (5) ~~00111 (6)~~	101111 (7)

6.12 Implementation Issues of Viterbi Algorithm

There are two major issues in using Viterbi decoding algorithm for convolutional codes. First, the arbitrarily small error probabilities promised by Shannon's 'noisy channel coding theorem' are not achievable in practice by Viterbi algorithm. This is because only small constraint lengths can be used due to limitations on the encoder and decoder memory.

The first memory issue is the Decoder memory requirements. Since, there are 2^K states in the state diagram of the encoder, the decoder must reserve 2^K words of storage for the survivors. Each word must be capable of storing the survivor path along with its metric. Since, the storage requirements increase exponentially with K, in practice it is not feasible to use codes with large K, and a value of about 8 is considered the practical limit for the Viterbi algorithm. This also limits the available free distance and hence the achievable error probability cannot be arbitrarily small with Viterbi algorithm. Bit error probabilities of around 10^{-5} and a coding gain of around 7 dB are considered the practical limit for the Viterbi algorithm in most cases.

The next memory issue is known as Path memory requirements. For an encoder of memory order 'm' and an information sequence of length kL, the factor $m/(L + m)$ represents the fractional loss in the effective rate of information transmission compared to the code rate R. Since energy per information bit (E_b) is inversely proportional to the code rate, (i.e., $E_b = E/R$), a lower effective rate means a larger 'E_b/N_0' to achieve a given performance. Hence it is desirable to have L as large as possible so that the effective rate is nearly R. The difficulty with large L is that each of the 2^K words of storage must be capable of storing a kL bit path plus its metric. For very large L, this is clearly not feasible.

Second, the fixed (large) number of complex computations required per decoded information block as mentioned earlier. This large amount of computations are not always required, particularly when the noise is not severe. For example, even if the entire encoded sequence is transmitted over a BSC without any error, the Viterbi algorithm will perform 2^K computations per decoded information block, all of which is a wasted effort. i.e., a decoding procedure whose decoding efforts are adaptable to the noise level may be required.

6.13 Sequential Decoding of Convolutional Codes

Sequential decoding technique was introduced by 'Wozencraft' in 1961 as an alternative to Viterbi algorithm for decoding Convolutional codes. In sequential decoding, the decoding efforts are adaptable to the noise level. Also, the decoding efforts of Sequential decoding is independent of K and hence large constraint lengths can be used to achieve low probability of decoding error.

In contrast to Viterbi decoding , where the trellis paths are maintained through all possible SR-states and the decision is made at a latter node; in sequential decoding, the improbable paths are discarded at each node and only one most probable path is chosen. This path is then carried forward to the next node and a new decision regarding the most probable path is made. However, a continuous measure of the accuracy of the decision is maintained and if necessary the decoder traces an ignored path and forms new decisions. This provision of back and forth search of the correct path is the special characteristics of sequential decoding. The number of such searches depends on the noise conditions of the channel. Hence the number of computations required in a sequential decoder is a random variable. For understanding the concept of sequential decoding, the 2^{kL} code words of length n $(m + L)$ for a (n, k, m) code and an information sequence of length kL can be considered as the paths through a tree diagram called the **Code Tree.** This code tree will have (L + m + 1) nodes or levels or time units. The code tree is an expanded form of the trellis diagram in which every path is distinct from every other path. The code tree for the (3,1,2) code of Fig. 6.8 is shown in Fig. 6.12.

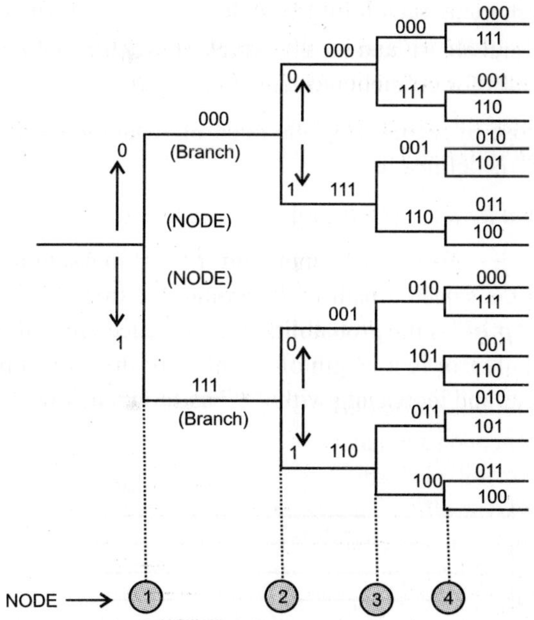

Fig. 6.12. Code Tree of (3, 1, 2) Sytematic Convolutional Encoder of Fig:(6.8)

The decoding strategy with reference to the tree of a code is: starting from the initial node of the tree, the decoder compares the 0th block (of n bits) of the received vector "r" with 2^k branches steming from the initial node and accepts the branch having minimum HD. The decoder then moves to the next node and decides on the message bit. The decoder now compares the first block of "r" and again accepts the nearest branch. For low error rate, the technique is quite successful and the decoder traces the correct path. For example, consider the code tree diagram for the (3, 1, 2) code of Fig. (6.8), as shown in Fig. (6.12) above. Assume, for this tree graph, that the transmitted sequence is an all-zero vector and the received vector 'r' is {001, 010, 000, 100, 000.......}. Since the HD between the two possible paths at each node is 3, and the error in each block is one only, then a correct decision will be made at each node by the decoder. If, however, 'r' is {000, 000, 011, 000......} then at the third node, the path decided upon by the decoder will be wrong. This node at which the wrong decision is made is known as the " **Node of Separation**".

The above branch-by-branch search strategy is not very efficient; because of the node of separation, the decoder will continue to make errors without returning to the stable state. It is thus necessary to have a decoding algorithm which will:

(*a*) recognise with high probability whenever the decoder moves to an incorrect path and,

(*b*) initiate a systematic search for the node of separation when "(a)" happens.

There are several algorithms for this, **stack algorithm** and **Fano's algorithms** are such sets of rules for efficient sequential decoding.

For the purpose of recognizing the node of separation a measure called the "**Tilted Distance**" is defined as :

$$\lambda\,(l) = p'\,n\,l - d\,[\,v'\,(l), r\,(l)\,] \tag{6.35}$$

where $p < p' < ½$, p = BER in channel, and d [ ] = measured H.D between the received $r(l)$ and the selected path v' (l) through the tree, " l " being number of blocks decoded; "$p'nl$" is the probability of correct decoding of 'n' bits and "pnl" is the probability of error in a length of 'n' bits. For the correct path $d\,[\,] \simeq p\,n\,l$ and λ (l) is positive and increasing with " l " as shown below :

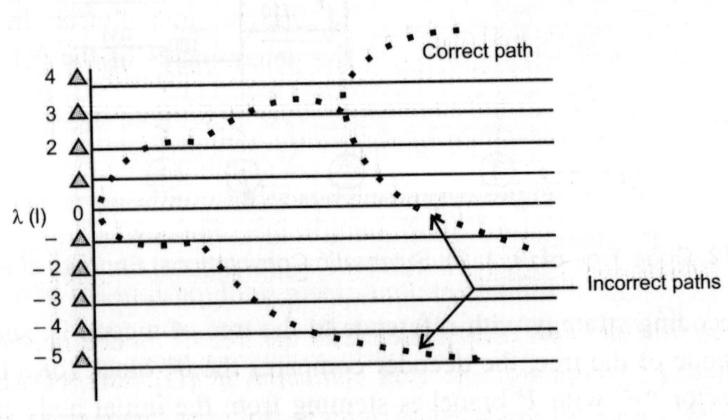

Fig. 6.13. Behaviour of "Tilted Distance λ (l) "with" l"

As long as $\lambda(l)$ is increasing, the decoder continues to move from node to node and follow the most probable path. If $\lambda(l)$ starts decreasing in value, then the decoder concludes that an incorrect path has been chosen, and as such it retraces its path testing alternate paths at earlier nodes. Once it is able to locate the node of seperation, then a new most probable path is followed with $\lambda(l)$ increasing with l. If $\lambda(l)$ starts decreasing again, then the 2nd node of separation is located through back and forth search, and this procedure continues till the "r" is completely decoded.

6.13.1. Drawback of Sequential Decoding

The major drawback of Sequential decoding is that noisy frames take large amounts of computation, and decoding times occasionally exceeds some upper limit, resulting in information loss or erasure.

6.14 Catastrophic Codes/Catastrophic Error Propagation

One of the advantages of a systematic code is that no inverting circuit is needed for recovering the information sequence from the code word. Non-systematic codes, on the other hand, require an inverter to recover the information sequence; that is an $(n \times k)$ matrix $G^{-1}(D)$ must exist such that,

$$G(D)G^{-1}(D) = I\,D^{l} \tag{6.36}$$

for some $l \geq 0$, and I is $k \times k$ identity matrix. As $v(D) = u(D)\,G(D)$, using the above expression we have,

$$v(D)G^{-1}(D) = u(D)\,G(D)G^{-1}(D) = u(D)\,I\,D^{l} = u(D)\,D^{l} \tag{6.37}$$

and the information sequence can be recovered with an l-time unit delay from the code word by letting v(D) as the input to the n-input, k-output linear sequential circuit whose transfer function matrix is $G^{-1}(D)$. For an $(n, 1, m)$ code, a transfer function *matrix* G(D) has feed forward inverse $G^{-1}(D)$ of delay l if and only if

$$\text{GCD}\,[g^{(1)}(D), g^{(2)}(D), \ldots\ldots\ldots, g^{(n)}(D)] = D^{l} \tag{6.38}$$

for some $l \geq 0$, where GCD is the greatest common divisor. In case of an (n, k, m) code with $k > 1$, let $\Delta_i(D)$ be the determinants of the $\binom{n}{k}$ distinct $k \times k$ submatrices of the transfer matrix G(D), for $i = 1, 2, 3, \ldots\ldots\ldots, \binom{n}{k}$. Then a feed forward inverse of delay l exists if and only if,

$$\text{GCD}\,[\,\Delta_i(D) : i = 1, 2, 3, \ldots\ldots\ldots, \binom{n}{k}] = D^{l} \tag{6.39}$$

for some $l \geq 0$. Procedures for constructing $G^{-1}(D)$ are available in the literature.

In order to understand what happens when a feed forward inverse does not exists, let us consider the following example.

Example* 6.16:** For the (2, 1, 2) code with $g^{(1)}(D) = 1 + D$ and $g^{(2)}(D) = 1 + D^2$, GCD $[1 + D, 1 + D^2] = 1 + D$ and a feed forward inverse does not exist. If the information sequence $u(D) = 1/(1+D) = 1 + D + D^2 + D^3 + \ldots\ldots\ldots\ldots\ldots\ldots$, the output sequences are $v^{(1)}(D) = 1$ and $v^{(2)}(D) = 1 + D$; that is, the code word contains only three non zero bits even though the information sequence has infinite weight. When this code word is transmitted over BSC, and if the three non -zero bits are changed to zeros by the channel noise, the received sequence will be all zeros. Then a MLD will produce the all zero code word as its estimate, since this is a valid code word and it agrees exactly with the received sequence. Thus the estimated information sequence will be $\tilde{u}(D) = 0$, implying an infinite number of decoding errors caused by a finite number (only three in this case) of channel errors. Clearly this is a deleterious situation, and the code is said to be subject to ***Catastrophic Error Propagation, and the code is called a ***Catastrophic Code***. *Hence a catastrophic error propagation can be defined as an event where by a finite number of code symbol (channel) error cause an infinite number of decoded data bit errors*. The codes which are responsible for this are called Catastrophic codes. Such codes are to be avoided for use in communication systems and can be done by proper selection of generator sequences for the code. Also Catastrophic codes can be identified by seeing their state diagram.

A code is catastrophic if and only if its state diagram contain a loop of zero weight, other than the self-loop around the zero state S_0. The state diagram of the catastrophic code discussed above is shown in Fig. 6.14. We can see that the self-loop around the state S_3 has zero weight (that is the transition around this loop produces all zero-(00) code word).

6.14.1. Measures to be taken in Selecting Non-catastrophic Codes

- The Transfer function matrix of the code should have a feed forward inverse $G^{-1}(D)$. That is equations (6.38) and (6.39) should be satisfied.
- The state diagram of the code should not have any loop with zero weight (other than the self-loop around the zero state S_0).
- Systematic codes are always non-catastrophic.

Literatures show that only a fraction $[1/(2^n-1)]$ of $(n, 1, m)$ non-systematic codes are catastrophic.

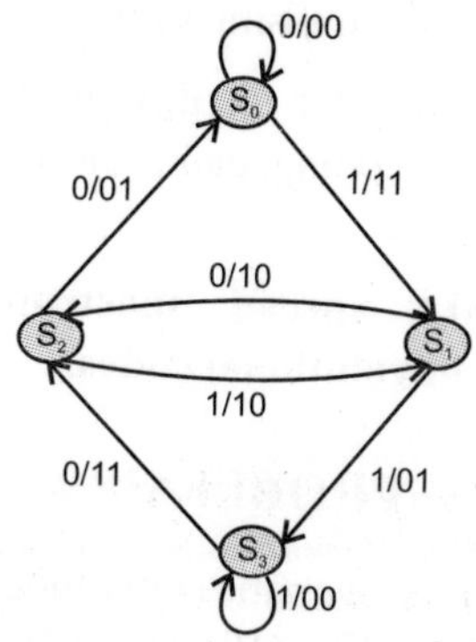

Fig. 6.14. State Diagram of a (2, 1; 2) Catastrophic Code of Example (6.16)

6.15 Applications of Convolutional Codes

Convolutional codes, using Viterbi decoding together with improved versions of sequential decoding, found lot of applications in deep-space and Satellite communication since early 1970's and in the modern communication systems. Some of them are listed below:

- **Satellite Communication: (with Vitrebi Decoding)**

(*a*) Voyager Space mission by NASA, 1977 (to Mars, Jupiter and Saturn) used

(*i*) (2, 1, 6) code $g^{(1)}(D) = 155$; $g^{(2)}(D) = 117$, $d_{free} = 10$

(*ii*) (3, 1, 6) code $g^{(1)}(D) = 155$; $g^{(2)}(D) = 117$; $g^{(3)}D = 127$, $d_{free} = 15$ with 3 bit soft decision; at 100 Kbps.

(*b*) Defence Science Communication System (DSCS) used 10 Mbps (2, 1, 6) code; Coding gain 5.1 dB to 5.6 dB.

(*c*) Tracking and Data Relay Satellite System (TDRSS) by NASA, used (2, 1, 6) code by Hariss with 10 Mbps speed; 16 parallel Decoders for 160 Mbps speed.

(*d*) Viterbi Decoded Convolutional codes were used with R.S. Codes in Concatenated Codes .

- ***Satellite Communication: (With Sequential Decoding)*** NASA Pioneer 9 Solar Mission- 1968, used,

(2, 1, 20) code with $g^{(1)} = 4000000$; $g^{(2)} = 7154737$ and $d_{min} = 10$,

(2, 1, 24) code with $d_{free} = 11$ (512 bps).

- **GSM cellular communication system** uses 1/2 rate (K = 5) convolutional code for the type I and Type II voice signals.

- **CDMA cellular system** uses convolutional codes in Concatenation with Reed-Solomon codes for data transmission.
- Convolutional codes are used for the **design of "Turbo codes"**, which can (in principle) achieve the coding gains close to the theoretic Shannon's limit.
- **Direct Sequence-CDMA spread spectrum systems use low-rate (R = 1/64) orthogonal convolutional codes.**

PROBLEMS

6.1. Consider the (3, 1, 2) non-systematic convolutional encoder whose generator sequences are $g^{(0)} = (110)$, $g^{(1)} = (101)$ and $g^{(2)} = (111)$.

(*a*) Draw the encoder block diagram.

(*b*) Find the time domain generator matrix G.

(*c*) Find the code word '*v*' corresponding to the message sequence $u = (11101)$.

6.2. For the (4, 3, 2) binary convolutional encoder of Fig. 6.3,

(*a*) Find the generator sequences.

(*b*) Find the time domain generator matrix G.

(*c*) Find the code word '*v*' for the message sequence $u = (110, 011, 101)$.

6.3. For the (3, 1, 2) encoder of problem 6.1,

(*a*) Find the transform domain generator matrix G(D).

(*b*) Find the set of output sequences V(D).

(*c*) Find the code word *v*(D) corresponding to the information sequence $u(D) = 1 + D^2 + D^3 + D^4$.

6.4. Consider the rate R = 2/3 convolutional encoder of Fig. 6.2,

(*a*) Find the composite generator polynomials $g_1(D)$ and $g_2(D)$.

(*b*) Find the code word *v*(D) corresponding to the information sequences $U(D) = [1 + D + D^3 \;\; 1 + D^2 + D^3]$.

6.5. For the (3, 2, 3) systematic convolutional encoder with $g_1^{(1)} = 1 + D^2 + D^3$ and $g_2^{(2)} = 1 + D + D^3$,

(*a*) Give the encoder diagram in the controller canonical form and specify how many shift register stages are required in this realization.

(*b*) Draw the modified observer canonical form of the encoder which requires only three stages of shift register.

6.6. Consider the (3, 1, 2) encoder of problem 6.1,

(*a*) Draw the state diagram of the encoder.

(*b*) Draw the modified state diagram of the encoder.

(*c*) Find the WEF for this code.

6.7. Determine which of the following rate R = 1/2 convolutional codes are catastrophic:

(*a*) $g_1 = D^2$ and $g_2 = 1 + D + D^3$.

(*b*) $g_1 = 1 + D^2 + D^4$ and $g_2 = 1 + D + D^3 + D^3$.

(*c*) $g_1 = 1 + D + D^2 + D^4$ and $g_2 = 1 + D^3 + D^3$.

(*d*) $g_1 = 1 + D^4 + D^5 + D^6$ and $g_2 = 1 + D + D^3 + D^5$.

6.8. Consider the (2, 1, 3) non-systematic encoder with $G(D) = [(1 + D^2)\ (1 + D + D^2 + D^3)]$.

(*a*) Find the GCD of the generator sequence polynomials given.

(*b*) Draw the encoder state diagram.

(*c*) Is the encoder catastrophic or non-catastrophic? Give reasons.

6.9. For the (3, 1, 2) encoder of problem 6.1, using the state diagram from problem 6.6,

(*a*) Draw the trellis diagram for an information sequence of length L = 5.

(*b*) Trace the trellis and find the code word for the information $u = 10011$.

6.10. For the (3, 1, 2) encoder of problem 6.1, using the state diagram from problem 6.6,

(*a*) Draw the code tree for an information sequence of length L = 4.

(*b*) Trace the code tree and find the code word for the information $u = 1011$.

CHAPTER 7

MULTIPLE AND BURST ERROR CORRECTING CODES

7.0 Introduction

In this chapter we will take up the study on well-known multiple and burst error correcting codes like Golay Codes, BCH Codes, RS Codes, Shortened Cyclic Codes, 'Random and Burst Error Correcting Codes' and Erasure Correcting Codes. We will also see the modifications that can be carried out to remodel known codes to fit a specific application.

7.1 The Golay Code

As it was mentioned in chapter 4 Golay code is the only multiple error correcting Perfect Code. It can be considered as an (n, k) Cyclic Code with block length $n = 23$ and $k = 12$ with an error correcting capability of '$t = 3$' or less errors in the block length of 23 digits. This code was discovered and proposed by Golay in 1949 and has large and unique algebraic properties. This made it to occupy an important place in the field of channel coding. Many work was carried out on the structure and decoding of this code. Its variants, (24, 12) extended Golay code and two types of Ternary Golay codes were also found out. Because of its excellent structure (23, 12) Golay code found variety of applications in the field of data communication. However, because of its limited block length it could not found place in the new modern applications.

The (23, 12) Golay code can be defined as a Binary Cyclic code which can be generated by any one of the two reciprocal polynomials $g_1(X)$ and $g_2(X)$ as the generator polynomial, where

$$g_1(X) = 1 + X^2 + X^4 + X^5 + X^6 + X^{10} + X^{11},$$

$$\text{and } g_2(X) = 1 + X + X^5 + X^6 + X^7 + X^9 + X^{11} \tag{7.1}$$

We can see that $(X^{23} + 1) = (1+X)g_1(X)g_2(X)$ and hence $g_1(X)$ and $g_2(X)$ are valid generator polynomials of the (23,12) cyclic code, as they are factors of $(X^n + 1)$ for $n = 23$. From the generator polynomials, it is clear that d_{min} of this code, which is equal to the number of terms in $g(X)$, is 7. Hence, the error correcting capability of this code is $t = 3$.The encoding of this code can be carried out using an $(n - k) = 11$ stage linear feedback shift register. The feedbacks will be based on the coefficients of either $g_1(X)$ or $g_2(X)$.

7.1.1. Error Trapping Technique for Decoding Golay Code

Many decoding methods are available in the literature for this code. The Error trapping technique proposed by Kasami and a refined error trapping technique namely, the Systematic Search method are the very efficient decoding methods for the Golay code. An error trapping decoder is a modification of Meggitt decoder (Fig. 5.9). They are very efficient for short, random and burst error correcting codes; but less efficient for high rate and long codes with large error correcting capability. The principle of error trapping decoder is as follows:

After forming the syndrome, the syndrome register must be shifted a certain number of times before its contents are identical to the error digits. This shifting of the syndrome register until its contents are identical to the error digits is called ***error trapping***. In order to detect the event of error trapping, we can test the weight of the syndrome after each shift of the 'n–k' stage syndrome register. If the number of errors in the received vector is 't' or less and if they are confined to 'n–k' consecutive positions, the errors are trapped in the syndrome register only when the weight of the syndrome in the register becomes 't' or less. Therefore, while we are shifting the syndrome register, when the weight of the syndrome becomes 't' or less, it is assumed that the errors are trapped in the syndrome register. Then the error correction can be accomplished by modulo-2 addition of the contents of the syndrome register with the received digits at the $n - k$ proper positions.

7.1.2. (23,12) Binary Golay Code Decoder using Table Look-up Method

The number of Correctable error patterns for the (23,12) Binary Golay code = Number of error patterns with three or fewer errors in the 23 bit block $= {}^{23}C_0 + {}^{23}C_1 + {}^{23}C_2 + {}^{23}C_3 = 1 + 23 + 253 + 1771 = 2048 = 2^{11}$ (7.2)

Assuming that the errors that occur in the channel are 3 or less, a decoder circuit based on the look-up table method for the (23,12) Golay code can be devised as shown in Fig. 7.1.

The Error pattern look-up table can be in the form of a Read-Only-Memory (ROM) which can store 2^{11} error patterns of 12 bit length each (i.e., error pattern in the message portion). The error patterns and the corresponding syndrome can be found out by considering 'all zero vector' as the transmitted code vector and the error patterns themselves as the received vectors. The syndrome can be computed for these received vectors (i.e., error patterns). That is the look-up table can be pre-computed for all the 2048 error patterns of 3 or fewer errors (by computer simulation), and stored in a read only memory. The syndrome register

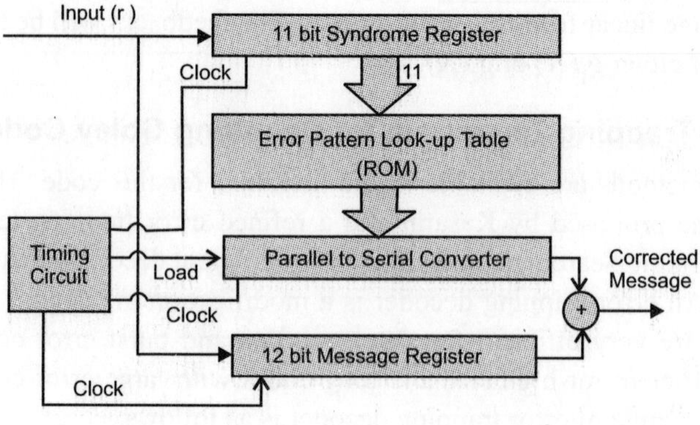

Fig. 7.1. (23,12) Golay Code Decoder

(Note: This scheme was implemented using embedded software in a processor based system. Additional facility can be included to indicate when more than 3 errors are encountered.)

can be realized as usual with an 11 bit shift register having feedbacks based on the generator polynomial $g(X)$ chosen. The operation of the decoder circuit is as follows:

***Step* 1:** The received code vector is fed into the Syndrome Register and into the 12 bit message buffer, simultaneously. The timing circuit takes care that all the 23 received bits are fed into the syndrome register and only the 12 message bits into the message buffer.

***Step* 2:** At the end of 23 clocks the syndrome is formed at the output of the syndrome register. This output acts as the address to the Error pattern look-up table and the corresponding error pattern is loaded into a parallel to serial converter.

***Step* 3:** Next the error pattern in the parallel to serial converter and the received message vector in the message buffer are read out bit by bit and added in the modulo-2 adder to get the corrected message.

If there are three or fewer errors in the received vector, the message read out of the decoder will be the transmitted message. If there are more than three errors the message read out will not be the transmitted message, i.e., the decoding will be erroneous.

7.1.3. Probability of Undetectable Error, $P_u(E)$

For the (23,12) Golay code $\mathbf{P_u(E)}$ can be computed as follows:

The probability of undetectable error = the probability of occurrence of number of error patterns resembling non-zero code vectors.

$$= P_u(E) = \sum_{i=0}^{n} A_i p^i (1-p)^{n-i} \tag{7.3}$$

Where A_i is the number of code words with weight 'i'. The number of code words with different weights are given by the ***Weight Enumerator Function (WEF) or Weight Distribution Function*** given by,

$$A(z) = A_0 + A_1 z + A_2 z^2 + \ldots\ldots\ldots\ldots + A_i z^i + \ldots\ldots\ldots\ldots + A_n z^n. \tag{7.4}$$

The weight enumerator for the (23,12) Golay code is given by,

$$A(z) = 1 + 253\, z^7 + 506\, z^8 + 1288\, z^{11} + 1288\, z^{12} + 506\, z^{15} + 253\, z^{16} + z^{23}.$$

In equation (7.4) above $A_0 = A_{23} = 1$, $A_7 = A_{16} = 253$, $A_8 = A_{15} = 506$, $A_{11} = A_{12} = 1288$.

Using these values and equation (7.3), we get,

$P_u(E) \le 0.258 \times 10^{-11}$ for channel transition probability of '$p = 10^{-2}$',

$P_u(E) \le 0.27 \times 10^{-10}$ for channel transition probability of '$p = 10^{-1}$'.

From this it is clear that the code satisfies the upper bound namely $P_u(E) \le 2^{-(n-k)}$, i.e., $P_u(E) \le 2^{-11}$.

7.1.4. The Probability of Decoding Error, P(E), for the (23,12) Golay Code

The Probability of Decoding Error P(E)

= [1 – probability of correct decoding]

= [1 – probability of occurrence of error patterns resembling coset leaders]

$$P(E) = [\,1 - \sum_{i=0}^{n} \alpha_i p^i (1-p)^{n-i}\,] \tag{7.5}$$

Where α_i is the weight distribution of the Coset Leaders in the standard array. It is seen that $\alpha_0 = 1$, $\alpha_1 = 23$, $\alpha_2 = 253$, $\alpha_3 = 1771$ and, $\alpha_i = 0$ for $(4 \le i \le 23)$.

Hence, for '$p = 10^{-2}$', from the above expression $P(E) = 0.362 \times 10^{-4}$.

7.1.5. Coding Gain for the (23, 12) Binary Golay Code

It is estimated that Golay code provides a coding gain of more than 4 dB when compared to an un-coded transmission system, for a required bit error probability of '1×10^{-5}' (refer to Fig. 7.10)

7.2 The BCH Codes

The main purpose of designing an error control code will be, for a required error correcting capability 't' and given message length 'k' the number of parity bits to be added should be as minimum as possible; and for a given value of message length and number of allowed parity check bits the code should correct as high a number of errors as possible. That is we will be requiring the best code possible. One method to achieve the requirement is to use Bose – Chaudhuri – Hocquenghem (BCH) codes. BCH codes are large class of powerful multiple error correcting Cyclic Codes that occupies an important place in the theory and practice of channel coding. They have well defined algebraic structure and knowledge on this will help in understanding many other classes of codes. The BCH codes are generalization of the Hamming codes for multiple error correction. The non-binary sub-class of the BCH code, namely the Reed – Solomon [RS] code is very popular for burst error correction.

The BCH codes were first introduced by Hocquenghem in 1959 and independently by Bose and Chaudhuri in 1960. The RS code was introduced by Reed and Solomon independently in 1960. Many decoding algorithms were proposed by different researchers and the most popular were the iterative algorithm by Berlekamp and the search algorithm by Chein, as they were the most efficient ones.

7.2.1. Definition of BCH Codes

Like Hamming codes for any positive integers 'm' ($m \geq 3$) and 't' ($t < 2^{m-1}$), there exists a binary BCH Code such that,

The Block length of the code $= n = 2^m - 1$

Number of Parity – check digits $= n - k \leq mt$

Minimum Distance $= d_{min} \geq 2t + 1.$

As $d_{min} \geq 2t + 1$, the code can correct any combination of 't' or fewer errors in a block length of $n = 2^m - 1$ digits. This code is called a 't' error correcting BCH Code.

7.2.2. Encoding of BCH Codes

Being a cyclic code, BCH code can be generated by generator polynomial '$g(X)$'. Hence, the first step in encoding a BCH code is to find out the generator polynomial $g(X)$. When the digits of the message are non-binary, the generator polynomial $g(X)$ for generating the code can be given in terms of its zeros (i.e., roots) from the Galois Field GF(2^m). Let 'α' be a primitive element in GF(2^m), ('α' is said to be primitive if the elements of the field has some powers of 'α'). Then the generator polynomial $g(X)$ of 't' error correcting BCH code is the lowest–degree polynomial over GF(2) which has,

$$\alpha, \alpha^2, \alpha^3 \ldots\ldots\ldots, \alpha^{2t} \quad (7.6)$$

as its roots [i.e., $g(\alpha^i) = 0$ for $1 \leq i \leq 2t$]. Let $\phi_i(X)$ be the minimal polynomial of α^i. Then g(X) must be the least common multiple of $\phi_1(X), \phi_2(X), \ldots\ldots\ldots, \phi_{2t}(X)$, that is,

$$g(X) = \text{LCM } \{\phi_1(X), \phi_2(X), \ldots\ldots\ldots, \phi_{2t}(X)\} \quad (7.7)$$

It can be proved that, every even power of α in the sequence of equation (7.6) has the same minimal polynomial as some preceding odd powers of α in the sequence. As a result, the generator polynomial $g(X)$ of the binary t-error-correcting BCH code given by equation (7.7) can be reduced to,

$$g(X) = \text{LCM } \{\phi_1(X), \phi_3(X), \ldots\ldots\ldots, \phi_{2t-1}(X)\} \quad (7.8)$$

Since the degree of each minimal polynomial is m or less, the degree of $g(X)$ is at most 'mt'. That is, the number of parity check digits, n–k, of the code is at most equal to 'mt'. The BCH codes defined above are called as Primitive BCH Codes. Apart from the required algebraic structure the main advantage of the BCH codes lie in the flexibility in selecting the code parameters like block length and code rate. The parameters of all binary BCH codes of length 2^m–1with $m \leq 5$ are given in table (7.1) below, $g(X)$ is given in octal representation.

Table 7.1: Parameters of Binary BCH Code

n	k	t	g(X) *
7	4	1	13
15	11	1	23
15	7	2	721
15	5	3	2467
31	26	1	45
31	21	2	3551
31	16	3	107657
31	11	5	5423325
31	6	7	313365047

***Example* 7.1:** 171 in octal = 001 111 001 in binary & $g(X) = X^6 + X^5 + X^4 + X^3 + 1$

From equation (7.8) it is clear that ***for the single error correcting BCH codes of length 2^m–1 is generated by $g(X) = \phi_1(X)$***. Since 'α' is a primitive element of GF(2^m), $\phi_1(X)$ is a primitive polynomial of degree m. Hence, the Single Error Correcting BCH code of length 2^m–1 is a Hamming code.

---

* The generator polynomial is given in octal form. Octal digit is first represented in binary form and then g(X) is written.

***Example* 7.2:** Let 'α' be a primitive element of the Galois Field GF(2^4), such that $1 + \alpha + \alpha^4 = 0$. Minimal polynomials of primitive element and its powers are available in literature. The minimal polynomials of α, α^3, and α^5 are,

$$\phi_1(X) = 1 + X + X^4;\ \phi_3(X) = 1 + X + X^2 + X^3 + X^4 \text{ and } \phi_5(X) = 1 + X + X^2$$

respectively. Then from equation (7.8) the double error correcting BCH code of length $n = 2^4 - 1 = 15$ is generated by,

$$g(X) = \text{LCM } \{\phi_1(X), \phi_3(X)\}$$

Since $\phi_1(X)$ and $\phi_3(X)$ are two distinct irreducible polynomials,

$$g(X) = \{\phi_1(X).\ \phi_3(X)\} = (1 + X + X^4).\ (1 + X + X^2 + X^3 + X^4)$$

$$= (1 + X^4 + X^6 + X^7 + X^8)$$

Therefore, the number of parity – check digits $(n-k)$ = the degree of $g(X)$ = 8 and hence $k = 7$. Thus, the code is a (15,7) cyclic code with d_{min} = the weight of $g(X) = 5$. Similarly the Triple Error correcting BCH code of length 15 is generated by,

$$g(X) = \text{LCM } \{\phi_1(X), \phi_3(X), \phi_5(X)\}.$$

$$= (1 + X + X^4).\ (1 + X + X^2 + X^3 + X^4).\ (1 + X + X^2)$$

$$= (1 + X + X^2 + X^4 + X^5 + X^8 + X^{10}).$$

This Triple error correcting BCH code is a (15, 5) cyclic code with minimum distance = 7.

Once $g(X)$ is known the encoding of the BCH code can be carried out using the shift register circuits with feedback as per $g(X)$.

7.2.3. Decoding of BCH Codes

Being a class of cyclic codes BCH codes can be decoded by any technique for decoding cyclic codes. Many better decoding algorithms were proposed by different researchers exclusively for BCH codes. The most popular among them are the ***iterative algorithm*** by Berlekamp and the ***search algorithm*** by Chein, as they are the most efficient ones. The major steps involved in decoding the BCH code are,

1. Computation of Syndrome (polynomial), 2. Computation of Error Locator Polynomial from the syndrome values, 3. Computation of Error Location and Error Value from the Error locator polynomial, 4. Error Correction.

The above steps are quite involved and cumbersome for most of the algorithms; less efficient due to large number of multiplication and division operations that must be performed which increases as square of the number of errors to be corrected. However, all these can be performed using the modern high speed processors which may be already existing as part of the system. The procedures in the above steps are explained in the following section (7.3) on Reed-Solomon codes.

7.3 The Reed–Solomon (RS) Codes

An important sub-class of non-binary, q-array BCH codes is called the Reed – Solomon (RS) code in honor of their discoverers. RS codes are known for their powerful burst error correcting capabilities, have high channel efficiency and are very versatile. A 't' error correcting RS code with digits (symbols) from GF(2^m), an extended binary field, has the following parameters:

Block Length: $n = 2^m - 1$
Number of parity check digits: $n - k = 2t$
Minimum Distance: $d_{min} = 2t + 1 = n - k + 1$

The length of the code 'n' is one less than the number of symbols in the code (i.e., 2^m). We know that for any (n, k) linear block code, by Singleton bound, the minimum distance $d_{min} \leq (n - k + 1)$; whereas for the RS code $d_{min} = n - k + 1$, that is the design distance of the code, that is one more than the number of parity-check digits. Hence it is called a "Maximum Distance Separable" code. Each element (digit) of the RS code is given by an m-bit binary word (where 'm' is a positive integer). RS codes are always referred as (n, k, t) codes.

The RS code is generated using a generator polynomial of the form,

$$g(X) = (X - a)(X - a^2)(X - a^3) \ldots\ldots\ldots\ldots (X - a^{2t}) \qquad (7.9)$$

where 'a' is the primitive element (i.e., the element whose powers form the field elements) of the field GF(2^m) and 't' is the length of the burst errors that the code can correct.

Example **7.3:** For the (255, 127, 64) RS code the code length is 255 symbols, each of 8 bits, over the field GF(2^8) containing 2^8 symbols. Number of message symbols, $k = 127$ symbols each of eight bits. The minimum distance is $n - k + 1 = 129$ and the code can correct $t = 64$ symbol error (each of 8 bits) over a block length of 255 symbols. That is $t = (d_{min} - 1)/2$. Now let us consider the design of the encoder and decoder for this (255, 127, 64) RS Code.

7.3.1. Generation of Galois Field, GF(2^8)

The first step in the design of the code is the generation of the field GF(2^8), whose elements are the code alphabets. This is done by using a primitive polynomial of degree 8, say, $P(X) = (1 + X^2 + X^3 + X^4 + X^8)$. The field elements are the '$2t$' consecutive powers of a primitive element say 'a' for which $P(a) = 0$. Normally the field elements are given as powers of 'a' as well as in the vector form [a binary word of m bits (= 8 in this case)]. The power form of the field element will be useful while doing multiplication operation and the vector form will be useful for addition operation of the symbols.

7.3.2. Generator Polynomial Computation

The next step in the design, involves the computation of generator polynomial $g(X)$. The generator polynomial for the (t =) 64 symbol error correcting (255, 127, 64) RS code is given by

$$g(X) = (X-a)(X-a^2)(X-a^3)\ldots\ldots\ldots(X-a^{2\times 64}).$$

This expression will be expanded to get the coefficients of the polynomial which will be used in the encoder as feedback weights.

7.3.3. Encoding of RS Codes

The formation of code words in the RS code over the field GF(2^m) is carried out using the generator polynomial $g(X)$ computed in the previous step. Basically the code words will be a multiple of $g(X)$ and the message polynomial $u(X)$ of degree $(k-1)$ or less. But the straight forward multiplication of $u(X)$ and $g(X)$ will result in a code polynomial in a non-systematic form. The code polynomial $v(X)$ in the systematic format can be obtained using the following algorithm.

$$v(X) = X^{n-k}.u(X) \,\big|\, g(X) \,\big| + X^{n-k}.u(X) \tag{7.10}$$

The above algorithm involves the following three steps.

- Pre-multiplication of the message polynomial $u(X)$ by X^{n-k}.
- Division of the above product by $g(X)$ to get the remainder $X^{n-k}.u(X) \,\big|\, g(X) \,\big|$
- Appending the product $X^{n-k}.u(X)$ with the remainder, $X^{n-k}.u(X) \,\big|\, g(X) \,\big|$, obtained.

All the above three steps can be implemented by a division circuit of the form shown in Fig. 7.2, where,

[b_i] Denotes a storage device that stores an element 'b_j' from GF(2^m)
(×) Denotes a multiplier that multiplies two elements from GF(2^m)
(+) Denotes an adder that adds two elements from GF(2^m).

This division circuit consists of a $(n-k)$ stage shift register with feedback based on the coefficients of $g(X)$. The operation of the circuit is as follows:

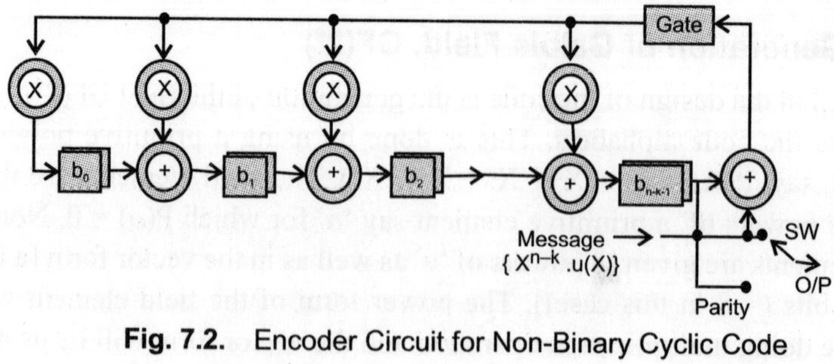

Fig. 7.2. Encoder Circuit for Non-Binary Cyclic Code

The message symbols are fed into the encoder circuit from the right end as well as into the channel as shown. Feeding the message symbols from the right end of the shift register accomplishes the pre-multiplication of $u(X)$ by X^{n-k}. Once all the 'k' (= 127) message symbols are shifted into the encoder circuit the division operation is over and the shift register contents are the $(n - k)$ parity symbols, which can be shifted out immediately following the message symbols. After shifting out the '$n - k$' (i.e. 128 symbols in this example) parity symbols, encoding of next message block can start.

7.3.4. Decoding of RS Codes

Like for any other error control code the decoding of RS code also involves the error detection and error correction. For the purpose of detecting the presence of the error in the received vector, the received code symbols are processed with the generator polynomial $g(X)$ to compute a set of $(n - k)$ symbols of the syndrome. When the syndrome symbols are not equal to zero, then it is decided that the received symbols contain error. There by error detection is accomplished.

The error correction process involves the computation of the most probable error pattern that produces the observed syndrome values. Therefore, error correction involves the solving of a set of simultaneous syndrome equations. In general the decoding of a non-binary code consists of the following operations:

1. Computation of the symbols (digits) of the Syndrome $(S_1, S_2, \ldots\ldots, S_{n-k})$ from the received code symbols.
2. Computation of error location which involves,
   - Determination of error locator polynomial C(X) from the syndrome values.
   - Computation of the inverse of the roots of C(X) which gives error locations.
3. Computation of error value using the inverse of the roots of the error locator polynomial.
4. Error correction.

The implementation of above operations is explained below:

7.3.4.1. Syndrome generation

For any cyclic code the code polynomial $v(X)$ is a multiple of the generator polynomial $g(X)$. If the received code polynomial $r(X)$ happens to be a multiple of $g(X)$ then it can be concluded that it is error free and equal to the corresponding code polynomial $v(X)$; but if $r(X)$ is not a multiple of $g(X)$ then it contains error. Therefore, the roots of $g(X)$ should also be roots of $r(X)$. In other words the syndrome values are obtained by evaluating $r(X)$ at the prescribed roots of $g(X)$, namely $(a, a^2, a^3, \ldots\ldots, a^{2t})$ .i.e $S_i = r(a^i)$ for $1 \leq i \leq (n - k)$. That is if all S_i's are not zero then error is detected.

7.3.4.2. Determination of error locator polynomial C(X)

We have the received polynomial $r(X) = v(X) + e(X)$ where $e(X)$ is the polynomial representation of the error pattern encountered during transmission. The error polynomial $e(X)$ has non zero terms only in those positions where errors have occurred.

Therefore, $S_i = r(a_i) = v(a^i) + e(a^i)$

$$= e(a^i), \text{ since } v(a^i) = 0, \text{ for } 1 \leq i \leq (n-k).$$

That is we have $(n–k)$ simultaneous equations relating to $(n - k)$ known syndrome digits and 'n' unknown error digits. Therefore, there are 2^k error patterns that give rise to a single syndrome value. To keep the probability of erroneous decoding as low as possible, a most probable error pattern which will result in the computed syndrome value can be selected as the error pattern that has occurred in the channel. The most probable error pattern will have the least number of non-zero elements. The following methods are generally used for finding the Error locator polynomial for a computed syndrome value:

1. Peterson's direct solution method.
2. Berlekamp's iterative algorithm.
3. Massey – Berlekamp's algorithm.

For error correction of moderate to large number of errors using RS code, the Peterson's direct method of solving for the coefficients of the error location polynomial from the syndrome values, is comparatively cumbersome and inefficient due to large number of multiplications and divisions that must be performed, in fact the number of multiplications required increases as square of the number of errors (i.e., 't') to be corrected.

For the second method, namely Berlekamp's iterative algorithm, the number of multiplication and division operations required increases as twice the number of errors to be corrected. Whereas for the third method, the number of operations required increases only as the actual value of the number of errors ('t') to be corrected. The third method namely the Massey-Berlekamp's algorithm which is more efficient than the other two is explained here.

7.3.4.3. Massey-Berlekamp's algorithm for calculating error locator polynomial C(X)

The parameters used in this algorithm are:

S = The syndrome value,

C(X) = Error Locator Polynomial,

L = Degree of the error locator polynomial, C(X)

D(X) = Correction term used to modify C(X)

= Discrepancy i.e., the difference between the newly entered syndrome value and the value generated in the corresponding sequence position.

The steps involved in the Algorithm for calculating the Error Locator Polynomial C(X) are explained below and is also given in the form of flow chart (Fig. (7.3)).

1. Compute Syndrome Values S_n, for $1 < n < 2t$.
2. Initialize Algorithm Variables C(X) = 1; D(X) = X; L = 0; $n = 1$;
3. Take new syndrome value and compute Discrepancy

$$\partial = S_n + \sum_{i=1}^{L} C_i S_{n-i}$$

4. Test Discrepancy: If $\partial = 0$ go to step 9, otherwise go to step 5.
5. Modify Error Locator Polynomial : Let , C*(X)= C(X) – ∂ D(X)
6. Test register length (i.e., degree of error locator polynomial): If $2L \geq n$, go to step 8, otherwise go to step 7.
7. Change Register length (i.e, degree of error locator polynomial) & update correction term: Let $L = n - L$ and D(X) = C(X) /∂.
8. Update Error Locator Polynomial: Let C(X) = C*(X).
9. Update Correction Term : Let D(X) = XD(X).
10. Update Syndrome Counter : Let $n = n + 1$.
11. Test Syndrome Count. If $n < 2t + 1$ go to step 3, otherwise STOP.

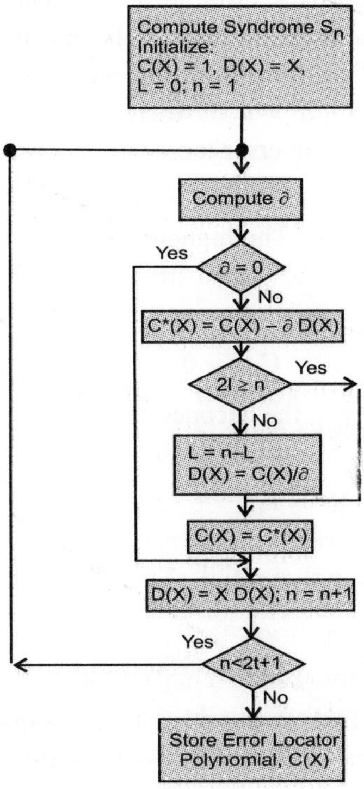

Fig. 7.3 Massey-Berlekamp's Algorithm to Compute Error Location Polynomial

The error locator polynomial is first initialized to its simplest possible form $C(X) = X^0 = 1$ and is subsequently modified as needed for correct reproduction of the syndrome values in sequence. The correction term D(X) is used to modify C(X) at each iteration, when a discrepancy is encountered between a generated value and the corresponding syndrome value. The syndromes are examined by the algorithm in sequence, one in each iteration. At each iteration the discrepancy ∂ is computed, using C(X) as it was structured at the end of the previous iteration. At each appearance of a non-zero value for ∂, the C(X) is modified using the computed value of ∂ and the correction term. The formation and use of the correction term is the most important part of the algorithm. One reason is that in addition to zeroing out the encountered discrepancy ∂, the modification of C(X) is such that the new C(X) also correctly generates all the previous syndrome values. This obviates the necessity of having to re-examine previous syndrome values each time C(X) is modified, and provides an algorithm in which the number of computations required per decoding is a linear function of the number of correctable errors 't', rather than some geometric function. Another important characteristic of the polynomial modification procedure is that it accomplishes the needed sequence modification with the smallest possible increase in the error locator polynomial. The computed $C(X) = C_1X + C_2X^2 + \ldots\ldots + C_vX^v$ where 'v' is the number of errors encountered.

7.3.4.4. Computation of error location and error correction

The inverse of the roots of the error locator polynomial gives the location of error and error value. The roots of the error locator polynomial are found out by substituting the values of the field elements, namely, $1, a, a^2, \ldots, a^{n-1}$ into C(X), where $n = 2^m - 1$. Since $a^n = 1$, $a^{-l} = a^{n-l}$, where l = error location number. Therefore, if 'a^1' is root of C(X), a^{n-1} is an error location. This procedure is described below:

The received polynomial $r\,(X) = r_0 + r_1X + r_2X^2 + \ldots\ldots\ldots + r_{n-1}X^{n-1}$, is decoded on a bit-by-bit basis. The higher order bits are decoded first. To decide r_{n-1}, the decoder tests whether a^{n-1} is an error location number; this is equivalent to testing whether it's inverse, 'a' is a root of C(X). If 'a' is a root then,

$$1 + C_1a + C_2a^2 + \ldots\ldots + C_va^v = 0 \quad (7.11)$$

Therefore, to decode r_{n-1}, the decoder computes $C_1a, C_2a^r, \ldots, C_va^v$ and if the sum $(1 + C_1a + C_2a^2 + \ldots\ldots + C_va^v) = 0$, then a^{n-1} is an error location number.

Let $v(X) = v_0 + v_1X + v_2X^2 + \ldots\ldots\ldots + v_{n-1}X^{n-1}$ be the transmitted code polynomial and let $r(X) = r_0 + r_1X + r_2X^2 + \ldots\ldots\ldots + r_{n-1}X^{n-1}$ be the corresponding received polynomial. Then the error pattern added by the channel is,

$$C(X) = r(X) - v(X) = C_0 + C_1X + C_2X_2 + \ldots\ldots + C_{n-1}X^{n-1} \quad (7.12)$$

where, $C_i = r_i - v_i$ is a symbol from $GF(2^m)$. So to determine the error pattern we need to know the error locations X_j and the error values e_j's. The procedure is as follows:

Let the error value polynomial be,

$$Z(X) = 1 + (S_1 + C_1)X + (S_2 + S_1C_1 + C_2)X^2 + (S_3 + S_2C_1 + S_1C_2 + C_3)X^3 + \\ + \dots\dots\dots\dots + (S_v + S_{v-1}C_1 + S_{v-2}C_2 + \dots\dots\dots + S_1C_{v-1} + C_v)X^v \quad (7.13)$$

Then the error value 'e' at location $\mathbf{b}_l = \mathbf{a}^{j_l}$ is given by

$$e_{j_l} = \left[\frac{Z(b_l^{-1})}{\prod_{i=1,\, i\neq l}^{v}(1 + b_i b_l^{-1})}\right] \quad (7.14)$$

Where $\Pi\ (X_i)$ refers to product of X_i for $i = 1$ to v and $i \neq l$. Finally the calculated error value is modulo added with the received value to get back the transmitted message.

It should be noted that RS codes are well known for their burst error correcting capabilities; however they are comparatively poor for random error correction. Reason being random errors, though they are less in number than the length of the burst which it can correct, when distributed over the length of the code, will cause many symbols in error. Hence, exceeding the symbol error correcting capability of the code.

As we have mentioned initially, RS codes are very versatile and the RS Code Encoder/Decoders are available in the form of VLSI devices. For example the **Advanced Hardware Architectures, Inc.,** has released AHA4011 / 4012 / 4013 RS code encoder/decoder in the form of VLSI. This uses a block length of 255 symbols of 8 bits each. The number of correctable errors 't' is a maximum of 10 per block. i.e., it is a (255, 235, 10) RS code encoder/decoder. These chips are fully programmable for use as encoders or decoders. The AHA4210 is an RS code decoder device that conforms to the Digital Video Broadcasting (DVB) specifications. Reed-Solomon Evaluation Software packages are also available in the industry.

7.4 Shortened Cyclic Codes

In real world situations there may not be a code available with a specific block length or information digits to meet a particular requirement. Also there may not be enough number of n, k pairs which will produce codes with good algebraic properties. For example the generator polynomials of cyclic codes should be factors of $X^n + 1$. For a given value of 'n', in general the polynomial $X^n + 1$ will have few factors which will produce good cyclic codes. In order to meet this situation and produce more (n, k) pairs which will produce good codes, the existing codes can be modified. The modification of codes can be used to remodel a known code to fit specific applications. They can also be used to device new classes of good codes. One such modification procedure is the shortening of a code. There are several other modifications possible and are dealt in section (7.7).

Shortening of a code means reducing the length of the code by dropping the information symbols. This corresponds to decreasing both the dimensions of the generator matrix by the same amount. One such popular shortened code is the Shortened Cyclic Code. Shortened cyclic codes are obtained by dropping, say, 'l' ($< k$) leading higher order information digits which are identical to zero. In that case there will be 2^{k-l} such code vectors and they form a linear sub-code of (n, k) code C. This sub-code will have (2^{k-l}) vectors each of length '$n-l$' and form the shortened cyclic code specified as **(n – *l*, k – *l*)** linear code. The l zero elements need not be transmitted. The decoder of the original cyclic code can be used for decoding the shortened cyclic code, by simply padding the received $(n-l)$ tuples with l zeros. Shortened cyclic codes are sometimes called a Polynomial Code and they are not cyclic codes. However, as they are a sub-class of the cyclic codes they will have the same d_{min} and error correcting capabilities from which they are derived. Using shortened cyclic codes (and other techniques like interleaving) a large collection of good burst error correcting codes can be created.

The encoding, syndrome computation and error correction techniques for the shortened cyclic codes are identical to that of the original cyclic codes (like Meggitt Decoding). Hence, the shortened cyclic codes have the same implementation advantages and structural properties as the original cyclic codes. However in decoding the shortened cyclic codes, after shifting the received '$n-l$' tuple in to the syndrome register, it has to be cyclically shifted 'l' times to generate proper syndrome. When 'l' is large these extra shifts of the syndrome register results in undesirable decoding delay. However, this decoding delay can be eliminated by modifying the feedback in the syndrome register or the error pattern detection circuit.

***Example* 7.4:** Examples of shortened cyclic codes and their original codes are:

- (28, 23) shortened cyclic code obtained by deleting 3 digits from the (31,26) Cyclic Hamming Code generated by $g(X) = 1 + X^2 + X^5$.
- (8, 4) shortened cyclic code obtained by deleting the seven leading high-order message digits from (15, 11) Cyclic Hamming code generated by $g(X) = 1 + X + X^4$.
- (64, 60) code over GF(2^4), derived from (85, 81) Cyclic Hamming Code generated by $g(X) = 1 + 3X + X^3 + X^4$.

7.4.1. Cyclic Redundancy Check (CRC) Codes

Shortened cyclic codes for error detection in conjunction with ARQ (Automatic Repeat Request) protocols are widely used for error control, particularly in computer communication. In these applications they are often called as ***Cyclic Redundancy Check (CRC) Codes***. A CRC code is, in general, generated by either a primitive polynomial $p(X)$ or a polynomial $g(X) = (X + 1)p(X)$. A number of CRC codes have become international standards for error detection in various contexts. A few standard CRC codes are:

1. CCITT X-25 (Consultative Committee for International Telegraphy and Telephony, Recommendation X-25). $g(X) = X^{16} + X^{12} + X^5 + 1.$
2. ANSI (American National Standards Institute).

   $g(X) = (X + 1)(X^{15} + X + 1) = X^{16} + X^{12} + X^5 + 1.$
3. IBM SDLC (IBM Synchronous Data Link Control).

   $g(X) = X^{16} + X^{15} + X^{13} + X^7 + X^4 + X^2 + X + 1.$
4. IEC TC57.

   $g(X) = X^{16} + X^{14} + X^{11} + X^8 + X^6 + X^5 + X^4 + 1.$
5. IEEE Standard 802.3.

   $g(X) = X^{32} + X^{26} + X^{23} + X^{22} + X^{16} + X^{12} + X^{11} + X^{10} + X^8 + X^7 + X^5 + X^4 + X^2 + X + 1.$

7.5 Burst Error Correcting Codes

Initially we have classified the errors, which are introduced to the information as they are transmitted over different channels, as random errors and burst errors. Random errors occur in the channels where in the transmitted symbols or digits are affected independently by the channel disturbances. All the error control techniques discussed so far are for this kind of channels. However, most of the transmission channels and data storage systems used encounter situations in which errors in neighbouring digits are dependent. For example the switching noise in a telephone circuit, motor ignition noise in an automobile, multipath reflections (Rayleigh fading) in a radio mobile communication channel, the noise encountered will be impulse in nature. For high rate transmission systems, the period of the impulsive noise will be high enough to affect large number of consecutive bits and will result in errors to cluster into bursts. Even in data storage systems like magnetic tapes and CDs the tape defects or dust accumulation and scratches result in burst errors when we readout the data.

In general the random error correcting codes are not efficient for correcting burst errors. Therefore, it is required to device special coding techniques for correcting burst errors. These special codes are called as the Burst Error Correcting Codes. Among the codes discussed so far, the cyclic codes are efficient both for detecting and correcting the burst errors. Many special codes have been proposed for burst error correction over the years by analytical approach as well as by computer search methods. A large class of Cyclic codes called **Fire codes** are very popular for burst error correction. This code was discovered by Fire and is the first class of cyclic codes constructed systematically for correcting burst errors and can be decoded using simple circuits.

A burst error of length 'l' can be defined as a vector whose non-zero components are confined to 'l' consecutive symbol or digit positions. For example the vector 00011011110010 has a burst of length 10. ***An 'l'– burst error correcting code can be defined as a code which can detect and correct all burst errors of length 'l' or less and is said to have a burst error correcting capability of 'l' bits or symbols.***

The design requirement for a 'l'– burst error correcting (n, k) linear code is that the number of parity bits (n–k) should be $\geq 2l$.

Or for a given value of 'n' and 'k' the burst error correcting capability of the code is given by $l \leq \lfloor (n-k)/2 \rfloor$ where, $\lfloor x \rfloor$ means the largest integer which is less than 'x'. This is an upper bound on the burst error correcting capability of a (n, k) code and also many times referred as ***Reiger bound***. The codes which meet this requirement are called ***optimal codes***.

The ***Burst Error Correcting Efficiency*** of a code is defined as twice the ratio of 'burst error correcting capability' to 'the number of parity bits'.

i.e.,
$$z = \frac{2l}{n-k} \tag{7.15}$$

An optimal code will have a burst error correcting efficiency of 100%.

Example **7.5:** The (15,9) cyclic code generated by $g(X) = 1 + X^3 + X^4 + X^5 + X^6$, which has a burst error correcting capability of $l = 3$ is an optimal code with $z = 1$.

The burst error correcting codes can be easily decoded using error-trapping techniques directly or with small modification to the error-trapping techniques. An optimum decoding technique for burst-error correcting codes was proposed by Gallager.

Apart from the efficient cyclic codes and shortened cyclic codes discussed above there are other efficient cyclic codes found in literature for correcting single short burst errors. Table (7.2) below lists some of them along with their generator polynomials, $g(X)$, in octal representation.

Table 7.2: Parameters of Burst Error Correcting Codes

n - k – 2l	Code (n,k)	Burst Error Correcting Capability (l)	g(X)*
0	(7,3)	2	35
	(15,9)	3	171
	(19,11)	4	1151
	(21,9)	6	14515
1	(15,10)	2	65
	(21,12)	4	1663
	(63,50)	6	22377

Contd...

n - k – 2l	Code (n,k)	Burst Error Correcting Capability (l)	g(X)*
2	(17,9)	3	471
	(35,23)	5	13627
	(51,35)	7	304251
3	(51,42)	3	1455
	(89,78)	4	4303
	(151,136)	6	114371
4	(43,29)	5	52225
	(93,83)	3	2065
	(255,239)	6	301565

*Octal digit is first represented in binary form and then g(X) is written.

***Example* 7.6:** 171 in octal = 001 111 001 in binary & $g(X) = X^6 + X^5 + X^4 + X^3 + 1$.

7.6 Burst and Random Error Correcting Codes

Consider a typical radio mobile channel. When the mobile is moving in a plain terrain where there is no high raised buildings or towers or vegetation the transmission errors will be due to thermal noise and will be random in nature. However, when the mobile moves into a hilly terrain or into urban area having high raised buildings and reflecting towers, multipath reflections of EM wave takes place and the resulting transmission errors will be busty in nature. Therefore, the communication system used in the mobile should be able to detect and correct both random and burst errors (Compound errors). In most such practical systems, errors occur neither independently, at random nor in well-defined bursts. Therefore, random error correcting code or single burst error correcting code alone will not be sufficient to protect the message in compound error channels. Therefore, there is a requirement for codes which can correct both random and burst errors. However, instead of going for new class of codes to meet this requirement, it will be easy and convenient to use the existing codes and at the same time achieve the requirement of devising a long powerful burst error correcting code from random and/or short optimal single burst error correcting codes. In this section we will see three such methods, namely **Interleaving of codes, Concatenation of codes** and **Product codes** to design a random and long powerful burst error correcting codes from known shorter codes.

7.6.1. Interleaved Codes

Consider an (n, k) linear code C and take 'λ' code vectors from C , and arrange them in the form of an array as shown in Fig. 7.4. Then the digits (symbol) are taken column wise one by one and transmitted as shown.

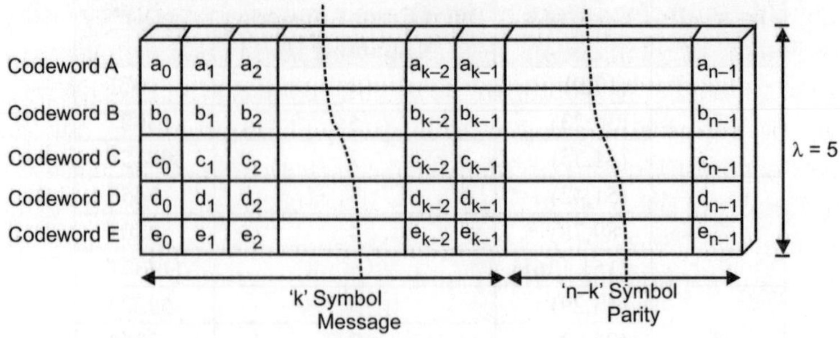

Fig. 7.4 Interleaver Table — Codewords are Fed Row-by-row and Output taken Column-by-column

a_0	b_0	c_0	d_0	e_0	–	–	–		a_{n-1}	b_{n-1}	c_{n-1}	d_{n-1}	e_{n-1}

Fig. 7.5 Output Sequence from the Interleaver.

As the digits are received at the receiver they are arranged back to get the original array. This method of arranging the λ code vectors of an (n, k) code, row wise in the form of an array and transmitting them column wise is called Interleaving of codes and the total code is called a **(λ_n, λ_k)** Interleaved or Interlaced code. That is the resulting code is **'λ'** time longer than the original code. The number of code vectors used for interleaving, that is **'λ'**, is called **'Degree of Interleaving'** and the total number of bits (digits), that is **'n.λ'**, used in the process of interleaving is called **'Depth of Interleaving'**.

Let us assume that the (n, k) code used is a single error correcting code and let a burst error of **'λ'** bit takes place during transmission. No matter where it starts, a burst of length **'λ'** will affect no more than one bit (digit) in each row. Thus if the original (n, k) code can correct single bit error in a length of 'n' digits, the interleaved code can correct a single burst of length **'λ'** or less over a length of **'n.λ'** bits (digits). If the original code can correct any single burst of length '*l*' or less, the interleaved code can correct any single burst of length **'λ*l*'**or less. By interleaving short codes with maximum possible burst error correction capability it is possible to devise codes of practically any length with maximum burst error correction capability. Thus this technique reduces the problem of devising long efficient burst error correcting codes to the selection of good short codes.

The requirement for the above method of implementing the interleaved code is that, the storage elements for the formation of the array and the transmission will have to start only after the array is formed. However, the delay can be reduced to one array formation time and while transmitting the first array the second array can be formed and the process can be continued. Only requirement is the additional array both at encoder and decoder. The array can be implemented using RAM (Random Access Memory) or other hardware.

A simplest implementation of interleaved code is possible by using cyclic codes. When the original code is cyclic the interleaved code will also be a cyclic code. If the generator polynomial of the original code is $g(X)$ then the interleaved code can be generated by a polynomial $g(X^{\lambda})$. The encoding and decoding (syndrome computation) can be done using shift registers, with each shift register stage of decoder of the original code replaced by 'λ' stages without changing the other connections. Thus, if the decoder of the original code is simple, the decoder of the interleaved code also will be simple.

7.6.1.1. Selection of 'degree of interleaving' ['λ']

***Example* 7.7:** Suppose we use the AHA4011 encoder/decoder chip to perform Reed-Solomon (255, 235) coding. This chip can correct up to '$t = 10$ ' bytes in error per message block of length up to 235 bytes. Let us assume that the maximum burst error size for the system is $l = 233$ bytes. Then we need to have '$\mathbf{l < t \cdot \lambda}$' bytes. Thus, the minimum required code vectors for interleaving will be,

$$\lambda > \frac{l}{t} = 233/10 = 23.3.$$

Thus the minimum value of λ will be 24. The array size for interleaving will be $n \times \lambda = 255 \times 24$ bytes.

7.6.2. Concatenated (Nested) Codes

Concatenation or Nesting of codes is a technique of generating long block codes using shorter codes. However the resulting codes called the Concatenated codes or Nested codes can be decoded with simple hardware than the complex hardware required for the usual long codes. A concatenated code is constructed using a (n, k) binary code C_1 and a (N, K) non-binary code C_2 with symbols from GF(2^k). The non-binary code can be a (N, K) Reed – Solomon code and binary code can be even the simple (7,4) Hamming code.

Consider a block consisting of kK symbols. This block can be divided into K sub-blocks each of 'k' bits. A sequence of 'K' such sub-blocks can be encoded with a (N, K) code over GF(2^k). Then each of the N symbols of length 'k' bits can be coded with a (n, k) code. Thus a concatenated code has two levels of coding.The digits are then transmitted one C_1 code vector at a time, in succession. Thus the length of the resulting concatenated code is nN bits and the resultant code is a (nN, kK) binary linear code. The (n, k) code C_1 is called the Inner Code and the (N, K) code C_2 is called the Outer Code as shown in Fig. (7.6). If the minimum distances of inner code C_1 and the outer code C_2 are d_1 and d_2 respectively then the minimum distance of their concatenated code will be d_1d_2.

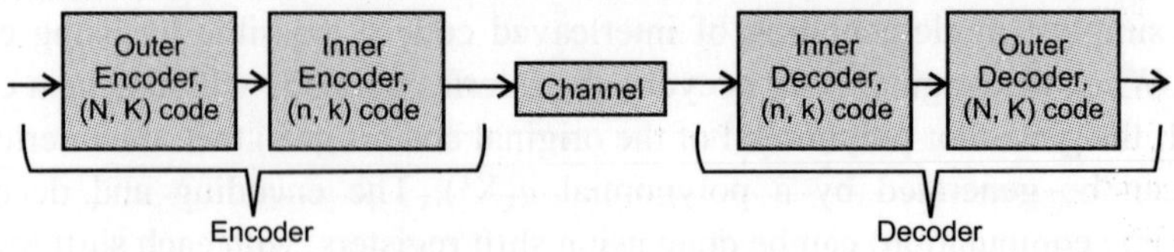

Fig. 7.6 Concatenated Code

The decoding of the concatenated code is done in two steps as the encoding was done and in the reverse order. As the digits are received decoding is done for each C_1 code words and the parity digits are removed. This results in a sequence of Nk digits. These digits are then decoded by the outer decoder as per C_2 to get the final corrected message. The scattered random errors are corrected by the inner code C_1. Burst errors in the channel may corrupt relatively few symbols, which C_1 cannot correct and these few errors can then be corrected by C_2.

***Example* 7.8:** Let us take the Single error correcting (7, 4) Hamming code as the inner code C_1 and the (15, 11) double-burst error correcting Reed-Solomon code, with symbols from $GF(2^4)$, as the outer code C_2. The code symbol of the RS code will be of 4 bits each. Each 4 bit symbol coming out of the outer encoder is encoded into a code vector in the (7,4) Hamming code. The resultant concatenated code will be a (105, 44) binary code. The d_{min} of the resultant code = $d_1 \times d_2$ = $3 \times 5 = 15$.

The concatenated codes can be constructed with single outer code and multiple inner codes. Another method available in the literature is [Hirasawa *et al*] to use an outer code obtained by interleaving 'λ' codes with symbols from $GF(2^k)$ and a single inner code. The amount of circuitry (hardware) required will be the circuitries required for both the codes C_1 and C_2.

7.6.3. Product Codes

Let us consider two linear codes, an (n_1, k_1) code C_1 and an (n_2, k_2) code C_2. Then we can form an $(n_1 n_2, k_1 k_2)$ linear code such that each code is a rectangular array of n_1 columns and n_2 rows in which every row is a code vector in C_1 and every column is a code in C_2, as shown in Fig. 7.7. This two dimensional code is called the **Product Code** of C_1 and C_2. The $k_1 k_2$ digits in the upper right corner of the array are the information symbols. The digits in the upper left corner of this array are computed from the parity-check equations of C_1 on rows, and the digits in the lower right corner are computed from the parity-check equations of C_2 on columns. The digits in the lower left corner of the array can be computed by using the parity check equations of C_2 on the columns or the parity check equations of C_1 on rows of the array. Either way we will get the same $(n_1–k_1) \times (n_2–k_2)$ check digits. Therefore, for encoding the product code $C_1 \times C_2$, we may first encode the k_2 rows of the information array based on the parity check equations of C_1 and then encode the n_1 resulting columns based on the rules of C_2, or vice versa.

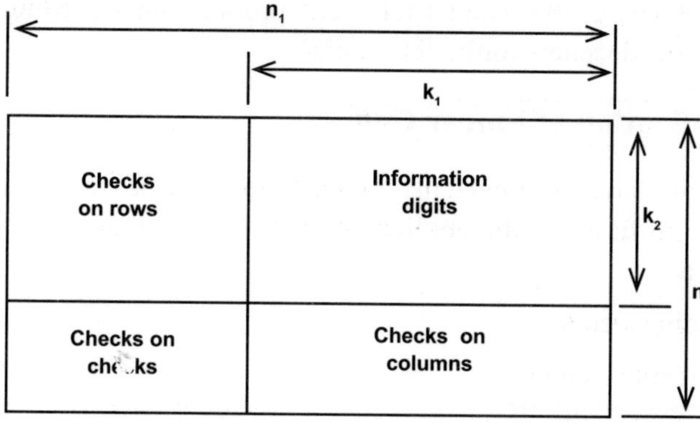

Fig. 7.7 Code Array for the Product Code $C_1 \times C_2$

If the code C_1 has minimum distance d_1 and the code C_2 has minimum distance d_2, then the minimum distance of the product code will be d_1d_2. Then the product code will be capable of correcting any combination of $[(d_1d_2-1)/2]$ errors. It is also proved that this product code $C_1 \times C_2$ is capable of correcting any combination of $t = [(d_1d_2-1)/2]$ or fewer random errors and simultaneously correcting any error bursts of length $l = \max(n_1t_2, n_2t_1)$ or less, where t_1 and t_2 are the error correcting capabilities of C_1 and C_2 respectively.

7.6.3.1. Implementing product codes using cyclic codes

The straight forward way to implement a product code is to set up the code array and operate on rows and then columns (or columns and then rows) in encoding and decoding. However, there is an alternative and extremely attractive way. In many cases the product codes of cyclic codes are also cyclic, and cyclic code implementation is much simpler. If the component codes, C_1 and C_2, are cyclic and if their lengths, n_1 and n_2 are relatively prime, the product code $C_1 \times C_2$ is cyclic if the code digits are transmitted in proper order. If $g_1(X)$ and $g_2(X)$ are the generator polynomials of such component codes C_1 and C_2 respectively then the generator polynomial of the cyclic product code is given by,

$$g(X) = \text{GCD}\,[\,X^{n_1n_2}-1,\ g_1(X^{bn_2})g_2(X^{an_1})\,] \tag{7.16}$$

where n_1 and n_2 relatively prime ; 'a' and 'b' are integer s such that $an_1 + bn_2 = 1$.

Suppose the cyclic code C_1 has random error correcting capability of 't_1' and burst error correcting capability of 'l_1' and the cyclic code C_2 has random error correcting capability of 't_2' and burst error correcting capability of 'l_2'. Then the burst error correcting capability 'l' of the cyclic product code of C_1 and C_2 is given by,

$$l \geq \max(n_1t_2 + l_1,\ n_2t_1 + l_2) \tag{7.17}$$

The complexity of the decoder for cyclic product codes is comparable to the complexity of the decoders for both C_1 and C_2.

7.7 Modification of Linear Codes

In section 7.4 we have mentioned that modifications can be carried out to remodel a known code to fit a specific application. The following are such modification used in practice.

- Expanding a code,
- Lengthening a code,
- Puncturing a code,
- Shortening a code,
- Augmenting a code,
- Expurgating a code.

7.7.1. Expanding a Code

Expanding a code means increasing the length of the code by adding more parity check symbols. This corresponds to increasing the larger dimension of the generator matrix.

Any binary (n, k) code whose minimum distance, d, is odd can be expanded to an $(n+1, k)$ code with minimum distance '$d+1$', by appending the sum of all other code word components as an overall parity check. This is because if the original code word has odd weight, the new bit will be a one. Hence, all the code words of weight 'd' become words of weight '$d+1$'. If 'H' is the parity check matrix of the original code, then all the expanded code has parity check matrix $\mathbf{H}'$ given below.

$$H' = \begin{pmatrix} 1 & 1\ 1 \ldots\ldots 1 \\ 0 & \\ 0 & \\ . & [\ \ H\ \ ] \\ . & \\ 0 & \end{pmatrix}$$

Specifically, every $(2^m-1,\ 2^m-1-m)$ Hamming code can be expanded into a $(2^m,\ 2^m-m)$ Single Error Correcting (SEC) and Double Error Detecting (DED) code usually, also, called a Hamming code. (24, 12) Golay Code is also an example of expanded code, of (23, 12) Golay code, by adding one extra overall parity bit.

7.7.2. Lengthening a Code

Lengthening a code means increasing the length of the code by adding more information symbols. This corresponds to increasing both the dimension of the generator matrix by the same amount.

The process of expanding or lengthening is generally called as **'Extending a code'**. That is by 'extending' we mean, increasing the block length either by expanding or by lengthening.

7.7.3. Puncturing a Code

Puncturing a code means reducing the length of the code by dropping parity check symbols. This corresponds to decreasing the larger dimension of the generator matrix.

7.7.3.1. Punctured convolutional codes

The Viterbi algorithm for convolutional code is suitable for low rate (e.g. 1/2 or 1/3) convolutional codes. In applications where a high rate convolutional code is required, the following method is used. Suppose we need a '3/4' rate convolutional code. Then, first, we encode the message sequence by a 1/2 rate convolutional encoder. Second we remove two symbols from each segment of six output symbols, and we send the remaining four symbols, as shown in Fig. 7.8 below:

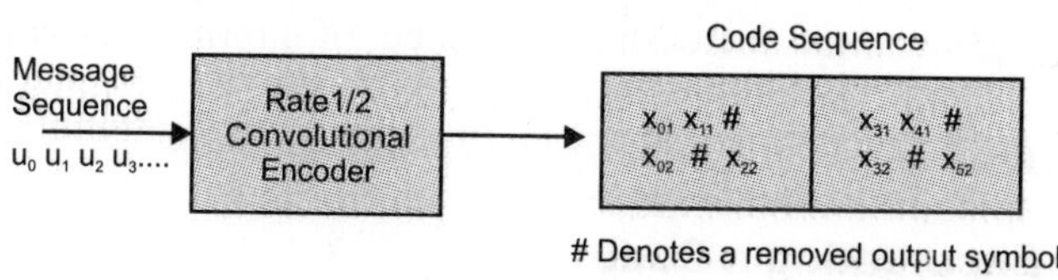

Fig. 7.8 A Punctured Convolutional Code of Rate 3/4

Thus we have a convolutional code of rate '3/4', which is called a punctured convolutional code. Maximum likelihood decoding of the code just described can be performed by using the Viterbi decoder for the original '1/2' rate code, where the computation of the distances for the removed symbols is omitted in the comparison of the paths at each node.

7.7.4. Shortening a Code

Shortening a code means reducing the length of the code by dropping the information symbols. This corresponds to decreasing both the dimensions of the generator matrix by the same amount.

7.7.4.1. Shortened hamming code: (SEC and DED Code)

We can delete any '*l*' columns from the H matrix to get a shortened Hamming code. By proper deletion of columns in H (all even weighted columns) we can get a code with $d_{min} = 4$ and hence, SEC and DED code.

***Example* 7.9:** The H matrix of the (7, 4) Hamming code is,

$$H = \begin{pmatrix} 1 & 0 & 0 & 1 & 0 & 1 & 1 \\ 0 & 1 & 0 & 1 & 1 & 1 & 0 \\ 0 & 0 & 1 & 0 & 1 & 1 & 1 \end{pmatrix}_{3\times 7}$$

The Parity check Matrix H′ for the shortened hamming code is obtained by removing even weighted columns in H, and is given below:

$$H' = \begin{pmatrix} 1 & 0 & 0 & 1 \\ 0 & 1 & 0 & 1 \\ 0 & 0 & 1 & 1 \end{pmatrix}_{3\times 4}$$

We can see from H′, that the minimum number of columns that add to zero is equal to four and hence, the minimum distance of the code generated by H′ is equal to four.

Examples of shortened cyclic codes were given in section (7.4).

7.7.5. Augmenting a Code

Augmenting a code means increasing the number of information symbols without changing the length. This corresponds to increasing the smaller dimension of the generator matrix.

7.7.6. Expurgating a Code

Expurgating a code means decreasing the number of information symbols without changing the length. This corresponds to decreasing the smaller dimension of the generator matrix.

In an expurgated Hamming code, the number of message bits, is reduced by one bit, but the number of parity bits is increased by one bit so that the total bit length remains at '*n*'. Such a Hamming code is called as "(*n*, *k*–1) Expurgated Hamming code". The principle is illustrated in the example (7.10) below:

***Example* 7.10:** A (7, 3) Expurgated Cyclic Hamming code can be devised using the (7, 4) Cyclic Hamming code whose generator polynomial is $g(X) = 1 + X + X^3$. The generator polynomial $g'(X)$ of the expurgated (7, 3) code is obtained by multiplying $g(X)$ by $(1 + X)$ which is a factor of $(X^n + 1)$. Hence, the product is also a factor of $(X^n + 1)$, which is the requirement of a polynomial to be a generator polynomial.

$$g'(X) = (1 + X)g(X)$$

$$= (1 + X)(1 + X + X^3) = 1 + X + X^3 + X + X^2 + X^4$$

$$g'(X) = 1 + X^2 + X^3 + X^4$$

From the properties of the generator polynomial, it is clear that the number of parity bits, $(n - k)$, is equal to the power of generator polynomial, $g'(X)$; equal to four in this case. Also the minimum distance (d_{min}) of the code is equal to the number of terms in the generator polynomial and is equal to four.

Using the generator polynomial the required encoder and decoder can be devised for the expurgated cyclic Hamming code by usual procedure for the regular cyclic code.

7.8 Binary Repetition Codes (0th-order Reed-Muller Code)

The simplest type of a block code allowing a variable amount of redundancy, and hence, the error detecting and correcting capability, is the Binary Repetition Code. With this code a single information bit is encoded into a block of 'n' identical bits, producing a $(n, 1)$ code.That is a binary bit is transmitted 'n' number of times. Generally 'n' is an odd value. The code, therefore, contains only two code words, namely, the all zeros code word and all ones code word. That is the binary '0' or binary '1' is repeated 'n' times. The repeated transmission of the information symbol is a form of time-diversity signaling. The received 'n' bit word, detected one bit at a time, may be operated on by the majority voting rule. That is a Maximum Likelihood decoding principle.

7.8.1. Decoding of Binary Repetition Codes

As in an $(n, 1)$ repetition code, 'n' is generally selected as an odd number, if the received code word contains more number of ones, it is decided that the transmitted bit is a binary one, else it is decided that the transmitted bit is a binary zero.The error detection is very simple. The 'n' received bits are fed to an 'n' input 'Exclusive OR' gate; and if the output of the gate is 'one', the syndrome is one and hence, the error is detected. Let us take the following example (7.11) to understand the concept.

***Example* 7.11:** Let us consider the transmission of a code word from the (5, 1) code. Suppose that the received word is '01001'. Obviously there are errors in the word, since only '00000' or '11111' could have been sent. Two interpretations are possible; either five zeros were transmitted and two errors occurred, or five ones were transmitted and three errors occurred. For a BSC with transition probability '$p < 1/2$', the conditional probability of the first event (with two errors) is $p^2(1-p)^3$ and the conditional probability of the second event (with three errors) is $p^3(1-p)^2$. Thus the event with two error pattern is the event of greater likelihood. It can be seen that, in this example, single and double error patterns are correctable with the majority voting rule.

Also it is clear that the Hamming distance between the two valid codes (namely '00000' and '11111') is five. That is for an $(n, 1)$ repetition code the minimum distance is 'n'. Therefore, considering the $(n, 1)$ repetition code as a linear block code the error correction capability is 't' $\leq (d_{min}-1)/2$. Hence, in this example $t = (5-1)/2 = 2$.

Therefore, in a repetition code, to correct 't' errors the number of repetitions required is given by 'n' ≥ 2t +1

The repetition code can be considered as a parity-check code. The $(n-1)$ parity bits can be generated by the following parity check equations

$$
\begin{aligned}
i_1 + p_1 &= 0 \\
i_1 + p_2 &= 0 \\
i_1 + p_3 &= 0 \\
\cdot \quad \cdot &= 0 \\
\cdot \quad \cdot &= 0 \\
i_1 + p_{n-1} &= 0
\end{aligned}
$$

Based on the above equations, the parity check Matrix for the (5, 1) repetition code is

$$
\text{H} = \begin{pmatrix} 1 & 1 & 0 & 0 & 0 \\ 1 & 0 & 1 & 0 & 0 \\ 1 & 0 & 0 & 1 & 0 \\ 1 & 0 & 0 & 0 & 1 \end{pmatrix}
$$

The Probability of correct decoding of an $(n, 1)$ Repetition Code is given by,

$$
\text{P}_c = \sum_{i=0}^{\frac{n-1}{2}} {}^n\text{C}_i p^i (1-p)^{n-i} \tag{7.18}
$$

The Probability of Erroneous Decoding {P(E)} = {1 – Probability of correcting decoding}.

$$
= \{1 - \text{P}_c\}
$$

Note: *The Binary repetition code is the* ***0th-order Reed-Muller code.*** *Reed-Muller codes are known for their simplicity. Reed-Muller codes are a class of binary linear codes that are easy to describe and can be decoded by a simple voting technique. For each integer 'm', and for each integer 'r' less than 'm', there is a Reed-Muller code of block length $n = 2^m$, called the rth-order Reed-Muller code of block length 2^m. The Reed-Muller code has a minimum distance of 2^m. The 0th–order Reed-Muller code is an (n, 1) code, which is the binary repetition code.*

7.9 Error and Erasure Correction Codes

In section 3.6.3 we have mentioned about the erasure channels. A receiver may be designed to declare a symbol erased when it is received ambiguously, or when the receiver recognizes the presence of interference or transient malfunction. Such a channel has an input alphabet of size 'q' and an output of size 'q + 1'. The extra symbol is called an erasure. For example, an erasure of symbol '*r*' from the message 'control' gives 'cont-ol'.

An error control code can be used with such a channel. If the code has minimum distance d_{min}, then any pattern of 'ρ' erasures can be filled if $d_{min} \geq$ 'ρ + 1'. Further any pattern of '*t*' errors and 'ρ'erasures can be decoded, provided,

$$d_{min} \geq 2t + \rho + 1 \tag{7.19}$$

Proof: Let us delete, from all code words, the **'ρ'** components where the receiver has specified erasures. This is a new code whose minimum distance is not smaller then **'$d_{min} - \rho$'**; hence, 't' errors can be corrected provided the above inequality in equation (7.19) is satisfied. Thus we can recover the shortened code word with **'ρ'** components erased. Finally, because $\boldsymbol{d_{min} \geq \rho + 1}$, there is only one code word that agrees with the un-erased components; thus the entire code word can be recovered.

Now let us see how a received code word with erasures and errors can be decoded. Let the largest 't' that satisfies the inequality of equation (7.19) be represented by 't_p'. In order to decode a received word with 'ρ' erasures, it is necessary to find a code word that differs from the un-erased portion of the received word in at most 't_p' places. Any such word will do because by the definition of 't_p', such a word is unique. So, we need a solution to find such a code word.

One method to decode is to guess the erased symbols and then apply any error correction procedure. That is filling the erasures with some possible symbols (on guess) and apply an error correction procedure. If the procedure finds that less than 't_p' of the un-erased symbols are in error, then these errors can be corrected, and the code word is recovered. If more than 't_p' errors are found, then a new guess is made for the erased symbols. In this way any error-correction algorithm suitable for a given code can be adapted to a procedure for correcting both errors and erasures. It is feasible, if the number of combinations that must be tried out for the erasures is small. Otherwise, such a trial and error procedure becomes impractical.

Special class of codes that are supported by a variety of strong decoding algorithms, like BCH codes, can be extended to algorithms for correcting both errors and erasures. The Massey – Berlekampalgorithm explained in section (7.3.4.3) can be modified for procedures correcting both errors and erasures. The procedure is explained below:

7.9.1. Massey—Berlekamp Algorithm for Error and Erasure Correction

Let 'r_i' be the received code vector, for i = 0 to (n–1). Suppose erasures have occurred in locations i_1, i_2,……,$i\rho$. At these known locations, the received word has blanks, which we will initially fill with zeros. Let us define the erasure vector as that n-tuple having component f_{i_l}, for l = 1, ….. , ρ equal to the erased symbol and in other components f_i equals zero. Then,

$$r_i = v_i + e_i + f_i \text{ for } i = 0 \text{ to } (n-1),$$

where 'v_i' is the transmitted vector and 'e_i' is the error vector.

Let Φ be any vector that is zero at every erased location and otherwise non-zero. Then,

$$\Phi_i r_i = \Phi(v_i + e_i + f_i) = \Phi_i v_i + \Phi_i e_i \tag{7.20}$$

Now let us define the modified received vector as $(r')_i = \Phi_i r_i$, the modified error vector as $(e')_i = \Phi_i e_i$, and the modified code word as $(v')_i = \Phi_i v_i$. The modified error vector will have errors in the same locations as 'e'. Then equation (7.20) can be written as,

$$(r')_i = (v')_i + (e')_i \tag{7.21}$$

Let $U_l = a^{i_l}$ for $l = 1, \ldots\ldots, \rho$ denote the erasure locations. Now let us define the erasure-locator polynomial as,

$$\Phi(X) = \sum_{k=0}^{n-1} \phi_k X^k = \prod_{l=1}^{\rho} (1 - X\,U_l) \tag{7.22}$$

This definition is done such that the inverse of the vector Φ has components Φ_i equal to zero whenever $f_i \neq 0$. Therefore, $\Phi_i f_i = 0$.

Now we have to modify the syndrome to take into account of both the erasure and error. This is done by initializing the error Locator Polynomial, C(X), with Φ(X) instead of with 1, in the Massey-Berlekamp algorithm for computing the error locator polynomial. By this initialization the modified syndromes are computed implicitly and need not explicitly appear. Now the error locator polynomial, C(X), will be called as the *error-and-erasure-locator polynomial.* The Massey-Berlekamp algorithm initialized with Φ(X) generates recursively the error-and-erasure-locator polynomial.

The Massey-Berlekamp algorithm, revised to handle erasures also, is shown in Fig. 7.9. This should be compared with error only case in Fig. 7.3. The only change from the decoder for errors only is the computation of the erasure-locator-polynomial, which is trivial compared to the other decoding computations. This is provided by a special loop for the first 'ρ' iterations, as shown in Fig. 7.9. The index 'n' is used to count the first 'ρ' iterations while the erasure polynomial is being computed and then continues to count the iterations of the Massey-Berlekamp algorithm, stopping when 'n' exceeds '$2t$'. The shift –register length L (degree of error locator polynomial) is increased once for each erasure, and therefore the lengthy changes according to the procedure of the Massey-Berlekamp algorithm. Since this algorithm was developed for the error only case, when we are including erasures also, the length test is modified by replacing 'n' by 'n–ρ' and 'L' by 'L–ρ', respectively.

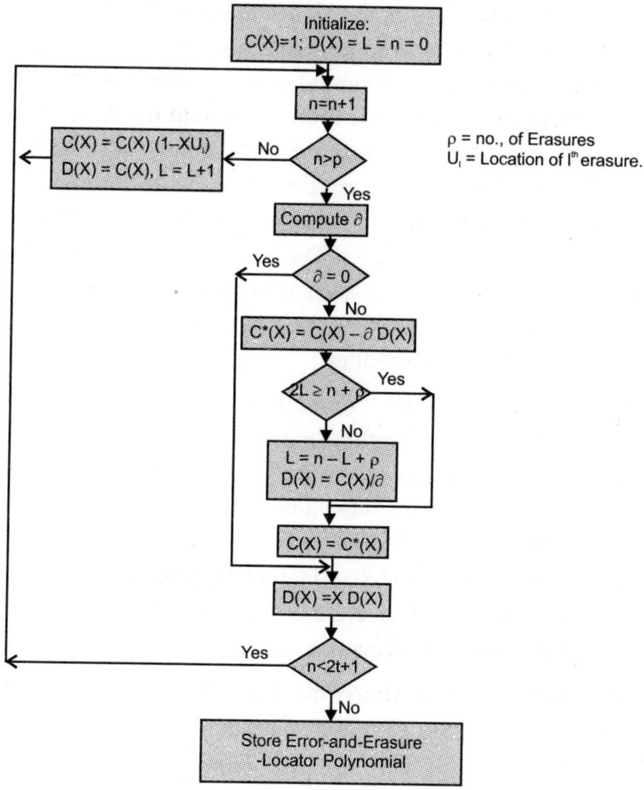

Fig. 7.9 Massey—Berlekamp Algorithm to Compute Error-and Erasure Location Polynomial

As a general note, instead of using a blank to denote an erased symbol, we can leave a best guess of a symbol in this location and specify that it is erased, by attaching a flag bit to this location. This is called **readable erasure.** Sometimes, outside the main decoder this readable erasure flag can be used by the user whether to accept (use) the information or not. One may refuse to accept a decoder output if too many of the readable erasures disagree with the decoder output word.

7.10 Block Codes Versus Convolutional Codes

We end this chapter with a discussion on the very frequently asked question, namely whether to choose block codes or convolutional codes. The system designer's choice between block code and convolutional code should depend on the application. The die hard block code supporters always advocate in favour of block codes, while their counter parts on the other hand claim that in almost all situations convolutional codes outperform block codes. As always, the "truth" is not only somewhere in between, but it also depend on the application.

The theory of block codes is much richer than the theory of convolutional codes. Many sophisticated finite field concepts have been used to design block codes with excellent mathematical structure that has simplified the development of efficient error correction algorithms. From a practical point of view, the Reed Solomon (RS) codes constitute the most important family of block codes. They are extremely well suited for digital implementation with the modern hardware technology. Berlekamp's bit serial RS encoders have been adopted as a NASA standard for deep space communication. The RS codes are particularly powerful when the channel errors occur in clusters-burst errors– which are the case in secondary memories such as magnetic tapes and disks. All compact disc players use RS codes with table-lookup decoding.

Considering a digitized voice communication system which requires a bit error rate of 10^{-5}, for satisfactory performance, convolutional codes in combination with Viterbi decoding appears to be an extremely good combination for communication over AWGN channels. For example, Qualcomm Inc. has on a single chip implemented a memory '$m = 6$' Viterbi decoder for rates R = 1/3, 1/2, 3/4 and 7/8. The rate R = 1/2 coding gain is 5.2 dB at $P_b = 10^{-5}$. This very powerful error-correcting system operates either with hard decisions or with eight-level quantized soft decisions.

The major drawback of RS codes is the difficulty of making full use of soft-decision information. The Viterbi algorithm can easily exploit the full soft-decision information provided at the decoder input and thus easily pick up the 2 dB gain over hard-decision. Further, code synchronization is in general much simpler for convolutional codes than for block codes.

If a combination of high level of data integrity of the order of $P_b = 10^{-10}$, and a larger coding gain which a Viterbi decoder cannot provide, then we could use either an RS code or a large memory (say, of the order of $m = 25$) convolutional encoder. The complexity of the Viterbi decoder, which is essentially 2^m, will be prohibitively larger in the latter case. Of course we could use sequential decoding of the convolutional code whose complexity is essentially independent of memory of the encoder. However, the major drawback of Sequential decoding is that noisy frames take large amounts of computation, and decoding times occasionally exceeds some upper limit, resulting in information loss or erasure.

In many applications where the noise is predominantly Gaussian, the best solution is obtained when block code and convolutional codes join forces and are used in series, that is in cascade. The convolutional code can be used as the inner code of the cascaded scheme explained in section 7.6.2, to clean up the channel. The Viterbi inner decoder will correct most channel errors but will occasionally output a burst of errors. This output then becomes the input to the outer decoder. Since an RS code is well suited to cope with burst of errors, RS code can be used as the outer code. Such a concatenated coding system can provide a high level of data integrity with large coding gain and low complexity. In order to have more strength an interleaver is often used between the outer and inner encoders and a corresponding de-interleaver between the inner and outer decoders. Then the output error burst from the inner decoder will be smeared out by the de-interleaver before the inner decoder has to cope with it. From the above discussion it is clear that depending on the application we can use an appropriate block code or convolutional code or a mix of both to strengthen each other and arrive at a system which may provide a high level of data integrity with large coding gain, low complexity and cost effectiveness.

7.11 Comparison of Uncoded and Coded Systems

Having discussed the different error control codes it is necessary to compare the coded and uncoded systems, on a quantitative basis, to understand the benefits and limitations of coded system over the uncoded system. In this section we will compare the two schemes, based on error probability under similar constraints of power and information rate.

The performance of a coded communication system is in general measured by its probability of decoding error and its coding gain over an uncoded system that transmits information at the same rate. When we talk about the error probability, there are two types of error probability, namely the probability of code word (or block) error and probability of bit error. The probability of code word error is the probability that a decoded code word (or block) at the output of the decoder is in error. This probability is also referred to as the **word-error rate (WER)** or **block-error rate (BLER)**. The probability of bit error, also known as the **bit-error rate (BER),** is the probability that a decoded information bit at the output of the decoder is in error. An efficient coded system is one which minimizes both these error probabilities under a given power, bandwidth and system complexity constraints.

Considering a '*t*' error correcting (n, k) code with code rate $R = (k/n)$, if the energy per transmitted code word is E_s, then the energy per information bit is given by,

$$E_b = E_s/R = (n/k).E_s \qquad (7.23)$$

In general the error probability of a system is expressed in terms of the ratio of energy per bit (E_b) to the one sided noise power spectral density (N_0) of the channel noise, expressed by

$$\lambda = E_b / N_0 \qquad (7.24)$$

For proper comparison, it can be assumed that '*k*' information digits are transmitted in the same time interval over the systems and that the transmitted power is E_s, is also maintained the same for both systems. Considering a '*k*' massage block, E_s energy is shared by k bits in the un-coded system, whereas it is shared by 'n' bits in the case of a coded system. Hence λ is higher for the uncoded system when compared to the coded system by an amount (n/k). Hence the bit error probability of the uncoded system will be lesser than for the coded one. Let P_{eu} and P_{ec} be the bit error probabilities of the uncoded and coded systems respectively. For the uncoded system , a word of 'k' bits will be received erroneously if any one of the 'k' bits is in error. Let P_{Eu} and P_{Ec} are the word error probabilities of the uncoded and coded systems respectively. Then,

P_{Eu} = 1 – Probability that all k bits are received correct $= 1-(1-P_{eu})^k$

$$P_{Eu} \cong kP_{eu} \quad \text{for } P_{eu} << 1 \tag{7.25}$$

For a 't' error correcting (n, k) code, the received code word will be in error if more than 't' errors occur in 'n' bits. If P(j, n) is the probability of 'j' errors in 'n' bits, then,

$$P_{Ec} = \sum_{j=t+1}^{n} P(j,n) \tag{7.26}$$

As there are ${}^nC_j$ ways in which 'j' errors can occur in 'n' bits we have

$$P(j, n) = {}^nC_j\,(P_{ec})^j(1-P_{ec})^{n-j} \tag{7.27}$$

And
$$P_{Ec} = \sum_{j=t+1}^{n} {}^nC_j\,(P_{ec})^j(1-P_{ec})^{n-j} \tag{7.28}$$

For $P_{ec} << 1$, the first term in the above summation dominates all the other terms, and hence can be ignored. Hence,

$$P_{Ec} = {}^nC_{t+1}\,(P_{ec})^{t+1}(1-P_{ec})^{n-(t+1)} \cong {}^nC_{t+1}\,(P_{ec})^{t+1} \text{ for } P_{ec} << 1. \tag{7.29}$$

Assuming a coherent PSK transmission scheme, for an AWGN channel,

The bit error probability $P_{eu} = Q(\sqrt{2\lambda})$ where Q(y) is related to 'erfc(y)', the error complementary function of 'y' (as explained in the note given below).

Since λ for the coded system is (k/n) times that of the uncoded system we have the bit error probability for the coded system given by,

$$P_{ec} = Q\left(\sqrt{\left(\frac{2k\lambda}{n}\right)}\right) \tag{7.30}$$

Hence, the word error probabilities are given by

$$P_{Eu} = kQ(\sqrt{2\lambda}) \tag{7.31}$$

$$P_{Ec} = {}^nC_{t+1}\left[Q\left(\sqrt{\left(\frac{2k\lambda}{n}\right)}\right)\right]^{t+1} \quad \text{for } P_{ec} << 1. \tag{7.32}$$

In order to compare the coded and uncoded systems we can plot P_{Eu} and P_{Ec} as functions of λ. For the case of a (7, 4) single error correcting code (t = 1, n = 7, k = 4), P_{Eu} and P_{Ec} are plotted in Fig. 7.10a as a function of λ. It can be seen from the plot that the performance of the coded system is better than that of the uncoded system; however the improvement is about 1dB only. For large value of n and

k the coded system will become significantly superior to the uncoded system. Fig. 7.10b shows the plots of P_{Eu} and P_{Ec} as a function of λ for the (23,12) Golay code ($n = 23$, $k = 12$ and $t = 3$). We can see that the improvement is of the order of 4 dB or more. For practical channels affected by fading and impulse noise, coding can yield substantial gains.

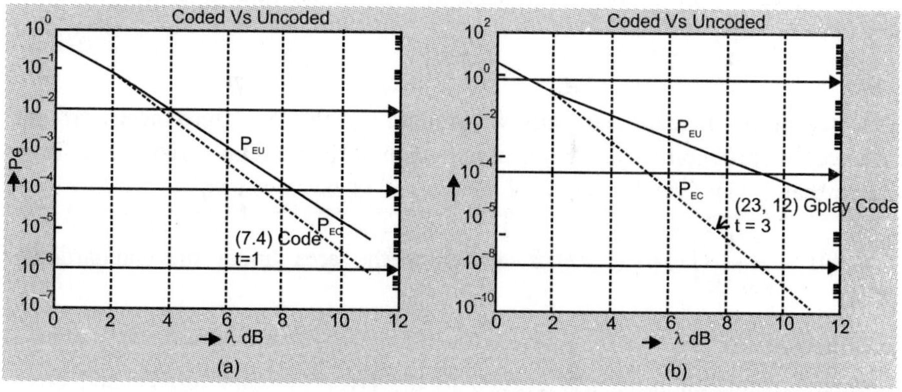

Fig. 7.10. Comparison of Coded and Un-coded Systems

***Example* 7.12:** For an AWGN – BSC compare the performance of an uncoded coherent PSK transmission scheme with that of a coded scheme using a single error correcting (15,11) code. Assume $\lambda = 9.12$.

Solution:

For the uncoded coherent PSK transmission scheme the word error probability

$$= P_{Eu} = kQ\,(\sqrt{2\lambda})$$

$$= 11\,Q(\sqrt{18.24}) = 1.1 \times 10^{-4}.$$

For the coded scheme the word error probability =

$$P_{Ec} = {}^{n}C_{t+1}\left[Q\left(\sqrt{\left(\frac{2k\lambda}{n}\right)}\right)\right]^{t+1}$$

$$= {}^{15}C_2\left[Q\left(\sqrt{\left(\frac{11(18.24)}{15}\right)}\right)\right]^2 = 105(1.37 \times 10^{-4})^2 = 1.96 \times 10^{-6}.$$

That is the word error probability of the coded system is reduced by a factor of $(1.1 \times 10^{-4}/1.96 \times 10^{-6} =)$ 56. Suppose, if we would like to have the error probability, 1.96×10^{-6}, of the coded system using an uncoded coherent PSK scheme, we must increase the transmitted power. If λ' is the new increased value of λ to obtain $P_{Eu} = 1.96 \times 10^{-6}$ then,

$$P_{Eu} = 11\, Q\left(\sqrt{2\lambda'}\right) = 1.96 \times 10^{-6}$$

This gives $\lambda' = 10.7$; this is an increase over the old value of 9.12 by a factor of 1.17 or 0.7 dB.

Note:

Considering the standardized Gaussian random variable X of zero mean and unit variance, the probability that an observed value of the random variable X will be greater than 'y' is given by the "Q" function and is expressed as,

$$Q(y) = \frac{1}{\sqrt{2\pi}} \int_y^{\infty} \exp(-z^2/2)\, dz;$$ which is the area under the standardized Gaussian tail.

The Q function is closely related to the error function 'erf(y)' and error complementary function 'erfc(y)'. The error function is given by,

$$\text{erf}(y) = \frac{2}{\sqrt{\pi}} \int_0^{y} \exp(-z^2)\, dz \text{ and when } y \to \infty, \text{erf}(y) = 1.$$

and the erfc is given by erfc(y) = 1 – erf(y)

From these $Q(y) = \frac{1}{2}\,\text{erfc}\left(y/\sqrt{2}\right)$.

This integral of Q(y) cannot be evaluated in a closed form and must be computed numerically. This function is widely tabulated and can be found in most of the standard mathematical tables.

7.12 Limitations of Error Control Coding Techniques

We may like to ask the question, what is the praise we have to pay for this saving in power and the improved performance? The error control codes combined with binary modulation, to obtain coding gain, results in channel bandwidth expansion. That is the coded systems require more bandwidth than the uncoded system. This is because of the addition of redundant bits, which are added to combat the channel noise. In order to maintain a transmission rate, even when we use an (n, k) error control code, the bandwidth expansion required is $1/R = n/k$. Therefore, the coding gain is achieved at the cost of increased bandwidth. For example when we use a ½ rate code, like an extended (24, 12) Golay code, coding gain is obtained at a cost of doubling the required bandwidth of the transmitted signal and the additional cost of decoder implementation. This type of error control codes are suitable for digital communication systems that are designed to operate on power-limited channels where (R/W) < 1 ('W' is the bandwidth), such as deep-space, satellite and other wideband communication systems. With the tremendous improvements in the microelectronics the hardware complexity and cost is insignificant for the benefit of decreased power requirement achieved.

7.13 Applications of Error Control Schemes

Applications of error control schemes in different areas are given in table 7.3 below.

Table 7.3: Applications of Error Correcting Codes

Communication Systems	Control Systems	Information Security and Retrieval Systems	Information Storage Systems	Signal Processing Systems
Broadcasting (VHF,HF etc.,), Line, Microwave (Civil, Military), Mobile network, Optical power line, Satellite, Sonar, Space, Telemetry, Spread Spectrum, Telex and Fax, Troposcatter.	Aeronautical, Aerospace, Automobile, Naval, Railway, RPV	Account & card codes, Bar and dot codes, Crypto, Product ID codes, ISBN.	Compact Discs, Magnetic tapes & discs, Punched cards and tapes, Semi- conductor Memories, RAID, ZFS, Digital Audio Tape (DAT).	Fault Tolerance, Image coding, Speech coding, Internet, TCP/IP

PROBLEMS

7.1 Consider the (23, 12) Golay code:

(*a*) Prove that this code is a Perfect code.

(*b*) What is the error detecting and correcting capabilities of its extended (24, 12) code?

7.2 Modify the (23, 12) Golay code decoder of Fig. 7.1 to indicate when the error encountered is more than three.

7.3 Consider a triple error correcting BCH code of length $n = 15$.

(*a*) Find out the generator polynomial g(X), given the minimal polynomials of the elements a, a^3, and a^5 are $\Phi_1(X) = 1 + X + X^4$; $\Phi_3(X) = 1 + X + X^2 + X^3 + X^4$ and $\Phi_5(X) = 1 + X + X^2$.

(*b*) Find the value of '*k*'.

(*c*) Draw the encoding circuit for this code.

(*d*) Mention the steps involved in decoding the received vector.

7.4 Draw a syndrome computation circuit for the binary double error correcting (31, 21) BCH Code.

7.5 Show that the 't' error correcting Reed-Solomon code generated by $g(X) = (X-a)(X-a^2)(X-a^3)\ldots\ldots\ldots\ldots(X-a^{2t})$ has minimum distance exactly $2t+1$, where 'a' is a primitive element in GF(2^m).

7.6 Consider the (31, 15) Reed-Solomon Code.

(*a*) How many bits are there in a symbol of the code?

(*b*) What is the block length in bits?

(*c*) What is the minimum distance of the code?

(*d*) How many symbol errors the code can correct?

(*e*) What is the length of an in-phase burst the code can correct?

7.7 Consider an (127, 65) Reed-Solomon code.

(*a*) What is the symbol error correcting capability of this code?

(*b*) If this code is used in an Interleaving scheme, what should be the degree of interleaving to be used, for a system burst error size of 300 symbols?

7.8 Consider a (19, 11) cyclic code whose generator polynomial in octal form is 1151.

(*a*) What is the burst error correcting capability of this code?

(*b*) Find the burst error correcting efficiency of this code.

7.9 Let C_1 be the (3, 1) cyclic code generated by $g_1(X) = 1 + X + X^2$ and let C_2 be the (7, 3) code generated by $g_2(X) = 1 + X + X^2 + X^4$.

(*a*) Find the generator and parity check polynomials of the product code of C_1 and C_2.

(*b*) Find the minimum distance of the product code.

(*c*) What is the error correcting capability of this product code?

7.10 Prove that the concatenation of an (n_1, k_1) inner code with minimum distance d_1 and an (n_2, k_2) outer code with minimum distance d_2 has minimum distance at least $d_1 d_2$.

APPENDIX

ALGEBRA

A.0 Introduction

The subject of Information Theory and Coding requires considerable amount of mathematical background, particularly algebra. Development of good codes has relied to a large extend on the useful structures of modern algebra. Different important codes based on the structural properties of polynomials and Galois Field has been developed. Also this algebraic structure provides the tools required for the design of encoders and decoders. This appendix provides an elementary knowledge of modern algebra that will aid in understanding the concepts of Information theory and coding. By-and-large a qualitative description is given without going for rigorous treatment like mathematical derivations etc.

Addition, subtraction, multiplication and division are the basic operations in general mathematics on real as well as complex numbers. Both of this real and complex number arithmetic systems are of fundamental importance in any mathematically oriented subject. However, there are less familiar arithmetic systems that are useful in the study of information theory and coding. The new system consists of sets together with operations on the elements or mathematical objects. In these systems the basic mathematical operations mentioned need not be the same as those of elementary arithmetic.

Modern algebraic theory classifies many arithmetic systems according to their mathematical strength. Brief definitions of these classifications are given below:

(1) *Group*: A set of elements or mathematical objects that can be "added" and "subtracted" is called a Group, G.

(2) *Field*: A set of elements or mathematical objects that can be "added", "subtracted", "multiplied" and "divided" is called a Field, F. (The elements of F are also called as *scalars*)

(3) *Vector space*: A set V is called a "vector space" and its elements are called "vectors", if there is defined an operation called "vector addition" on pairs of elements from V, and an operation called "scalar multiplication on an element from F and an element from V to produce an element in V.

A.1 Modulo Operations

Let us define the following addition and multiplication on the basic elements '0' and '1'.

$$0 + 0 = 0 \qquad 0 . 0 = 0$$
$$0 + 1 = 1 \qquad 0 . 1 = 0$$
$$1 + 0 = 1 \qquad 1 . 0 = 0$$
$$1 + 1 = 0 \qquad 1 . 1 = 1$$

The addition and multiplication defined here are called *Modulo-2 addition* and *Modulo-2 multiplication.*

Here we see that 1 + 1 = 0; or in general $x + x = 0$. That is "$-x = x$" and hence, we can say that in modulo arithmetic ***"subtraction = addition".***

Also we see that 1.1 = 1; or in general $x \cdot x = x$. That is "$x^{-1} = x$" and hence, we can say that in modulo arithmetic, except for division by zero, ***"division = multiplication".***

The binary elements '0' and '1', together with modulo-2 addition and modulo-2 multiplication is called the field of two elements or binary field and is represented by GF(2). A field consisting of 'q' elements is represented by GF(q). Here GF means Galois Field named after Evariste Galois (1811–1832) who proposed this concept. In a field of 'q' elements modulo-q operations are used. For example the field with 5 elements are represented as GF(5) = {0, 1, 2, 3, 4} and the modulo-5 addition and multiplication are given in table A1 below:

Table A1: (a) Modulo-5 Addition (b) Modulo-5 Multiplication.

+	0	1	2	3	4
0	0	1	2	3	4
1	1	2	3	4	0
2	2	3	4	0	1
3	3	4	0	1	2
4	4	0	1	2	3

.	1	2	3	4
1	1	2	3	4
2	2	4	1	3
3	3	1	4	2
4	4	3	2	1

A.2 Groups

A group is a mathematical abstraction of an algebraic structure. A Group G is a set of elements together with an operation on pairs of elements in the set (denoted by '$*$') satisfying the following four properties.

1. ***Closure:*** For every element a, b in the set, '$c = a * b$' is in the set.
2. ***Associativity:*** For every element a, b, c in the set $\boldsymbol{a * (b * c) = (a * b) * c}$.

3. ***Identity:*** There is an element 'e' called the identity element such that $\boldsymbol{a * e = e * a = a}$.

4. ***Inverse:*** If 'a' is in the set, then there is some element 'b' in the set called the inverse of 'a' such that $\boldsymbol{a * b = b * a = e}$.

If G has finite number of elements then it is called a ***finite group***, and the number of elements in G is called the ***order of the group*** G. Some groups satisfy the additional property that for all 'a, b' in the group, $\boldsymbol{a * b = b * a}$. This is called the ***commutative property***. Groups with this additional property are called the ***Commutative Groups*** or ***Abelian Groups***.

If H is a subset of group G, then H is called the **sub-group** of G, provided H satisfies all the four properties specified above. However, it is sufficient for H to satisfy the closure and inverses property as the other properties will get automatically satisfied, as the elements belong to G. Also if G is a finite group then H needs to satisfy only the closure property and all other properties will get satisfied automatically. As an example, in the Group formed by the set of integers (positive, negative and zero) under addition, the set of even integers form a sub-group.

One way to form a sub-group is to take an element, say 'p' from G and let H be the set of elements obtained by taking powers of 'p'. That is the elements of H will be $p, p^2, p^3, p^4, \ldots$. If G is a finite group, only finite number of elements can be distinct, and the sequence will repeat. The first element repeated will be 'p' itself. The set H is called the sub-group generated by 'p'. The number of elements, c, in H is called the order of the element 'p'. The set of elements namely $p, p^2, p^3, p^4, \ldots p^c$ is called a cycle. *A group that consists of all the powers of one of its elements is called a* ***Cyclic Group.***

A.3 Fields

A field is a more powerful algebraic structure, which is a set in which we can do all the four operations namely, addition, subtraction, multiplication and division.

Definition: A field F is a set that has two operations defined on it, namely, addition and multiplication, such that the following axioms are satisfied:

1. The set is an abelian group under addition.
2. The field is closed under multiplication, and the set of non-zero elements is an abelian group under multiplication.
3. The distributive law, $(a + b)c = ac + ab$, holds for all a, b, c in the field.

It is customary to denote the identity element under addition by '0' and call as 'zero'; the additive inverse of 'a' by '$-a$'; the identity element under multiplication by '1' and call as 'one' and the multiplicative inverse of 'a' by 'a^{-1}'. By subtraction $(a - b)$ we mean '$a + (-b)$'; by division (a/b), we mean $b^{-1}a$.

Popular examples of fields are,

1. The set of real numbers.
2. The set of complex numbers.
3. The set of rational numbers.

These fields all have infinite number of elements. However, we will be interested in fields with finite number of elements. The field with 'q' elements, if it exists, is called a finite field, or a Galois Field, and is denoted as GF(q), as mentioned in section A.1. The smallest field will consist of the binary elements '0'and '1'.

A subset of a field F is called a subfield, if it is a field under the inherited addition and multiplication. The original field F is then called an ***extension field*** or ***extended field*** of the sub-field. For example using the binary field GF(2) as the ground field we can form fields having 2^m elements. Then the field with 2^m elements or symbols are called the Extended Galois Field, GF(2^m) .

A.4 Polynomials

Polynomials play a very vital role in coding theory. They are used to generate Galois Fields, to design encoder and decoder circuits etc., A polynomial can be defined as an expression consisting of powers of one or more variables connected using algebraic symbols ['+' or '–'].

Examples of polynomials: $aX^2 + bY + c$; $pX^3 + qY^2 + Y + m$; $aX^2 + bX + c$;

$$pX^n + qX^{n-1} + ... + aX^2 + bX + c.$$

Here X and Y are the variables, a, b, c, p, q, m are the coefficients and 'n' is the power of the variable(s) and the highest value of 'n' is called the degree of the polynomial. Polynomials with binary digits as their coefficients can be manipulated using modulo operations.

A polynomial f(X) with one variable X and with coefficients from GF(2) is of the form:

$$f(X) = f_0 + f_1X + f_2X^2 + \cdots + f_nX^n$$

where $f_i = 0$ or 1 for $0 \le i \le n$

For the polynomial above, if $f_n=1$, f (X) is a polynomial of degree 'n' and is known as a ***monic polynomial***; if $f_n = 0$, f (X) is a polynomial of degree less than 'n'. The degree of the polynomial $f(X) = f_0$ is zero, and in this case f (X) is known as a ***zero polynomial.*** A polynomial with coefficients from GF(2) is called "a polynomial over GF(2)".

There are two polynomials over GF(2) with degree1, namely X and 1 + X ; there are four polynomials over GF(2) with degree 2, namely X^2, $1+X^2$, $X + X^2$, $1+ X + X^2$. In general, there are 2^n polynomials over GF(2) with degree

'n'. Polynomials over GF(2) can be manipulated using modulo operations; they can be added, multiplied and divided.

Example **A1:** Consider the product of the polynomials $f_1(X) = X + 1$ and $f_2(X) = X^3 + X + 1$.

$f_1(X)\ f_2(X) = X^4 + X^2 + X + X^3 + X + 1 = X^4 + X^3 + X^2 + 1$ (because $X + X = 0$)

Example **A2:** We can divide one polynomial, say $f_2(X) = X^6 + X^5 + X^3$ by another polynomial, say $f_1(X) = X^3 + X + 1$ using long-division method as shown:

$X^3 + X + 1)X^6 + X^5 + X^3(X^3 + X^2 + X + 1$ ← Quotient

$X^6 + X^4 + X^3$

\- - -

$X^5 + X^4$

$X^5 + X^3 + X^2$

\- -

$X^4 + X^3 + X^2$

$X^4 + X^2 + X$

\- - -

$X^3 + X$

$X^3 + X + 1$

\- - -

----

1 ⟵ Remainder

Example **A3:**

Let $f(X) = X^4 + X + 1$, then,

$[f(X)]^2 = X^8 + X^2 + 1 + 2X^5 + 2X^4 + 2X,$

$= X^8 + X^2 + 1$, (because $2X^5 = 2X^4 = 2X = 0$).

$= f(X^2)$.

In general for any $n \geq 0$, $[f(X)]^{2n} = f(X^{2n})$

Roots of polynomials can also be calculated by using the condition that if $f(1) = 0$, then $(X+1)$ is a factor of the polynomial $f(X)$. That is $f(X)$ is divisible by $(X+1)$. The factors can be calculated in the normal way by factorization. From example A1, (X+1) is a factor of $f(X) = X^4 + X^3 + X^2 + 1$, since $f(1) = 0$. For a polynomial $f(X)$ over GF(2), if it has an even number of terms, it is divisible by X+1.

Another kind of polynomial frequently used is the *irreducible polynomial*. ***A polynomial f(X) of degree 'm' over GF(2), is said to be irreducible over GF(2) if f(X) does not have a factor of degree less than 'm' and greater than zero.***

Among the four polynomials of degree 2 over GF(2), mentioned above, X^2, $1+X^2$ and $X + X^2$ are not irreducible as they are either divisible by X or X+1. However, $1+X+X^2$ does not have either '0' or '1' as a root and so is not divisible by any polynomial of degree one. Therefore, $1+X+X^2$ is an irreducible polynomial of degree two.

***Example* A4:** Examples of irreducible polynomials are X^3+X+1; X^4+X+1; X^5+X^2+1.

Another general definition of an irreducible polynomial is, ***a polynomial p(X) that is divisible only by 'a.p(X)' or 'a', where 'a' is an arbitrary field element in GF(q), is called an irreducible polynomial.***

In general for any $m \geq 1$, there exists an irreducible polynomial of degree '*m*'. Also an important property of irreducible polynomials is that, ***any irreducible polynomial over GF(2) of degree 'm' divides*** $X^{(2^m-1)}+1$. For example the irreducible polynomial X^3+X+1 divides $X^{(2^3-1)}+1 = X^7+1$ (can be verified by long division method).

Typical applications of irreducible polynomials are generation of random numbers in computers; generation of pseudo random binary sequences (PRBS) in cryptography, CDMA and Frequency Hopping communication systems; construction of Galois Field and implementation of encoders and decoders in coding of information (source coding as well as error control coding).

An irreducible polynomial p(X) of degree 'm' is said to be a ***primitive polynomial****, if the smallest positive integer 'n' for which p(X) divides* X^n+1 *is* $n = 2^m-1$.

***Example* A5:** $p_1(X)=X^5+X^2+1$ divides $X^{31}+1$, but does not divide any X^n+1, for $1 \leq n < 31$, and $p_2(X) = X^4+X+1$ divides $X^{15}+1$, but does not divide any X^n+1, for $1 \leq n < 15$. Hence, both $p_1(X)$ and $p_2(X)$ are primitive. However, all irreducible polynomials are not primitive polynomials. For example the polynomial $(X^4 + X^3 + X^2 + X + 1)$ is irreducible but not primitive since it divides X^5+1.

It is difficult to identify primitive polynomials. However, there are tables of irreducible polynomials available in the literature in which primitive polynomials are marked. Also for a given value of '*m*' there may be more than one primitive polynomial. A list of primitive polynomials with smallest number of terms for each degree '*m*' is given in table A2 below:

Table A2: List of Primitive Polynomials.

m	Primitive Polynomials	*m*	Primitive Polynomials
3	$X^3 + X + 1$	10	$X^{10} + X^3 + 1$
4	$X^4 + X + 1$	11	$X^{11} + X^2 + 1$
5	$X^5 + X^2 + 1$	12	$X^{12} + X^6 + X^4 + X + 1$
6	$X^6 + X + 1$	13	$X^{13} + X^4 + X^3 + X + 1$
7	$X^7 + X^3 + 1$	14	$X^{14} + X^{10} + X^6 + X + 1$
8	$X^8 + X^4 + X^3 + X^2 + 1$	15	$X^{15} + X + 1$
9	$X^9 + X^4 + 1$	16	$X^{16} + X^{12} + X^3 + X + 1$

A monic irreducible polynomial of degree of at least one is called a ***prime polynomial.*** The ***greatest common divisor*** of two polynomials $p(X)$ and $q(X)$, denoted by GCD[$p(X)$, $q(X)$], is the monic polynomial of largest degree that divides both of them. The ***least common multiple*** of two polynomials $p(X)$ and $q(X)$, denoted by LCM[$p(X)$, $q(X)$], is the monic polynomial of smallest degree divisible by both of them. It can be proved that GCD and LCM of two polynomials are unique. If the GCD of two polynomials is '1', then they are said to be ***relatively prime.***

If β is an element in GF(2^m) and $\Phi(X)$ is the polynomial of smallest degree over GF(2) such that $\Phi(\beta) = 0$, then the polynomial $\Phi(X)$ is called the ***Minimal polynomial*** of the element β. The minimal polynomial of the zero element, '0', of GF(2^m) is 'X' and that of unit element '1' is 'X + 1'. Minimal polynomials are irreducible polynomials. The typical uses of minimal polynomials are in forming the generator polynomials of BCH error correcting codes and in the construction of Galois Fields.

A.5 Construction of Galois Field GF(2^m)

We have mentioned previously in section A.1 and A.3, that GF(2^m) is an extended field of GF(2). Hence, GF(2^m), for $m > 0$, can be constructed using the binary field GF(2).

Consider a polynomial $p(X)$ of degree 'm', over GF(2). Using a new variable 'α', such that $p(\alpha) = 0$, a set of elements which are powers of 'α' is developed. That is

$$F = \{0, 1, \alpha^1, \alpha^2, \ldots\ldots\ldots, \alpha^{q-2}, \ldots\}$$

From the definition of the primitive polynomial $p(X)$, it divides X^n+1.

That is $X^n+1 = p(X).\, q(X)$,

Therefore, $\alpha^n+1 = p(\alpha)q(\alpha) = 0$ (because $p(\alpha) = 0$).

i.e., $\alpha^{2m-1}+1=0$.

i.e., $\alpha^{2m-1}=1=\alpha^0$;

i.e., $\alpha^{q-1}=1$, (because $q=2^m$)

Therefore, $\alpha^q=\alpha$, $\alpha^{q+1}=\alpha^2$, ...

Therefore, under the condition $p(X)$ is a primitive polynomial and $p(\alpha)=0$, F becomes finite and contains the following elements,

$F=\{0,1,\alpha^1,\alpha^2,\ldots\ldots,\alpha^{q-2}\}$

The element 'α' is known as the primitive element of the field GF(2^m). Also the non-zero elements of F are closed under the multiplication operation.

Therefore, if p(X) is a primitive polynomial, then 0, α^0, α^1, α^2, ..., α^{q-2} for q = 2^m, will have distinct values and are called the set of 'q' (= 2^m) field elements. That is using a primitive polynomial and the binary GF(2) we can construct a Galois Field GF(2^m).

The field elements can be represented in power form, polynomial form, integer form and in an ordered sequence of '*m*' components called an *m*-tuple. The '*m*' components are the '*m*' coefficients of the polynomial representation of the field element. The power representation is convenient for multiplication and the polynomial representation is convenient for addition. The construction of a Galois Field GF(2^3) and different representation of the field element are illustrated in the following example A6.

***Example* A6:** Let us construct GF(2^3) using the binary ground field GF(2) and a third degree primitive polynomial $p(X)=X^3+X+1$.

$p(X)=X^3+X+1$

$\therefore p(\alpha)=\alpha^3+\alpha+1=0,$

$\therefore \alpha^3=\alpha+1;$

$\alpha^4=\alpha^3\alpha=\alpha+\alpha^2;\ \alpha^5=\alpha^4\alpha=\alpha^2+\alpha^3=1+\alpha+\alpha^2;\ \alpha^6=1+\alpha^2;$

$\alpha^7=\alpha^6\alpha=\alpha+\alpha^3=\alpha+\alpha+1=1;\ \alpha^8=\alpha^7\alpha=1.\alpha=\alpha;\ldots\ldots\ldots$

The field elements of GF(2^3) are given in table A3, in power (exponential), polynomial, 3-tuple and integer forms.

Table A3: Field Elements of GF(2^3) Generated by $p(X)=1+X+X^3$.

Power Representation	Polynomial Representation	3-tuple Representation	Integer Representation
0	0	0 0 0	0
1	1	1 0 0	1
α	α	0 1 0	2
α^2	α^2	0 0 1	4
α^3	$1+\alpha$	1 1 0	3
α^4	$\alpha+\alpha^2$	0 1 1	6
α^5	$1+\alpha+\alpha^2$	1 1 1	7
α^6	$1+\alpha^2$	1 0 1	5

Another primitive polynomial which can generate $GF(2^3)$ is $p(X) = X^3 + X^2 + 1$. We can see that $p(X) = X^3 + X + 1$ is the minimal polynomial of the field element 'α' and $p(X)=X^3+ X^2 + 1$ is the minimal polynomial of the field element 'α^3' of $GF(2^3)$.

A.6 Computations using Galois Field Arithmetic

We can solve equations using arithmetic over $GF(2^m)$. In this section we will consider two examples, one solving two given linear equations and the other factorizing a given equation.

***Example* A7:** Let us consider the following linear equations over $GF(2^3)$:

$$X + \alpha^3 Y = \alpha^4$$

$$\alpha^6 X + \alpha^5 Y = 1$$

Referring to table A3 we can solve these equations. Multiplying the second equation by 'α' we have,

$$X + \alpha^3 Y = \alpha^4$$

$$X + \alpha^6 Y = \alpha$$

By adding the above two equations we have, $(\alpha^6 + \alpha^3)Y = \alpha^4 + \alpha$

$$\alpha^4 Y = \alpha^2$$

Therefore, $\quad Y = \alpha^2/\alpha^4 = \alpha^2.\alpha^3 = \alpha^5$

(That is division by an element is equal to multiplication by its inverse)

Substituting $Y = \alpha^5$, in the first equation we have,

$$X = \alpha^4 + \alpha^3 Y = \alpha^4 + \alpha^3\alpha^5 = \alpha^4 + \alpha = \alpha^2$$

Thus, the solution for the given equations are $X = \alpha^2$ and $Y = \alpha^5$.

***Example* A8:** Let us find out the roots of the following equation over $GF(2^3)$,

$$f(X) = X^2 + \alpha^4 X + \alpha^2 = 0.$$

One of the straight forward methods of solving this equation is by substituting all the elements of table A3 for X in the given equation. Accordingly we see that $f(\alpha^3) = 0$ and $f(\alpha^6) = 0$, since,

$$f(\alpha^3) = (\alpha^3)^2 + \alpha^4.\alpha^3 + \alpha^2 = \alpha^6 + \alpha^7 + \alpha^2 = 0$$

$$f(\alpha^6) = (\alpha^6)^2 + \alpha^4.\alpha^6 + \alpha^2 = \alpha^{12} + \alpha^{10} + \alpha^2 = 0$$

Thus, α^3 and α^6 are the roots of $f(X)$ and $f(X) = (X+\alpha^3)(X+\alpha^6)$

Such computations are typical and required for decoding of codes such as BCH and Reed-Solomon error correcting codes and the procedure can be programmed on a general purpose computer or embedded processor, easily.

A.7 Properties of Galois Fields

The basic properties of Galois Fields are listed below:

i. In any Galois Field, the number of elements is a power of a prime.

ii. If '*p* 'is a prime and '*m*' is a positive integer, the smallest subfield of GF(p^m) is GF(p). The elements of GF(p) are called the integers of GF(p^m), and '*p*' is called its characteristic.

iii. In a Galois Field of characteristic 2, $-\beta = \beta$ for every field element β in the field.

iv. If '*p*' is a prime and '*m*' is an integer then there is a Galois Field with p^m elements.

v. Every Galois field GF(q) has at least one primitive element.

vi. Every Galois field GF(q) has at least one primitive polynomial over it of every positive degree.

vii. Every primitive element has over any subfield a minimal polynomial that is a prime polynomial.

viii. Two Galois Fields with the same number of elements are isomorphic.

ix. For any prime power q and positive integer '*m*', GF(q) is a subfield of GF(q^m), and GF(q^m) is an extension field of GF(q).

x. GF(q^n) is not a subfield of GF(q^m) if '*n*' does not divide '*m*'.

xi. The degree of minimal polynomial over GF(q) of any element of GF(q^m) is a divisor of '*m*'.

A.8 Vector Spaces

The concept of an *n*-dimensional vector space is closely related to the ideas of linear algebra and matrix theory and is important in many applications. The polynomial f(X) over GF(2) can be represented as an ordered sequence of binary symbols, such as:

$$(1+X+X^2+X^5) \leftrightarrow (1\ 1\ 1\ 0\ 0\ 1),$$

with the highest power coefficient at the right most (since in actual practice, the higher order coefficients are transmitted first). This ordered sequence of *n* binary digits is generally called an "*n*-tuple" over GF(2) and is considered as a vector. ***The set of all binary n-tuples is called an n-dimensional Vector Space over GF(2),*** and is denoted by $\{V_n\}$.

For '*n*', binary digits in each vector, the vector space consists of 2^n vectors. For example, for $n = 3$, $\{V_n\} = \{V_3\} = \{000, 001, 010, 011, 100, 101, 110, 111\}$.

General vector spaces can be defined with respect to any field as follows:

Let F be a field. The elements of F are called *scalars*. A set V is called a *vector space* and its elements are called *vectors* if there is an defined operation called *vector addition* (denoted by '+') on pairs of elements from V, and an operation called *scalar multiplication* (denoted by '.') on an element from F and an element from V to produce an element also belonging to V, provided the following postulates are satisfied.

(*a*) V is an abelian or commutative group under vector addition.

(*b*) For an element 'a' in F and an element 'v' in V, the product '$a.v$' is an element in V.

(*c*) Distributive Law: For any vectors v_1, v_2 in V and any scalars 'c_1' and 'c_2' in F,

$$c_1(v_1 + v_2) = c_1v_1 + c_1v_2 \text{ and } v_1(c_1 + c_2) = c_1v_1 + c_2v_1.$$

(*d*) Associative Law: For any vector v in V and any scalars 'c_1' and 'c_2' in F,

$$(c_1 \cdot c_2) \cdot v = c_1 \cdot (c_2 \cdot v)$$

(*e*) For the unit element '1' of F and 'v' in V, $1 \cdot v = v$

(*f*) There is an all – zero vector '0' in V called the origin of V, such that for every v in V,

$$0 + v = v + 0 = v.$$

(*g*) For every v in V, there is a vector '$-v$' called the additive inverse, such that

$$v + (-v) = (-v) + v = 0.$$

(*h*) For all scalars C from GF(2), and for all v in V, the scalar product Cv is a vector in V, for example

$$\{v\} = \{ C_1v_1 + C_2v_2 + C_3v_3 + \ldots\ldots\ldots\ldots\ldots + C_k v_k\} \in V$$

where C_i's are from GF(2) and are called the coefficients of v_i and v is called the Linear Combination of the vectors $v_1, v_2, v_3, \ldots\ldots\ldots\ldots, v_k$.

A8.1. Subspaces

A sub-set S of the vector space V is called a subspace if

(*a*) S is a non-empty subset.

(*b*) The all-zero vector is in S.

(*c*) The sum of any two vectors in S is also in S.

A8.2. Dot Product of Vectors

The dot product, also known as the inner product, of two vectors is defined as,

If $v_1 = \{x_1, x_2, x_3\}$ a triple

$v_2 = \{y_1, y_2, y_3\}$ a triple then,

$$v_1 \cdot v_2 = \{x_1y_1 + x_2y_2 + x_3y_3\}$$

where '+' and '.' are modulo-2 operations. Hence, the inner product of two vectors is a scalar in GF(2).

If the inner product of two vectors, $v_1.v_2$, is zero then the two vector v_1 and v_2 are said to be ***orthogonal*** to each other. Any non-zero vector in a field can be orthogonal to itself. The concept of inner product or dot product can be generalized to any Galois field.

The following example illustrates the concepts of vector space discussed above.

***Example* A9:** For $n = 4$, the vector space, $\{V_4\}$, consists of 2^4 vectors as shown in table A4.

TableA4: Vector Space $\{V_4\}$

0000	1000
0 0 0 1	1 0 0 1
0 0 1 0	1 0 1 0
0 0 1 1	1 0 1 1
0 1 0 0	1 1 0 0
0 1 0 1	1 1 0 1
0 1 1 0	1 1 1 0
0 1 1 1	1 1 1 1

We can verify easily from the table that $v_i + v_j = v_k$

i.e., (0011) + (0100) = (0111); all the three vectors belonging to V_4.

Similarly the scalar products: $C_1v_1 + C_2v_2 + C_3v_3 = v_j$

$$1.(0011) + 0.(0100) + 1.(0101) = (0110)$$

Out of the 2^4 (=16) vectors in $\{V_4\}$, let us consider the subset S_1 = {0000, 1100, 0011, 1111}.

We see that S_1 is a non-empty subset; contain the zero-vector and sum of any two non-zero vector in S_1, is the third non-zero vector in the subset. Hence S_1 satisfy the requirement to be a subspace and hence S_1 is a subspace of $\{V_4\}$. Similarly S_2 = {0000, 1010, 0101, 1111} also form a subspace of $\{V_4\}$.

A8.3. Linear Combination of Vectors

Let us consider $v_1, v_2, v_3, \ldots\ldots, v_k$ be 'k' vectors in a vector space $\{V_n\}$ over a field F. Also let us consider $a_1, a_2, a_3, \ldots\ldots, a_k$ be scalars from F. Then the sum $\{a_1v_1 + a_2v_2 + a_3v_3 + \ldots\ldots + a_kv_k\}$ is called the ***Linear Combination*** of the vectors $v_1, v_2, v_3, \ldots\ldots, v_k$.

It is apparent that the sum of any two linear combination of the vectors v_1, v_2, v_3, ..., v_k namely

$$(a_1v_1 + a_2v_2 + a_3v_3 + \ldots\ldots\ldots\ldots\ldots + a_kv_k) + (b_1v_1 + b_2v_2 + b_3v_3 + \ldots\ldots\ldots\ldots\ldots + b_kv_k)$$

$$= (a_1 + b_1)v_1 + (a_2 + b_2)\, v_2 + \ldots\ldots\ldots\ldots + (a_k + b_k)v_k.$$

is also a linear combination of v_1, v_2, v_3,................, v_k.

Similarly the product of a scalar '*c*' in F and a linear combination of v_1, v_2, v_3 ,................, v_k namely

$$c.(a_1v_1 + a_2v_2 + \ldots\ldots\ldots\ldots\ldots + a_kv_k) = (c.a_1)v_1 + (c.a_2)v_2 + \ldots\ldots\ldots\ldots\ldots + (c.\, a_k)v_k$$

is also a linear combination of v_1, v_2, v_3,................, v_k.

From the subspace principle, a set of linear combinations of the '*k*' vectors, v_1, v_2, v_3,................, v_k, in the vector space $\{V_n\}$, forms a subspace of $\{V_n\}$.

A8.4. Linearly Dependent and Linearly Independent Vector Sets

A set of '*k*' vectors v_1, v_2, v_3, ..., v_k in a vector space $\{V_n\}$ over the field F is said to be ***linearly dependent*** if and only if there exist '*k*' scalars a_1, a_2,..., a_k from GF(2), not all zero, such that the linear combination "$a_1v_1 + a_2v_2 + \ldots\ldots\ldots\ldots\ldots + a_kv_k = 0$".

A set of '*k*' vectors v_1, v_2, v_3, ..., v_k in a vector space $\{V_n\}$ over the field F is said to be ***linearly independent*** if it is not linearly dependent, that is when the linear combination "$a_1v_1 + a_2v_2 + \ldots\ldots\ldots\ldots\ldots + a_kv_k \neq 0$".

***Example* A10:** In $\{V_4\}$ of example A9, the set of vectors (1001), (0110) and (1111) are linearly dependent, since

$$1.(1001) + 1.(0110) + 1.(1111) = (0000)$$

The set of vectors (1011), (0100) and (1101) are linearly independent, since

$$1.(1011) + 1.(0100) + 1.(1101) = (0010) \neq 0.$$

All eight linear combination of these three vectors are as follows:

$$0.(1011) + 0.(0100) + 0.(1101) = (0000)$$
$$0.(1011) + 0.(0100) + 1.(1101) = (1101)$$
$$0.(1011) + 1.(0100) + 0.(1101) = (0100)$$
$$0.(1011) + 1.(0100) + 1.(1101) = (1001)$$
$$1.(1011) + 0.(0100) + 0.(1101) = (1011)$$
$$1.(1011) + 0.(0100) + 1.(1101) = (0110)$$
$$1.(1011) + 1.(0100) + 0.(1101) = (1111)$$
$$1.(1011) + 1.(0100) + 1.(1101) = (0010)$$

A8.5. Generating System and Dimension of Vector Space

A subset of vectors is said to span or generate a vector space V, if every vector in the vector space V is equal to at least one linear combination of the vectors in this subset and the vectors in the subset are called a ***generating system*** of the vector space V. In any vector space or subspace, there is at least one set of linearly independent vectors that span the vector space; this set is called the ***basis of the vector space***. The number of vectors in the basis is called the ***dimension of the vector space***. A vector space that is spanned by a finite set of vectors is called a ***finite-dimensional vector space.*** In general we will be using finite-dimensional vector spaces.

Let us consider the following set of 'n' linearly independent n-tuples,

$$g_0 = (1, 0, 0, \ldots\ldots\ldots, 0)$$
$$g_1 = (0, 1, 0, \ldots\ldots\ldots, 0)$$
$$\ldots\ldots$$
$$\ldots\ldots$$
$$g_{n-1} = (0, 0, 0, \ldots\ldots\ldots, 1),$$

where the ith n-tuple g_i contains only one non-zero element in the ith location. Then every n-tuple in vector space V_n can be formed as a linear combination of the above 'n' number of n-tuples. Therefore, $g_0, g_1, g_2, \ldots\ldots, g_{n-1}$ span the vector space V_n of all binary n-tuples over GF(2). Hence they form a basis for V_n and the dimension of V_n is 'n'. Thus the vector space V_n of all binary n-tuples is an "n-dimensional" vector space and is spanned by a basis of 'n' linearly independent vectors.

If $\{v_1, v_2, v_3, \ldots, v_k\}$ are 'k' linearly independent vectors and $k < n$, then all linear combinations of $\{v_1, v_2, v_3, \ldots, v_k\}$ of the form

$$q = a_1v_1 + a_2v_2 + \ldots\ldots\ldots\ldots + a_kv_k$$

form a k-dimensional subspace S of V_n. There are 2^k possible combinations of these vectors $\{v_1, v_2, v_3, \ldots, v_k\}$. Thus S has 2^k distinct vectors and is a k-dimensional subspace of V_n.

***Example* A11:** Consider the set of four linearly independent vectors $v_1 = 0111$; $v_2 = 1011$, $v_3 = 1101$ and $v_4 = 1110$ in $\{V_4\}$ of example A9. Any vector in $\{V_4\}$ is a linear combination of these vectors. Then v_1, v_2, v_3 and v_4, span the vector space $\{V_4\}$.

A8.6. Dual Space or Null Space or Orthogonal Complement Space

We have seen the concept of orthogonal vectors in section (A8.2). A vector orthogonal to every vector in a set is said to be orthogonal to the set. Let us consider the vector space V_n of all n-tuples. Let 'S' be a k-dimensional subspace in V_n. A set of vectors, which are orthogonal to S will form a subspace. This can be proved as follows:

Let 'S_d' be the set of all vectors orthogonal to S. Hence zero vector is in S_d, and hence S_d is not empty. Let **w** be any vector in S, and let **v** and **u** be any vector in S_d. Then w.v = w.u = 0 and w.v + w.u = 0 = w (v + u); thus $v + u$ is in S_d. Also w.(c.v) = c(w.v) = 0, and thus cv is in S_d. Therefore S_d satisfies all the requirements for subspace. Hence S_d is also a subspace in V_n.

This subspace S_d is called the orthogonal complement of S. That is ***a subspace S_d in a vector space is said to be an orthogonal complement of another subspace S in the same vector space if the vectors in S_d are orthogonal to the vectors in S.*** S_d is also known as Dual space or null space of S. Conversely S is also null space of S_d.

For a k-dimensional subspace S in the vector space V_n of all n-tuples over GF(2), the dimension of its null space S_d is '$n - k$'. Also, sum of the dimensions of S and S_d is equal to the dimension of the vector space V_n. i.e dim(S) + dim(S_d) = dim(V_n) = n.

***Example* A12:** Consider the vector space V_4 of all 4-tuples over GF(2) given in example A9. The following eight vectors form a 3-dimensional subspace S of V_4.

(0000), (1110), (0101), (1000), (1011), (0110), (1101), (0011).

The null space S_d of S consists of the following four vectors:

(0000), (1010), (0111), (1101).

The basis of S_d are (1010) and (0111) which are linearly independent. Therefore the dimension of S_d is 2.

All the concepts discussed on vector space here can be straight away used to the vector space of all n-tuples over GF(q), where q is a power of a prime number.

A.9 Matrices

The next mathematical structure which we are going to discuss is the Matrices. The vectors discussed in examples A9 and A11 can be arranged in the form of a matrix and the usual matrix manipulations can be carried out. We will use the same notations used during the discussions of codes in the main chapters.

A 'k × n' matrix consists of 'kn' elements arranged in a rectangular array of 'k' rows and 'n' columns.

Consider a '$k \times n$' matrix G, with elements g_{ij}, as shown below:

$$G = \begin{pmatrix} g_{00} & g_{01} & g_{02} & \cdots & g_{0,n-1} \\ g_{10} & g_{11} & g_{12} & \cdots & g_{1,n-1} \\ . & . & . & \cdots & . \\ . & . & . & \cdots & . \\ g_{k-1,0} & g_{k-1,1} & g_{k-1,2} & \cdots & g_{k-1,n-1} \end{pmatrix}$$

where each element of the matrix, g_{ij}, with $0 \leq i < k$ and $0 \leq j < n$, is from the binary field GF(2). The first index, '*i*' refer to the row containing the element, and the second index '*j*' refers to the column in which the element is there. Each row of G is an *n*-tuple and each column is a *k*-tuple over GF(2).The matrix G can be abbreviated as $[g_{ij}]$. The set of elements for which the row number is equal to the column number is called the ***main diagonal***. If $k = n$, the matrix is called a ***square matrix***. A '$k \times k$' matrix with the field element '1' in every entry of the main diagonal and all other elements of the matrix being the field element 'zero' is known as a '$k \times k$' ***identity matrix***. This identity matrix is denoted by $\mathbf{I}_k$. The ***transpose*** of G, denoted by G^T, is an '$n \times k$' matrix whose rows are columns of G and whose columns are rows of G. A ***submatrix*** of the matrix G is a matrix obtained by deleting given rows or columns of G.

The matrix G can also be represented by its '*k*' rows $g_0, g_1, \cdots, g_{k-1}$ as follows:

$$G = \begin{pmatrix} g_0 \\ g_1 \\ \cdot \\ \cdot \\ g_{k-1} \end{pmatrix}$$

The following are the useful properties of the Matrix:

(*a*) The rows and columns are known as Row and Column Vectors.

(*b*) For $k \leq n$, if the '*k*' row vectors are linearly independent, then the 2^k linear combinations of these rows form a *k*-dimensional subspace of the vector space V_n of all the *n*-tuples over GF(2). This subspace is called the row space of G. Similarly the column space is defined.

(*c*) The row rank of the matrix is the dimension of its row space and is equal to the maximum number of linearly independent rows of the matrix. Similarly the column rank is defined.

(*d*) The row rank and row space of a matrix are invariant under elementary operations like.

(*i*). Interchange of any two rows.

(*ii*). Multiplication of any row by a non-zero field element.

(*iii*). Addition of any scalar multiple of one row to another.

(*e*) Two matrices can be added if they have same number of rows (*k*) and columns (*n*). The resultant matrix will also be a '$k \times n$' matrix. In order to add two '$k \times n$' matrices $X = [x_{ij}]$ and $Y = [y_{ij}]$ we have to add their corresponding entries x_{ij} and y_{ij} as $[x_{ij}] + [y_{ij}] = [x_{ij} + y_{ij}]$.

(*f*) Two matrices can be multiplied if the number of columns in the first matrix is equal to the number of rows in the second matrix. Multiplying a '$k \times n$' matrix, X, with an '$n \times m$' matrix, Y, the resultant matrix, Z, will be '$k \times m$' matrix. The elements of the matrix Z, is equal to the inner product of the ith row in X and the jth column in Y; that is

$$z_{ij} = x_{ij} \cdot y_{ij} = \sum_{t=0}^{n-1} x_{it} y_{tj}$$

Let S be the row space of the '$k \times n$' matrix G over GF(2) whose 'k' rows $g_0, g_1, \ldots, g_{k-1}$ are linearly independent. Let S_d be the null space of S. Then, the dimension of S_d is '$n-k$'. Let vectors $h_0, h_1, \ldots, h_{n-k-1}$ be the $n-k$ linearly independent vectors in S_d. Clearly, these vectors span S_d. We can form an $(n-k) \times n$ matrix H using $h_0, h_1, \ldots, h_{n-k-1}$ as rows:

$$H = \begin{pmatrix} h_0 \\ h_1 \\ \cdot \\ \cdot \\ h_{n-k-1} \end{pmatrix} = \begin{pmatrix} h_{00} & h_{01} & h_{02} & \cdots & h_{0,n-1} \\ h_{10} & h_{11} & h_{12} & \cdots & h_{1,n-1} \\ \cdot & \cdot & \cdot & \cdot & \cdot \\ \cdot & \cdot & \cdot & \cdot & \cdot \\ h_{n-k-1,0} & h_{n-k-1,1} & h_{n-k-1,2} & \cdot & h_{n-k-1,n-1} \end{pmatrix}$$

The row space of **H** is S_d. Since each row $\mathbf{g}_i$ of G is a vector in S, and each row h_j of H is a vector of S_d, the inner product of $\mathbf{g}_i$ and h_j must be zero (i.e., $\mathbf{g}_i \cdot h_j = 0$). Since the row space S of G is the null space of the row space S_d of H, S is the null space of H.

Therefore, for any '$k \times n$' matrix G over GF(2) with 'k' linearly independent rows, there exists an '$(n-k) \times n$' matrix H over GF(2) with $(n-k)$ linearly independent rows such that for any row $\mathbf{g}_i$ in G and any h_j in H, $\mathbf{g}_i \cdot h_j = 0$. The row space of G is the null space of H, and vice versa. This implies that $G.H^T = 0$

The concepts and results discussed on matrices over GF(2) can be readily used with **entries from GF(q) with 'q' as a power of prime.**

PROBLEMS

1. Using modulo-2 arithmetic solve the following simultaneous equations of A, B, C, and D:

$$A + B + D = 1,$$
$$A + C + D = 0,$$
$$A + B + C + D = 1,$$
$$B + C + D = 0.$$

2. Construct a group under modulo – 3 multiplication.

3. Construct a group under modulo – 5 addition.
4. Show that $X^6 + X + 1$ is primitive over GF(2).
5. Find all irreducible polynomials of degree 5 over GF(2).
6. Construct a Galois Field table GF(2^4) using the primitive polynomial $p(X) = 1 + X + X^4$, and display the power, polynomial, vector and integer representation of each field element.
7. Let 'α' be primitive element in GF(2^4). Using the table constructed in problem 6 above find the roots of $f(X) = X^3 + \alpha^6 X^2 + \alpha^9 X + \alpha^9$.
8. *a, b* and *c* are three linearly independent vectors over GF(*q*). Show that the vectors $(a+b)$, $(b+c)$, $(c+a)$ are also linearly independent.
9. Let S be a subset of the vector space V_n of all *n*-tuples over GF(2). Prove that S is a subspace, if for any '*a*' and '*b*' in S, $a + b$ also is in S.
10. Construct the vector space V_5 of all 5-tuples over GF(2). Find the 3-dimensional subspace and determine its null-space.
11. Construct the vector space of all 3-tuples over GF(3). Form the two-dimensional subspace and its dual space.
12. For the matrices G and H given below, show that the row space of G is the null space of H, and vice versa.

$$G = \begin{pmatrix} 1101100 \\ 1110010 \\ 0111001 \end{pmatrix}, \quad H = \begin{pmatrix} 1000110 \\ 0100111 \\ 0010011 \\ 0001101 \end{pmatrix}.$$

REFERENCES

1. Shannon CE, ***"A Mathematical Theory of Communication"***, Bell Systems Technical Journal, 27 (1948), pp: 379–423.
2. Reza FM, ***"Introduction to Information Theory"***, McGraw Hill Book Company, NY 1961.
3. Abramson N, ***"Information and Coding",*** McGraw Hill Book Company, NY 1963.
4. McEliece RJ, ***"The Theory of Information & Coding",*** Addison Wesley, Mass., 1977.
5. Carlson, Bruce A, ***"Communication Systems",*** 2nd Edn., McGraw Hill International Book Company, Singapore, 1984.
6. B.P Lathi, ***"Modern Digital and Analog Communication systems***", 2nd Edn., Holt, Rinehart & Winston, Inc., USA, 1993.
7. Hancock, ***"An Introduction to Communication Theory"***, Tata McGraw Hill Book Company, New Delhi,1968.
8. Taub H and Schilling D L, ***"Principles of Communication Systems"***, McGraw Hill Book Company, NY, 1986.
9. Mischa, Schwartz, ***"Information Transmission, Modulation and Noise"***, McGraw Hill Book Company, NY, 1970.
10. Das J, Mullik.SK and Chaterjee. PK, ***"Principles of Digital Communications",*** Wiley Eastern Ltd., New Delhi, 1986.
11. Simon Haykin, "***Digital Communication***", John Wiley, 2003.
12. K. SamShanmugam, "***Digital & Analog Communication Systems***", John Wiley, 1996.
13. Glover and Grant, ***Digital Communications*** , 2nd Edn. Pearson, 2008.

14. Proakis JG, ***"Digital Communications",*** McGraw Hill Book Company, NY, 1983.
15. Bernard Sklar, ***"Digital Communications- Fundamentals and Applications"***, Printice Hall, Englewood Cliffs, New Jersey, 1988.
16. Hamming RW, "***Error Detecting and Correcting Codes***", Bell Systems Technical Journal, 29 (1950), pp: 147–160.
17. W.W. Peterson and E.J. Weldon, Jr., ***Error Correcting Codes***, 2nd Edn., Wiley, NY, 1972.
18. S Lin and D.Costello, "***Error Control Coding: Fundamentals and Applications***" Printice Hall, Englewood Cliffs, New Jersey, 1983.
19. P Elias, ***Coding for Noisy Channels***, IRE National Conference Rec.,Vol.3, Part 4, 1955.
20. A.J.Viterbi, ***Convolutional Codes and their Performance in communication Systems***, IEEE Trans., Communication Technology, vol. CT-19, pp. 751–771, Oct.1971.
21. J L. Massey, "***Threshold Decoding***" M I T Press, Cambridge, Mass., 1963.
22. Blahut RE, ***"Theory and Practice of Error Control Codes",*** Addison Wesley , 1984.
23. Arnold M. Michelson, Allen H. Levesque, ***"Error Control Techniques for Digital Communication"***, John Wiley & sons, Inc., 1985.
24. George C Clark, Jr and J.Bibb Cain ***"Error Correction Coding for Digital Communication"***, Plenum Press, 1981.
25. Stephen B. Wicker, Vijay K Bhargava, ***"Reed Solomon Codes and their Applications",*** IEEE Press 1994.
26. Rao, T.R.N., and Fujiwara, E., ***"Error Control Coding for Computer Systems",*** Printice Hall, 1989.
27. Berlekamp, E.R., "***The Technology of Error Correcting Codes***", Proceedings of IEEE, May,1980, 68,(5), pp 564–593.
28. J M Wozencraft and B.Reiffen, ***"Sequential Decoding",*** MIT Press, Cambridge, 1961.
29. L H C Lee, ***"Convolutional Coding: Fundamentals and Applications",*** Norwood, Mass., Artech House, 1997.
30. Stephen B. Wicker, "***Error Control Systems for Digital Communication*** and Storage", *Printice Hall, Englewood Cliffs, New Jersey, 1995.*

INDEX

K

L

M

N

O

P

Q

R